Green Chemical Syntheses and Processes

Dedication

A founding co-chair, of the Green Chemistry and Engineering Conference, from which this book is derived, was named after Dr. Joseph J. Breen. Joe Breen was an inspiration to people around the world because of the passion and enthusiasm with which he articulated the need and the goals of green chemistry and engineering for sustainability. Joe's premature death in the summer of 1999 is a tragedy to all who knew him and a loss to the causes he championed. This book is dedicated to the memory of Dr. Joseph J. Breen.

ACS SYMPOSIUM SERIES **767**

Green Chemical Syntheses and Processes

Paul T. Anastas, EDITOR
White House Office of Science, Technology and Policy

Lauren G. Heine, EDITOR
International Sustainable Development Foundation

Tracy C. Williamson, EDITOR
U.S. Environmental Protection Agency

American Chemical Society, Washington, DC

Library of Congress Cataloging-in-Publication Data

Green chemical syntheses and processes / Paul T. Anastas, Lauren G. Heine, Tracy C. Williamson [editors].

    p. cm.—(ACS symposium series ; 767)

    Includes bibliographical references and index.

    ISBN 0–8412–3678–X

    1. Environmental chemistry—Industrial applications—Congresses. 2. Environmental management—Congresses.

    I. Anastas, Paul T.., 1962- II. Heine, Lauren G., 1957– III. Williamson, Tracy C., 1963- IV. Title. V. Series.

TP155.2.E58 G75   2000
660′.28′6—dc21   00–41638

The paper used in this publication meets the minimum requirements of American National Standard for Information Sciences—Permanence of Paper for Printed Library Materials, ANSI Z39.48–1984.

Copyright © 2000 American Chemical Society

Distributed by Oxford University Press

All Rights Reserved. Reprographic copying beyond that permitted by Sections 107 or 108 of the U.S. Copyright Act is allowed for internal use only, provided that a per-chapter fee of $20.00 plus $0.50 per page is paid to the Copyright Clearance Center, Inc., 222 Rosewood Drive, Danvers, MA 01923, USA. Republication or reproduction for sale of pages in this book is permitted only under license from ACS. Direct these and other permission requests to ACS Copyright Office, Publications Division, 1155 16th St., N.W., Washington, DC 20036.

The citation of trade names and/or names of manufacturers in this publication is not to be construed as an endorsement or as approval by ACS of the commercial products or services referenced herein; nor should the mere reference herein to any drawing, specification, chemical process, or other data be regarded as a license or as a conveyance of any right or permission to the holder, reader, or any other person or corporation, to manufacture, reproduce, use, or sell any patented invention or copyrighted work that may in any way be related thereto. Registered names, trademarks, etc., used in this publication, even without specific indication thereof, are not to be considered unprotected by law.

PRINTED IN THE UNITED STATES OF AMERICA

Errata Sheet

ACS Symposium Series 767

Green Chemical Syntheses and Processes
Paul T. Anastas, Lauren G. Heine, and Tracy C. Williamson

On page ii, the Dedication should read as follows: This book is dedicated to the memory of Dr. Joseph J. Breen. Joe Breen was a founding co-chair of the Green Chemistry and Engineering Conference from which this book is derived. He was an inspiration to people around the world because of the passion and enthusiasm with which he articulated the needs and goals of green chemistry and engineering for sustainability. Joe's premature death in the summer of 1999 is a tragedy to all who knew him and a loss to the causes he championed.

On pages xi and 1, the correct addresses for Paul Anastas and Lauren Heine should be the following: Dr. Paul Anastas, White House Office of Science and Technology Policy, Old Executive Office Building, Washington, DC 20502 and Lauren Heine, Zero Waste Alliance, International Sustainable Development Foundation, 121 SW Salmon Street, Suite 210, Portland, OR 97204. Footnote 4 on page 1 also has a typo: It should be University of Nottingham.

ACS apologizes for mistakenly publishing the draft introductory chapter. Literature citations should read:

Literature Cited

1. Raber, Linda R. (1998). "Green Chemistry Here to Stay." *Chemical & Engineering News,* July 6, pp. 25–26.
2. Miller, J. P. (1993). "Firms Give Chemistry Green Light." *The Wall Street Journal,* December 7, B6.
3. Anastas, P. T. and Williamson, T. C. (1998). *Green Chemistry: Theory and Practice,* Oxford University Press, Oxford.
4. U.S. Environmental Protection Agency. (1998). *The Presidential Green Chemistry Challenge Awards Program: Summary of 1998 Award Entries and Recipients.* EPA744-R-98-001.
5. Anastas, P. T. and Williamson, T. C. (1996). "Green Chemistry: An Overview." In *Green Chemistry: Designing Chemistry for the Environment,* American Chemical Society Symposium Series, No. 626, (ed. P. T. Anastas and T. C. Williamson), pp. 1–17. American Chemical Society.

Foreword

THE ACS SYMPOSIUM SERIES was first published in 1974 to provide a mechanism for publishing symposia quickly in book form. The purpose of the series is to publish timely, comprehensive books developed from ACS sponsored symposia based on current scientific research. Occasionally, books are developed from symposia sponsored by other organizations when the topic is of keen interest to the chemistry audience.

Before agreeing to publish a book, the proposed table of contents is reviewed for appropriate and comprehensive coverage and for interest to the audience. Some papers may be excluded in order to better focus the book; others may be added to provide comprehensiveness. When appropriate, overview or introductory chapters are added. Drafts of chapters are peer-reviewed prior to final acceptance or rejection, and manuscripts are prepared in camera-ready format.

As a rule, only original research papers and original review papers are included in the volumes. Verbatim reproductions of previously published papers are not accepted.

ACS BOOKS DEPARTMENT

Contents

Preface xi

1. Green Chemical Syntheses and Processes: Introduction 1
Paul T. Anastas, Lauren G. Heine, and Tracy C. Williamson

Designing Safer Chemicals

2. Tebufenozide: A Novel Caterpillar Control Agent with Unusually High Target Selectivity 8
Glenn R. Carlson

3. Iron-Complexed Dyes: Colorants in Green Chemistry 18
Harold S. Freeman and Laura C. Edwards

4. In Vivo Synthesis in Yeast of Insect Sex Pheromone Precursors: An Alternative Synthetic Route for the Production of Environmentally Benign Insect Control Agents 33
Douglas C. Knipple, W. L. Roelofs, C.-L. Rosenfield, P. Marsella-Herrick, and S. J. Miller

5. Chrome-Free Single-Step In-Situ Phosphatizing Coatings 43
Tao Yu, Mary C. Whitten, Carmen L. Muñoz, and Chhiu-Tsu Lin

Green Chemical Synthesis

6. Water as Solvent for Organic and Material Synthesis 62
Chao-Jun Li

7. The Greening of a Fundamental Reaction: Metal-Mediated Reactions in Water 74
Chao-Jun Li

8. Dimethylcarbonate as a Green Reagent 87
Pietro Tundo, Maurizio Selva, and Sofia Memoli

9. Indium-Promoted Coupling Reactions in Water 100
   Leo A. Paquette

Biocatalysis and Biosynthesis

10. Biocatalytic Production of 5-Cyanovaleramide from Adiponitrile 114
    Robert DiCosimo, Eugenia C. Hann, Amy Eisenberg,
    Susan K. Fager, Neal E. Perkins, F. Glenn Gallagher, Susan M. Cooper,
    John E. Gavagan, Barry Stieglitz, and Susan M. Hennessey

11. Bioconversion of Toluene to *p*-Hydroxybenzoate Using a
    Recombinant *Pseudomonas putida* 126
    Edward S. Miller, Jr., and Steven W. Peretti

12. Genetically Engineered *Saccharomyces* Yeasts for Conversion
    of Cellulosic Biomass to Environmentally Friendly Transportation
    Fuel Ethanol 143
    Nancy W. Y. Ho, Zhendao Chen, Adam P. Brainard,
    and Miroslav Sedlak

13. Bioconversion of Sugar Cane Vinasse into Microbial Biomass
    by Recombinant Strains of *Aspergillus nidulans* 160
    André O. S. Lima and Aline A. Pizzirani-Kleiner

Environmentally Benign Catalysis

14. Fluorous Biphasic Catalysis: A Green Chemistry Concept
    for Alkane and Alkene Oxidation Reactions 172
    Jean-Marc Vincent, Alain Rabion, and Richard H. Fish

15. Polymer-Facilitated Biphasic Catalysis 182
    David E. Bergbreiter

16. Environmentally-Benign Liquid-Phase Acetone Condensation
    Process Using Novel Heterogeneous Catalysts 194
    A. A. Nikolopoulos, B. W-L. Jang, R. Subramanian, J. J. Spivey,
    D. J. Olsen, T. J. Devon, and R. D. Culp

17. Photooxidation of Toluene in Cation-Exchanged Zeolites 206
    A. G. Panov, K. B. Myli, Y. Xiang, V. H. Grassian, and S. C. Larsen

18. Oxygenation of Hydrocarbons Using Nanostructured TiO_2 as a Photocatalyst: A Green Alternative 217
    Endalkachew Sahle-Demessie and Michael A. Gonzalez

Green Solvent Systems

19. The Design of Technologically Effective and Environmentally Benign Solvent Substitutes 230
    Renhong Zhao, Heriberto Cabezas, and Subba R. Nishtala

20. Volatile Methyl Siloxanes: Environmentally Sound Solvent Systems 244
    Dwight E. Williams

21. Green Chemistry Through the Use of Supercritical Fluids and Free Radicals 258
    J. M. Tanko, B. Fletcher, M. Sadeghipour, and N. K. Suleman

22. Supercritical Fluids as Solvent Replacements in Chemical Synthesis 270
    Jefferson W. Tester, Rick L. Danheiser, Randy D. Weintstein, Adam Renslo, Joshua D. Taylor, and Jeffrey I. Steinfeld

23. Expeditious Solvent-Free Organic Syntheses Using Microwave Irradiation 292
    Rajender S. Varma

24. Choosing Solvents That Promote Green Chemistry 313
    William M. Nelson

Author Index 330

Subject Index 331

Preface

Scientific innovation is at the heart of Green Chemistry. The same flame of innovation that powered industry and economic growth since science's beginnings is now also powering the protection of human health and the environment. This book contains chapters by individuals from different disciplines, different sectors (industry, academia, and government), and different countries. In many cases, they worked together as teams toward the unified goal of sustainability through science. Through this past century, many goals have been set out as challenges to humanity including conquering many terrible diseases, combating hunger, and even going to the moon. Just as it has been that science and technology was the engine to achieve those noble goals, it too will provide the engine that takes us down the road toward sustainability.

PAUL T. ANASTAS[1]
White House Office of Science, Technology and Policy
1600 Pennsylvania Avenue
Washington, DC 20500

LAUREN G. HEINE
Zero Waste Alliance
International Sustainable Development Foundation
3624 S.W. 10th Avenue
Portland, OR 97201

TRACY C. WILLIAMSON
U.S. Environmental Protection Agency
Waterside Mall, Mail Code 7406
401 M. Street, SW
Washington, DC 20460

---

[1]Visiting Professor at Department of Chemistry, The University of Nottingham, University Park, Nottingham NG7 2RD, United Kingdom.

Chapter 1

Green Chemical Syntheses and Processes: Introduction

Paul T. Anastas[1,4], Lauren G. Heine[2], and Tracy C. Williamson[3]

[1]White House Office of Science, Technology and Policy, 1600 Pennsylvania Avenue, Washington, DC 20500
[2]Zero Waste Alliance, International Sustainable Development Foundation, 3624 S.W. 10th Avenue, Portland, OR 97201
[3]Industrial Chemicals Branch, U.S. Environmental Protection Agency, Waterside Mall, Mail Code 7406, 401 M Street, SW, Washington, DC 20460

Green chemistry is enjoying significant adoption by industry around the world and widespread activity from the research community (1). One reason for this is that not only does green chemistry address the fundamental scientific challenges of protecting human health and the environment at the molecular level, it accomplishes this in an economically beneficial way for industry (2). One measure of this is the fact that while there is not a single regulation requiring industry to engage in the specific practices or methodologies of green chemistry (3), there are nevertheless plentiful examples of excellent green chemistry techniques being commercialized. This can be seen by both the number and quality of the nominations and winners of the Presidential Green Chemistry Challenge Awards given annually at the National Academy of Sciences (4).

Green Chemistry, defined as the design of chemical products and processes that reduce or eliminate the use and generation of hazardous substances (5), has been referred to as pollution prevention at the molecular level. This emerging area recognizes that during the design phase of any chemical synthesis, product, or process, minimized hazard must be viewed as a performance criterion. Moreover, hazard must also be viewed as a physical/chemical property that is possible to manipulate and control at the molecular level.

By using the same skills, techniques, and expertise that is central to traditional chemistry, the practices of green chemistry are realizing some notable and in some cases, dramatic, results in the protection of human health and the environment.

This book presents a number of the innovations that have been developed recently in the emerging area of green chemistry. The chapters of the book are derived from presentations made at the Green Chemistry and Engineering Conference at the National Academy of Sciences, Washington, D.C. This conference was established to

[4]Visiting Professor at Department of Chemistry, The University of Knottingham, University Park, Nottingham NG7 2RD, United Kingdom.

© 2000 American Chemical Society

highlight the cutting edge science and engineering in this field being conducted in industry, academia and government.

Over the years, the Green Chemistry and Engineering Conference has had different themes. One of those themes was "Implementing Vision 2020 for the Environment". This theme was based on the document produced by the chemical industry to provide goals for the next twenty years, entitled "Vision 2020 for the Chemical Industry". Throughout this document, there was a common thread that was woven throughout the goals ranging from new chemical technologies to supply chain issues and this thread was the environment. Throughout all aspects of the chemical enterprise and all of the goals outlined in the report was the recurring and overarching recognition that the environmental impact of all activities must be among the fundamental considerations.

It is because of the ubiquitous need to consider environmental issues in all parts of the chemical industry that Vision 2020 was viewed as a good theme for the conference. It was recognized that the science and technology of Green Chemistry and Engineering was, in real time, implementing or beginning to implement many of the stated goals of 'Vision 2020'. By developing new separation techniques, new syntheses, new reaction processes, etc., and doing it while engaging in fundamental protection of human health and the environment, Green Chemistry and Engineering was accomplishing, or at least beginning to accomplish, those challenges outlined in 'Vision 2020'.

Another theme of a Green Chemistry and Engineering Conference was "Global Perspectives". This conference emphasized the fact that the science and engineering that are the topic of the conference know no boundaries. By featuring presentations from industry, academia, and government, the conference illustrated that Green Chemistry and Engineering is being actively pursued in both the public and private sectors. By featuring talks from disciplines ranging from organic synthesis to biology to electrical engineering and many others, it was demonstrated that this area is not limited by disciplinary bounds. The industrial sectors represented as well spanned well beyond the traditional chemical industry into electronics, pulp and paper, and pharmaceuticals, to name a few. Finally, it was shown that green chemistry and engineering know no national boundaries by representatives from eleven nations around the world taking part in the conference. Therefore, in several ways, one can see that a global perspective can be achieved through the practice of green chemistry and engineering.

Each year an organizing committee with representatives from all parts of the chemical enterprise convene to construct a conference that will represent some of the most recent, innovative and topical advances in green chemistry and engineering. Those organizations, American Chemical Society, American Institute of Chemical Engineers, Chemical Manufacturers Association, Council for Chemical Research, U.S. Department of Energy, U.S. Environmental Protection Agency, National Academy of Sciences, National Research Council, National Institute of Standards and Technology, and Organization for Economic Cooperation and Development, represent a breadth of essential elements of the discovery, demonstration, and commercialization of chemical technology. The fine work presented in this volume is a tribute to the work these organizations do in promoting, encouraging, funding, supporting and catalyzing the emerging area of Green Chemistry and Engineering.

Discussion of Book Sections

There are multiple ways to categorize approaches to green chemistry research. When considering chemicals that serve important social or industrial needs, some work focuses on their synthesis via environmentally friendly synthetic pathways or processes. Other work focuses on developing new benign replacements capable of achieving the desired performance without negative human or ecological impacts. Each section in this book contains chapters that are related by the principles that underlie the authors' work in green chemistry.

Designing Safer Chemicals

In this section, four chapters present work that shares the common goal of designing new products with safer and more benign impacts over their life-cycle. The topics range from the design and synthesis of environmentally friendly metal-complexed textile dyes that have the potential to eliminate the source of wastewater containing toxic metals, to a superior replacement for the current energy intensive and hazardous waste producing process of organic coating on metals. In the area of pest control, a chemically and mechanistically novel caterpillar control agent is described that has high target selectivity but poses minimal hazard to non-target organisms and the environment. There is also a chapter on a new methodology to produce highly selective, nontoxic insect sex pheromones. The purpose of this research is to facilitate the environmentally benign practice of insect control based on mating disruption by making it more economical. While these topics are disparate in their subject matter, they demonstrate the breadth of the application of green chemistry principles to the design of benign products and processes.

Green Chemical Synthesis

The chapters in this section focus on chemical reactions which utilize more environmentally benign reagents or are conducted in more benign reaction media. This approach has attracted more and more attention over the last decade, due in part to the desire for the development of more environmentally safe reaction processes. Dimethylcarbonate is discussed as an environmentally friendly substitute for dimethylsulfate and methyl halides in methylation reactions. In addition, the chapters by Paquette and Li illustrate that many reactions that are traditionally carried out in organic solvents can be carried out in water with additional interesting features. The focus of this work is on organometallic reactions in water, with a particular emphasis on indium-promoted coupling reactions by Paquette.

Biocatalysis and Biosynthesis

Biocatalysis and biosynthesis of chemicals has the potential to allow synthesis under environmentally friendly conditions while achieving high specificity and yield. Such reaction systems may also use renewable feedstocks and transform materials with low or negative economic value into fuels or chemicals with higher value. A biocatalytic process for the hydration of adiponitrile to 5-cyanovaleramide has been demonstrated which can be run to higher conversion, produce more product per weight of catalyst, and generate significantly less waste products than alternate chemical processes. Ho describes how her research successfully developed a genetically engineered recombinant *Saccharomyces* yeast that can effectively ferment both glucose and xylose to ethanol. This work has the potential to make the conversion of cellulosic biomass to ethanol technically viable. The chapter by Lima discusses the bioconversion of sugar cane vinnasse into microbial biomass by recombinant strains of Aspergillus nidulans transforms a waste product with ecologically detrimental impacts to a product with value as a fuel resource.

Environmentally Benign Catalysis

A principle of green chemistry contends that all other factors being equal, catalytic systems (as selective as possible) are superior to stoichiometric reagents (*3*). Using catalytic reagents creates opportunities for increased selectivity and the use of alternative reaction conditions. In this section, the effectiveness of both homogeneous and heterogeneous catalysis is expanded in some intriguing and practical ways.

Nikolopoulos *et al.* promote the condensation/hydrogenation of acetone to methyl isobutyl ketone (MIBK) using novel multi-functional heterogeneous catalysts. Bergbreiter and Vincent *et al.* focus on techniques to facilitate recovery and reuse of homogeneous catalysts. Some of the separation strategies described include soluble polymers that precipitate on cooling, soluble polymers that precipitate on heating, and polymers that dissolve selectively in fluorous or aqueous phases. A novel two phase homogeneous process called fluorous biphasic catalysis is discussed where a lower phase fluorocarbon solvent solubilizes the homogeneous catalyst, and a second upper solvent phase solubilizes the substrate alkane/alkene, the oxidant, and the products of alkane and alkene oxidation.

Photochemical oxidation is a potentially environmentally benign method for selective oxidation of hydrocarbons. Panov *et al.* compare the photooxidation of toluene with molecular oxygen using zeolites, BaX, BaY, CaY, BaZSM-5, and NaZSM-5. Sahle-Demessie and Gonzalez synthesize high-value organic compounds from linear and cyclic hydrocarbons by photocatalytic oxidation under mild conditions using the semiconductor material, titanium dioxide (TiO_2). TiO_2 photooxidation using gaseous phase reaction conditions eliminates the need for a separation step involving liquid solvents and minimizes the adsorption of products to the catalyst.

Green Solvent Systems

There is presently considerable interest in the development and utilization of more environmentally benign solvents and solventless systems that reduce or eliminate the use of toxic or environmentally hazardous solvents. In this section, concepts and tools for finding "green" solvent systems are described. The advantages of green solvents can be broadened to include the development and optimization of novel reaction conditions that can help to maximize product yield and minimize energy usage.

Zhao *et al.* describe the concepts and mathematical foundations for PARIS II software that designs solvent mixtures based on optimal solvent properties. The solvent property categories include performance requirements and environmental requirements (a VOC index and an environmental impacts index), and are based on general, dynamic, and equilibrium properties. The chapter by Nelson contains perspectives on the defining characteristics of green solvents, their desired characteristics (ecologically and economically) and a methodological approach to their selection.

Williams, of Dow Corning Corporation, discusses the properties of linear volatile methyl siloxanes, a class of mild solvents having an unusual combination of environmentally benign qualities, that have the potential to replace less benign solvents in applications such as coating formulations or the removal of particulates, oils, fluxes, and aqueous contaminants. The chapter details data showing that their solvency can be tailored to specific applications by using azeotropes, co-solvents and surfactants.

The use of supercritical carbon dioxide (sc-CO_2) is further expanded as an environmentally benign solvent for chemical synthesis and processing. Tanko *et al.* describe sc-CO_2 solvent effects on chemical reactivity for free radical reactions, providing a unique alternative to many conventional solvents for these reactions which are either carcinogenic or damaging to the environment. Tester *et al.* describe the results of fruitful collaboration between chemists and chemical engineers in understanding carbon-carbon bond forming reactions in supercritical fluids and in optimizing them to enable scaled-up designs for economically competitive processes.

Varma moves away from the use of solvents altogether by demonstrating microwave expedited solvent-free synthetic processes. He exposes neat reactants to microwave (MW) irradiation in the presence of supported reagents or catalysts on mineral oxides resulting in enhanced reaction rates, greater selectivity and experimental ease of manipulation.

Conclusion

The chapters in the book serve to illustrate some of the examples of green chemistry that are being researched and developed in academia, as well as being implemented in industry. Also recommended to the reader is an accompanying volume entitled, "Green Engineering and Processing" that provides a complementary treatment of the area of green engineering that works together integrally with green chemistry.

Literature Cited

1. Raber, Linda. 1999. "GC is here to stay" Chemical & Engineering News. July??**

2. **Firms give green chemistry the green light- find reference**GC Scrap book with collections of articles.

3. Anastas, P.T. and Warner, J.C. (1998). *Green Chemistry: Theory and Practice*. Oxford University Press, Oxford.

4. U.S. Environmental Protection Agency. 1998. *The Presidential Green Chemistry Challenge Awards Program: Summary of 1998 Award Entries and Recipients.* EPA744-R-98-001

5. Anastas, P.T. and Williamson, T.C. (1996). Green chemistry: an overview. In *Green Chemistry: Designing Chemistry for the Environment*, American Chemical Society Symposium Series, No. 626, (ed. P.T. Anastas and T.C. Williamson), pp. 1-17. American Chemical Society.

Designing Safer Chemicals

Chapter 2

Tebufenozide: A Novel Caterpillar Control Agent with Unusually High Target Selectivity

Glenn R. Carlson

Insecticide Discovery Research, Rohm and Haas Research Laboratories, 727 Norristown Road, Spring House, PA 19477

Tebufenozide is a chemically and mechanistically novel caterpillar control agent that poses very minimal hazard to non-target organisms and the environment. It can be used to selectively control a wide range of agriculturally important caterpillar pests, effectively replacing the use of many older more toxic and/or more environmentally persistent broad spectrum insecticides. Rohm and Haas Company was granted a Presidential Green Chemistry Award for the discovery and development of tebufenozide.

Introduction

A target selective insect control agent is one that is capable of controlling a specific insect pest, or small group of related insect pests, without causing significant harm to other non-target organisms in the ecosystem, including other non-target insects. Intuitively such agents would seem highly desirable, yet surprisingly few have actually reached the marketplace. Most highly target selective insect control agents currently in use are of biological, rather than synthetic, origin (Bt toxins, insecticidal viruses/ fungi, etc.) Unfortunately, such biologicals tend to have certain major deficiencies (lack of storage stability, lack of residuality in the field, poor formulation characteristics, very slow speed of kill, mediocre cost-efficacy, etc.) that have severely limited their success in the marketplace.

Tebufenozide (RH-5992) is one of the first examples of a highly target selective insecticide that has been produced by rational design and chemical synthesis (1,2,3). This chemically novel insecticide is the first commercial product to act as a potent mimic of 20-hydroxyecdysone (20E) in caterpillars (larval members of the insect order Lepidoptera). It very effectively controls a wide range of agriculturally important caterpillar pests at low use rate, yet it is almost as safe as the biologicals to a wide range of non-target organisms, including an impressive list of important beneficial and predatory insects. The following paper briefly describes the insecticidal and toxicological properties of tebufenozide and discusses the probable mechanistic basis for the remarkable target selectivity of this compound.

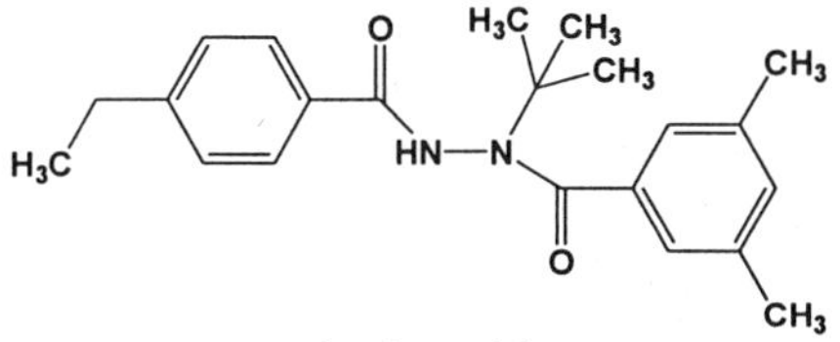

tebufenozide

20-Hydroxyecdysone and the Molting Process in Insects

Insects possess a characteristic laminar exoskeleton which gives them both great mechanical strength and impressive protection from physical and environmental stress. Insects certainly owe much of their evolutionary success to this distinctive body design. However, there is a serious drawback. The exoskeleton (cuticle) of insects is composed of a non-living polymeric material (chitin) which is relatively non-elastic and constrictive. Thus, in order to grow, the insect must periodically synthesize an entirely new and larger cuticle and then cast off the overlying old one. This reoccurring and highly complex process, called molting, is fraught with potential danger. Each time the insect molts, it must completely cease feeding, restrict its mobility, and initiate an intricate series of energy consuming physiological processes. This can make the insect temporarily more vulnerable to predation, mechanical injury, energy depletion, and/or desiccation.

Insects employ an unusual polyhydroxylated, cis-A/B ring-fused sterol to regulate molting. This substance, called 20-hydroxyecdysone (20E) or insect molting hormone, is biosynthesized by the insect from other simpler sterols extracted from the diet. It is released into, and removed from, the insect's blood during certain discrete stages of the insect's development. Precisely timed pulses of 20E concentration in the blood and tissues govern the key events of the molting cycle.

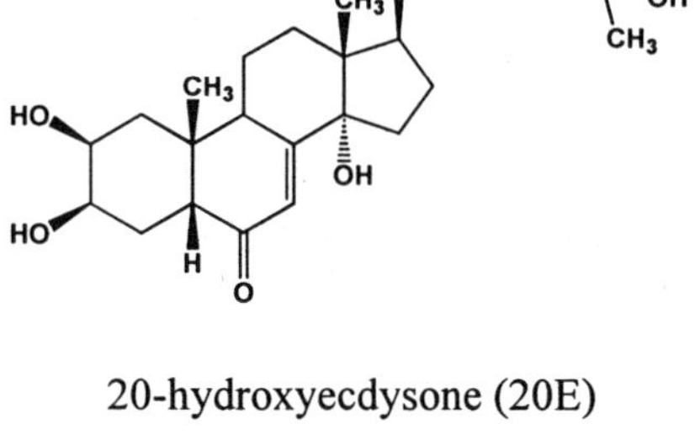

20-hydroxyecdysone (20E)

All the molting actions of 20E occur through its interaction with a specific receptor complex formed from the union of two smaller proteins called ecdysone receptor protein (EcR) and ultraspiracle protein (USP). As shown in Figure 1 below, neither EcR nor USP alone can bind 20E. However, when the EcR:USP heterodimer is formed, a conformational change in the EcR component of the heterodimer occurs, which allows the complex to bind 20E. Once 20E binding occurs, a second conformational change in the heterodimer takes place. This enables the heterodimer to interact with certain discrete segments of the insect's DNA known as ecdysone response elements (EcREs). The formation of a quaternary complex between 20E, EcR, USP, and EcRE results in the regulation of specific ecdysone responsive genes coding for proteins important to the molting process.

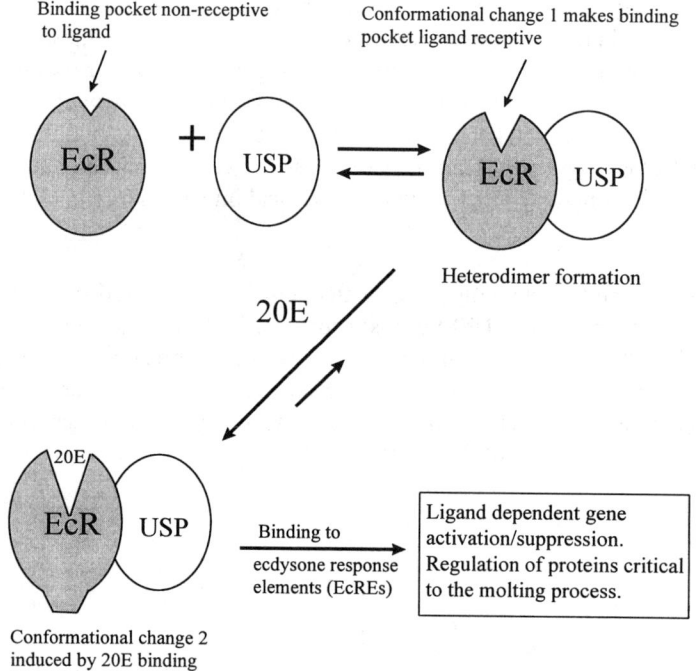

Figure 1. A schematic representation of the interaction of 20E with its target site in insects, the EcR:USP heterodimer.

As depicted in Figure 2, 20E is typically found in very low basal levels in the insect's blood during the normal intermolt period (the interval between molts in which the insect is feeding and gaining weight). Because the concentration of 20E is so low, the EcR:USP complex is essentially unoccupied during this time. As the molt attempt nears, however, 20E concentrations in the insect's blood begin to rise abruptly. Enough 20E is now present to occupy and activate the heterodimer. This causes the insect to abruptly cease feeding. A subsequent further increase in 20E concentration stimulates the epidermal cells to produce and release chitinolytic enzymes, which begin to erode the inner surface of the old cuticle. Just before 20E reaches peak concentration, the old partially digested cuticle detaches from the underlying epidermis (apolysis). The concentration of 20E then typically begins a rapid metabolic decline. This decline triggers the re-absorption of any remaining chitinolytic enzymes and the synthesis of new cuticle. Finally, after 20E blood levels have returned to near-basal levels and the synthesis of the new cuticle is complete, the insect begins a series of strenuous stereotypical escape behaviors which enable it to completely emerge from the remnants of its old partially digested cuticle (ecdysis). The new larger cuticle then hardens and feeding resumes.

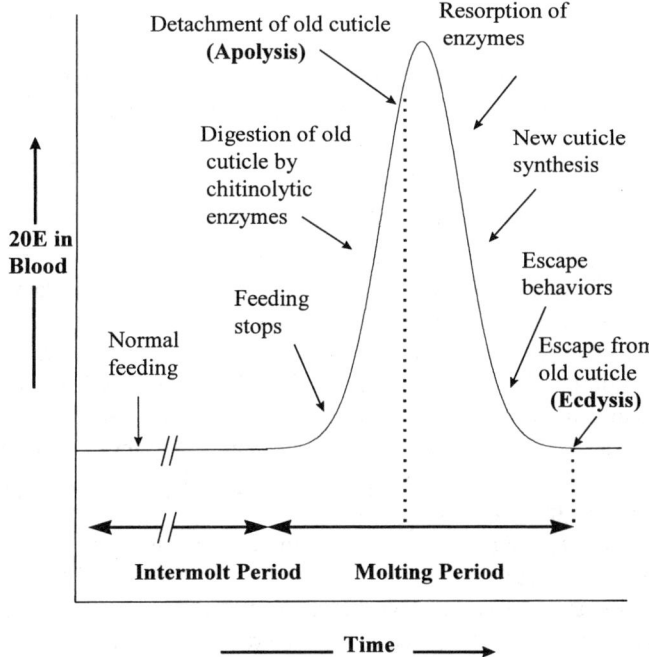

Figure 2. Critical events in the normal insect molting process relative to changes in 20E concentration.

The Insecticidal Mode of Action of Tebufenozide

It has been known for many years that insects treated with large exogenous doses of 20E, or any of a number of other closely related natural products derived from plants called phytoecdysteroids, experience severe, ultimately lethal, molting disruption. This toxic syndrome, characterized by premature apolysis, coupled with much impaired synthesis of new cuticle and complete (and permanent) inhibition of ecdysis, is commonly called hyperecdysonism. Apparently, a large overdose of 20E (or a substance that acts like 20E), administered during the intermolt period, can prematurely stimulate all the "early" events in the normal molting process (feeding inhibition, cuticle digestion, apolysis, etc,) but *inhibit* all the "late" ones (cuticle synthesis, ecdysis, etc.).

Tebufenozide is the first commercial insecticide to selectively mimic the actions of 20-hydroxyecdysone in Lepidoptera. It behaves like a potent analog of 20E (ecdysonoid) in such insects, binding directly to the 20E binding site on the lepidopteran EcR:USP heterodimer and acting as a full agonist at that site. As a consequence, caterpillar pests treated with tebufenozide express all the classic symptoms of an untimely and severe ecdysone overdose, including rapid inhibition of feeding (usually within 3 to 12 hours after exposure), premature apolysis, severe cuticle disruption, and permanent inhibition of ecdysis. The end result is quite effective caterpillar control at relatively low use rates (0.03 -0.3 lb/A). Tebufenozide, sold under the tradenames CONFIRM® and MIMIC™, is currently marketed around the world for control of such important pests as beet armyworm (*Spodoptera exigua*), cabbage looper (*Trichoplusia ni*), codling moth (*Cydia pomenella*), grape berry moth (*Lobesia botrana*), spruce budworm (*Choristoneura fumiferana*), rice leafroller (*Cnaphalocrocis medinalis*), and many others (*1,4*).

The Safety of Tebufenozide to Non-Target Organisms

Although tebufenozide has very high inherent toxicity to caterpillars, it has satisfyingly low toxicity to a wide range of vertebrates, such as mammals, birds, amphibians, and fish (*4*). As illustrated in Table I, representative mammal and bird species (rat and quail) are, at minimum, four orders of magnitude less susceptible to tebufenozide than is a representative caterpillar species, southern armyworm (*Spodoptera eridania*). Older broad spectrum neurotoxic insecticides such as the organophosphates or synthetic pyrethroids would typically have much less favorable insect/vertebrate selectivity ratios.

Tebufenozide has very low acute toxicity to range of laboratory mammals when administered orally, dermally, or by inhalation (*1,4*). It is also well tolerated at high dose in both sub-chronic and chronic dietary exposure studies, and is non-irritating, non-mutagenic, non-carcinogenic, and free of adverse reproductive and growth effects (*4*). It is one of the safest insecticides on the market today.

In retrospect, tebufenozide's relative lack of vertebrate toxicity might have been anticipated since such organisms do not biosynthesize or utilize 20E, EcR, USP, or any

Table I: The Relative Acute Oral Toxicity of Tebufenozide to Representative Caterpillar, Mammalian, and Avian Species

Organism	Acute LD$_{50}$ (mg/K)	Selectivity Ratio[a]
Southern Armyworm[b]	0.125	---
Rat	>5,000[c,d]	>40,000 X
Quail	>2,250[c]	>18,000 X

a) Selectivity ratio = rat or quail LD$_{50}$/armyworm LD$_{50}$, b) 2nd instar *Spodoptera eridania* (J. R. Ramsay, unpublished observations) c) No mortality observed at this maximum tested dose. d) For comparison, the acute LD$_{50}$ of aspirin in rats is 1,100 mg/K (Merck Index).

other closely homologous substances. Likewise, vertebrates are known to be relatively unaffected by large exogenous doses of 20E or related phytoecdysteroids. Tebufenozide's high safety to certain non-arthropod invertebrates such as earthworms (4) and nematodes can be explained similarly.

Another of tebufenozide's major attributes is its extraordinary safety to non-target insects/arthropods. This unusual property makes tebufenozide an ideal tool for integrated pest management (IPM) programs and control of caterpillar pests in ecologically sensitive areas. The extent of tebufenozide's caterpillar selectivity is really quite remarkable. It is extremely safe to pollinators such as honeybees (1) and bumblebees (order Hymenoptera); a wide range of valuable aphid and mite predators, such as lacewings (5) (order Neuroptera) and ladybird beetles (1,6) (order Coleoptera); various caterpillar endoparasites (4,7,8) (order Hymenoptera); certain generalist insect predators, such as spiders (4) (Arachnida), predatory bugs (4, 9) (order Hemiptera) and dragonflies (10) (order Odonata); ecologically important aquatic insects, such as stoneflies (10) (order Plecoptera), caddisflies (10) (order Trichoptera), and mayflies (10) (order Ephemeroptera); and a number of predatory mites (1,4) (Acarina) important to IPM. The only insects, other than caterpillars, that are known to be significantly intoxicated by tebufenozide are the larvae of certain mosquitoes (11,12) and midges (order, Diptera). Even here, however, the level of toxicity observed is generally low and other closely related dipteran species, like housefly and fruit fly, are relatively tebufenozide-insensitive (12).

All insects are thought to biosynthesize ecdysteroids as well as proteins that are homologous to the EcR and USP proteins found in Lepidoptera. They are also presumed to actively utilize these substances to regulate molting. How then can tebufenozide be so toxic to one kind of insect and yet so safe to another?

The Probable Mechanism(s) for Tebufenozide's High Order of Caterpillar Selectivity

In order for tebufenozide to cause the symptoms of hyperecdysonism in any insect, it must first reach the EcR:USP target site in that insect and then successfully activate it, as previously described. To do this, tebufenozide must conceptually pass through a series of "gates", represented in Figure 3. First it must penetrate through the insect's cuticle or gut membrane and move into the blood (pass through gate 1). There it must resist rapid excretion and/or rapid metabolic detoxification (pass through gates 2 and 3). Then it must successfully diffuse from the blood into cells containing the EcR:USP target site (gate 4). Finally, tebufenozide must have the correct molecular features (exactly the right size, shape, conformational flexibility, and spatial distribution of electron density) to potently bind and activate the particular EcR:USP heterodimer present in these cells (gate 5).

The high caterpillar toxicity of tebufenozide strongly suggests that this compound can pass through gates 1- 5 in such insects with high efficiency. Likewise, the very low toxicity of tebufenozide to non-caterpillar insects suggests a relative inability to pass through one or more of these same gates, probably due to some inherent insect-to-insect differences in body design, body chemistry, and/or target site structure.

Although much work remains to be done in this area, the evidence collected to date strongly suggests that target site discrimination (success or failure to pass through gate 5) may play an important, and perhaps dominant, role in determining tebufenozide's high degree of insect selectivity. As illustrated in Table II, tebufenozide binds to the EcR:USP heterodimer derived from a representative lepidopteran species *(Plodia interpunctella)* with very high potency (actually, 70-fold higher potency than 20E itself) *(13)*. Thus, it is no surprise that this compound is highly toxic to most members of this insect order. However, tebufenozide binds with significantly lower potency relative to 20E to receptor preparations derived from the fruit fly *(Drosophila melanogaster)*, which is a marginally tebufenozide-susceptible dipteran species, and even lower relative potency to receptors derived from completely non-susceptible species such as boll weevil *(Anthonomus grandis)* *(13)* and an ixodid tick *(14)* *(Amblyomma hebraeum)*.

Differences in the transport and metabolism do not seem to play an important role in tebufenozide's lack of toxicity to coleopteran species. Studies in which a highly susceptible caterpillar species *(Spodoptera eridania)* and a non-susceptible larval beetle species (Mexican bean beetle, *Epilachna varivestis*) were fed equal amounts of ^{14}C-labelled tebufenozide showed surprisingly similar patterns of tebufenozide uptake, metabolism, and overall tissue distribution *(13)*. The minor differences observed in this study were clearly insufficient to account for the greater than 2,800-fold difference in tebufenozide susceptibility between these two species (LC_{50} = 0.7 ppm for *Spodoptera*, >2,000 ppm for *Epilachna*) *(12)*. Studies comparing beet armyworm (Spodoptera exigua) with Colorado potato beetle (*Leptinotarsa decemlineata*) yielded similar results *(15)*.

Other factors may well contribute to tebufenozide's safety to other non-target insects. For instance, there is some recent evidence to suggest that enhanced cellular export (poor cellular uptake, Gate 4) may additionally limit the toxicity of tebufenozide to *Drosophila* cells in culture *(16)*.

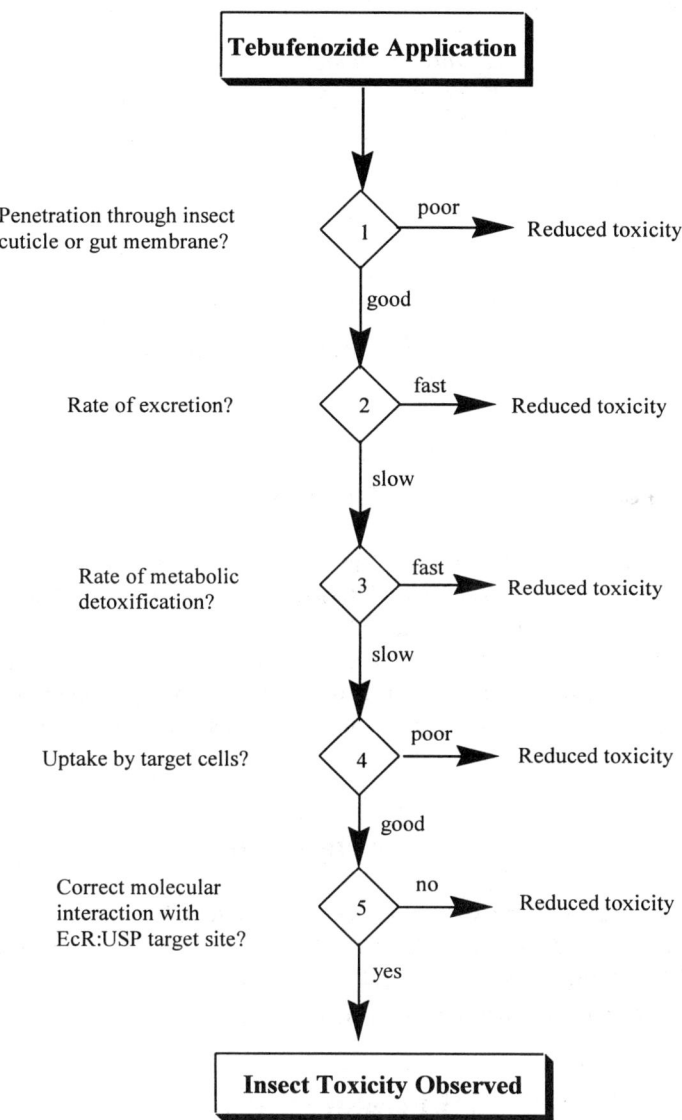

Figure 3. Conceptual "gates" determining the degree of tebufenozide-induced toxicity observed in a particular insect species.

Table II: The Binding Affinity of Tebufenozide Relative to 20E for EcR:USP Heterodimers Derived from Various Sources

Source of Receptor	Approx. X-Fold Binding Affinity of Tebufenozide Relative to 20E[a]	Relative Susceptibility of the Source Organism to Tebufenozide Intoxication
Caterpillar (*Plodia*)	70	+++
Fruit fly (*Drosophila*)[b]	0.25	+/0
Boll Weevil (*Anthonomus*)	0.02	0
Tick (*Amblyomma*)[c]	0.007	0

a) K_d of 20E/K_d of tebufenozide for the receptor in question, as determined by the relative ability of these ligands to displace tritiated Ponasterone A from that receptor. b) Kc cell extract. c) G. R. Charrois et al (1996)

Conclusion

Most of the unique and highly useful insecticidal and toxicological properties of tebufenozide, including its unusually high target specificity, seem to flow directly from its novel "ecdysonoid" mode of action. Other compounds in this mechanistic class, such as methoxyfenozide (*2,17*) and halofenozide (*2*), are currently being introduced into the marketplace. This mode of action will undoubtedly be further exploited in the future to create other highly target selective insecticides with outstanding performance characteristics.

Acknowledgments

Thanks to T. S. Dhadialla (Rohm and Haas Co) for tebufenozide receptor binding, metabolism, and tissue distribution studies. Thanks also to J. R. Ramsay (deceased), H. E. Aller, and M. Thirugnanam (Rohm and Haas Co) for information on the relative toxicity of tebufenozide to members of different insect orders.

Literature Cited

1: Heller J., Mattioda H., Klein E., Sagenmuller A., Brighton Crop Prot. Conf. Pests. Dis., 1: 59-65, 1992.
2: Dhadialla T., Carlson G. R., Le D. P., Ann. Rev. Entomol., 43: 545-69, 1998.
3: Hsu A. C. T., Fujimoto T. T., Dhadialla T. S., ACS Symposium Series 658, Phytochemicals for Pest Control, 206-219, 1997.
4: Long J. and Oakes R. L., CONFIRM 2F Technical Information Bulletin, Rohm and Haas Co, 1994.

5: Rumpf S., Frampton C., Chapman B., <u>J. Econ. Entomology,</u> 90(6), 1493-1499, 1997.
6: Hull L. A., <u>Arthropod Management Tests</u>, 20, 20-24, 1995.
7: Brown J. J., <u>Arch. Insect. Biochem. Physiol.</u>, 26, 235-48, 1994.
8: M. Thirugnanam, unpublished observations.
9: Smagghe G., Degheele D., <u>J. Econ. Entomology,</u> 88(1), 40-45, 1995.
10: Kreutzweiser D. P., Capell S. C., Waino-Keizer K. L., Eichenberg D. C., <u>Ecotox. and Envir. Safety</u>, 28, 14-24, 1994.
11: Song M. Y., Stark J. D., Brown J. J., <u>Environ. Toxicol. and Chem.</u>, 16(12) 2494-2500, 1995.
12: J. R. Ramsay, unpublished observations.
13: T. S. Dhadialla, unpublished observations.
14: Charrois J. R., Mao H., Kaufman R. W., <u>Pestic. Biochem. Physiol.</u>, 55(2), 40-49, 1996.
15: Smagghe G., Degheele D., <u>Pest. Biochem. and Physiol.</u>, 49, 224-234, 1994.
16: Sundaram M., Palli S. R., Krell P. J., Sohi S. S., Dhadialla T. S., Retnakaran A., <u>Insect Biochem. and Mol. Biol.</u>, 28, 693-704, 1998.
17: Le D. P., Thirugnanam M., Lidert Z., Carlson G. R., Ryan J. B., <u>Brighton Crop Prot. Conf. Pests Dis.</u>, 2:481-486, 1996.

Chapter 3

Iron-Complexed Dyes: Colorants in Green Chemistry

Harold S. Freeman and Laura C. Edwards

Department of Textile Engineering, Chemistry, and Science, North Carolina State University, Box 8301, Raleigh, NC 27695

Few textile and dye chemists would argue against the suggestion that textile wastewater containing toxic metal ions is a matter of considerable concern, and that a pollution prevention/source reduction approach to addressing this concern would be better than currently available waste treatment methods. With that point in mind, this chapter reflects work pertaining to the design and synthesis of metal-complexed dyes that contain potential replacements for metals now designated as "priority pollutants". The focus of the present report is on dye structures and their properties rather than the synthetic reactions employed. The goal of this research was the development of environmentally friendly metal complexed dyes. It was hoped that this would provide a green chemistry approach to minimizing the need to treat wastewater after the dyes are manufactured and/or applied to textiles, by eliminating the source of dye wastewater containing toxic metals. Our strategy was to identify alternative metal complexes that could replace chromium-, cobalt-, and copper-based synthetic dyes. This chapter includes discussion of the metals that are used to prepare metal-complexed dyes and the properties of key intermediates (ligands) from which the dyes are made.

Metal complex dyes are used in applications requiring very high stability in the presence of prolonged and repeated exposure to UV light and water (*1*). They are important colorants for carpet fibers that are used in automobiles or in other environments involving exposure to sunlight for extended periods. In the case of cotton fibers, the dyes employed also need to stay in place even when the substrate undergoes laundering. Examples include 1) CI Direct Blue 218 (**1**) and CI Reactive Violet 5 (**2**), both of which are dyes for cellulosic fibers, 2) CI Acid Blue 356 (**3**) and CI Acid Black 99 (**4**), acid dyes for polyamide and protein fibers, and 3) CI Acid Brown 98 (**5**) - a colorant for leather.

19

1

2

3

4

5

6 (CI Mordant Yellow 8)   7 (CI Mordant Blue 13)

While metal complex dyes are divided into roughly four classes (direct, reactive, acid, and mordant dyes) (2), this chapter deals with metal-complexed acid dyes for polyamide and protein substrates. Acid dyes derive their name from the fact that they are applied to polyamides and protein substrates from an acidic medium. Commercial dyers may apply acid dyes for wool fibers in the metal-complexed form (cf. **3-4**) or as unmetallized ligands such as **6-7**. In the latter case, metal complex formation (metallization) is conducted inside the fiber matrix, in what is commonly known as the mordanting process. Mordanting can be especially problematic because many of the dyes employed are metallized using chromium, which is quite toxic in the 6^+ oxidation state (3). In the mordanting process, an excess of chromium ions is required to obtain efficient conversion of dye ligands to the metal-complexed form. This leads to a significant level of toxic metal ions in the wastewater generated. Transferring the metallizing step from the hands of the dyer to the dye manufacturer, to produce "pre-metallized" dyes (4), reduces the toxicity of dyehouse wastewater but does not eliminate the problem. There remains the potential for unbound metal ions in effluents following the use of pre-metallized dyes, and effluents from dye manufacturing will also contain excess metal ions from the metallization step. Therefore, prudence dictates a search for potential alternatives to toxic metal ions and dyes derived therefrom.

Approach

The first step was to identify the structures of commercial dyes that were used in high volume. While it was deemed critical to reduce/eliminate toxicity, it was also important to maintain dye properties such as photostability and stability to exposure to water, particularly under laundering conditions. Consequently, the task of preserving the very good technical properties of the prototype dyes was an important design constraint. In view of the benign nature of iron and the fact that it has been used to make metal-complexed dyes for leather (5), the experimental work initiated by examining Fe(II) and Fe(III) analogs of well-known Cr and Co-complexed acid dyes (**8-13**).

Dyes **8-12** belong to the family of azo compounds, while **13** is a formazan dye. Formazan dyes are more commonly used as dyes for cellulose-based fibers, in which case they would contain a reactive moiety, and are 1:1 Cu-complexed dyes (1 metal per dye molecule) (6). Metal complexes of the 1:2 type (1 metal per 2 dye molecules) are used for polyamide and protein fibers. In the latter case, Cr and Co complexes predominate, although the metals employed have been designated as priority pollutants (7).

8 (CI Acid Yellow 151) **9** (CI Acid Orange 60)

10 (X = H; CI Acid Red 182) **12** (CI Acid Black 172)
11 (X = NHAc; CI Acid Blue 151)

Toxicity was evaluated in part as mutagenicity. The standard *Salmonella* mammalian mutagenicity assay (Ames test) (8-9) was conducted along with the Prival preincubation modification (10). The Ames assay was conducted with and without enzyme (S9) activation because azo dyes may undergo reductive cleavage by reductases to produce genotoxic aromatic amines (11), as illustrated in Figure 1 for the azo compound used to make dye **12**. The Prival modification was employed because it is more effective at carrying out the reductive cleavage of azo bonds to give a pair of aromatic amines, one of which was used in the synthesis of the starting dye. This insures that the genotoxicity of the parent dye and its potential metabolites (reductive-cleavage products) is assessed.

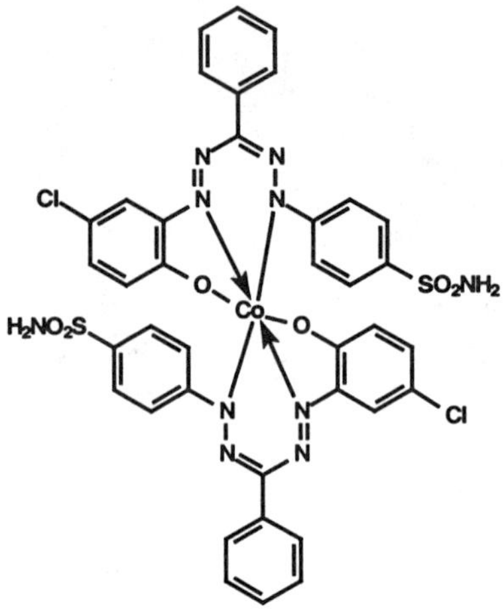

13 (CI Acid Black 180)

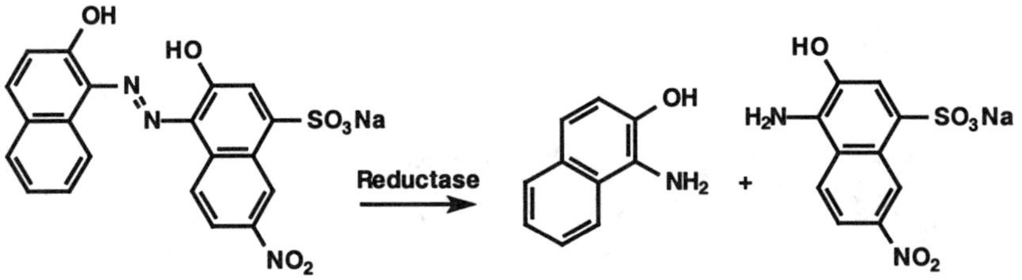

Figure 1. Reductive-cleavage of azo dye 14.

There was also testing of aquatic toxicity, using a plant protocol involving *lemna minor* (*12*), commonly known as "duckweed", instead of an animal protocol such as *Ceriodaphnia dubia* (*13*). The plant protocol was used because of its sensitivity to metal ions. The desired endpoint was an EC_{50} concentration $\geq$ 300mg/L. Aquatic toxicity testing was performed on 1) the metal complexed dyes and their unmetallized ligands, 2) the metal ions commonly employed in the metallization step, and 3) ozone decolorized dye solutions. The third set of analyses was conducted to address the concern that there might be a problem if metal ions were released as a result of ozone treatment of dye wastewater containing metallized dyes.

Results

Structure-Color-Fastness Relationships

Following the development of synthetic routes to Fe-analogs of dyes **8-13**, colors produced, stability to UV light (lightfastness), and laundering (washfastness) were assessed (*14-15*). It was clear that metallizing ligands **14** and **17-20** led to a significant broadening of the absorption spectra (*16*). While it was not possible to retain the yellow, orange, red, and blue colors of the prototypes (**8-11**) following Fe-complex formation, we found that the Fe-analogs of dyes **10** and **11** gave black colors on wool and nylon. This was an important finding, since Cr-free lightstable black dyes for wool were widely sought at the time. Interestingly, it was not possible to retain the black color of **12** when the corresponding Fe complex was made.

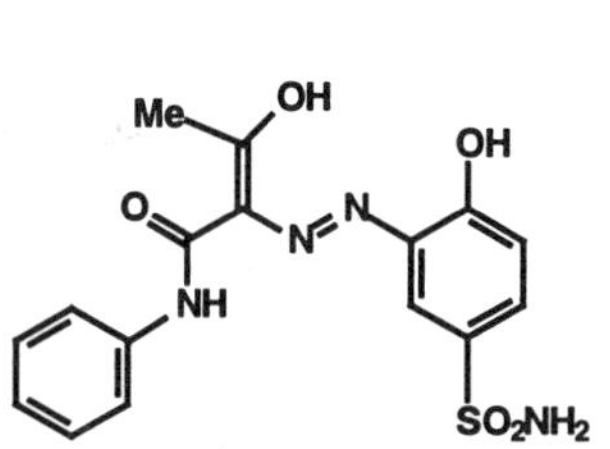

17

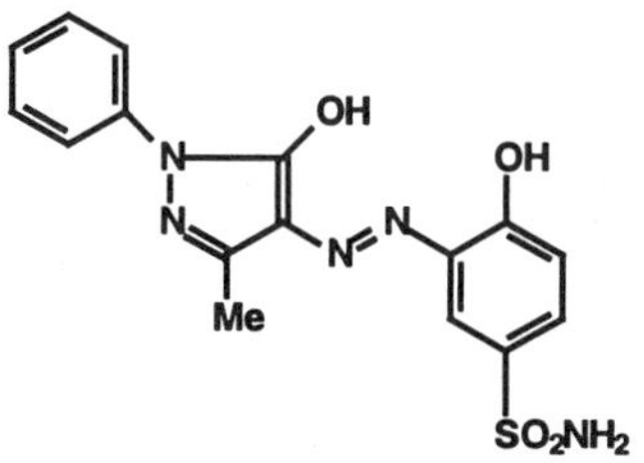

18

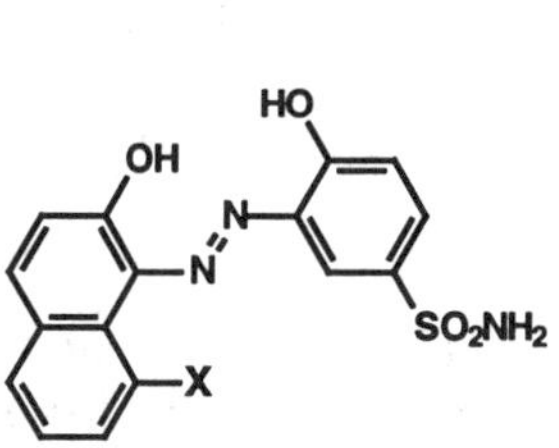

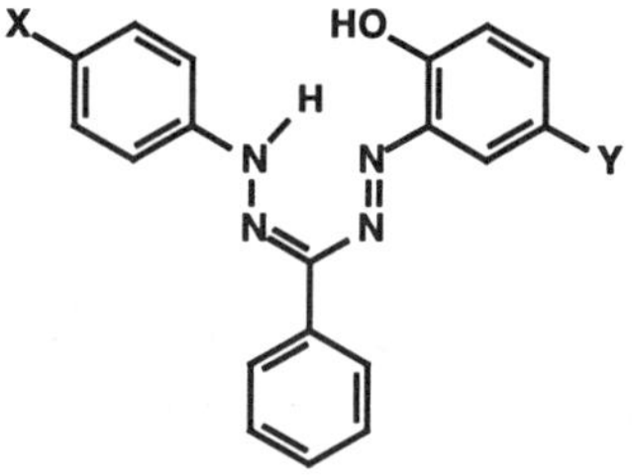

19 (R=H)
20 (R=NHAc)

21
X = SO$_2$NH$_2$, H, SO$_3$Na
Y = Cl, NO$_2$, H, SO$_2$NH$_2$

Using the lightfastness protocol employed to evaluate dyes for automotive applications (*17*) and a standard test for washfastness (*18*), we observed a reduction in lightfastness for the Fe-analogs of **8**, **9**, and **12**. However, **10** and **11** produced lightfast black colors on wool fibers. All of the Fe-complexed azo dyes exhibited very good washfastness on nylon and wool. With regard to Fe-analogs of formazan **13**, it was found that the ring substituents in the prototype dye could be varied to give Fe-complexed dyes having a variety of interesting properties. The substituents employed are listed with structure **21**, while the unmetallized formazans used most often in our work were **22-25** (Figure 2).

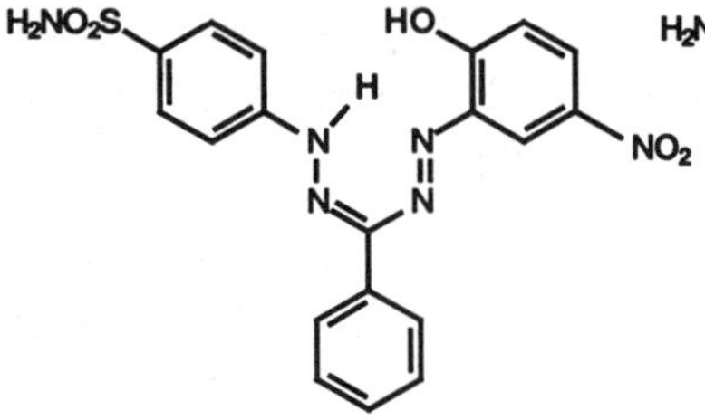

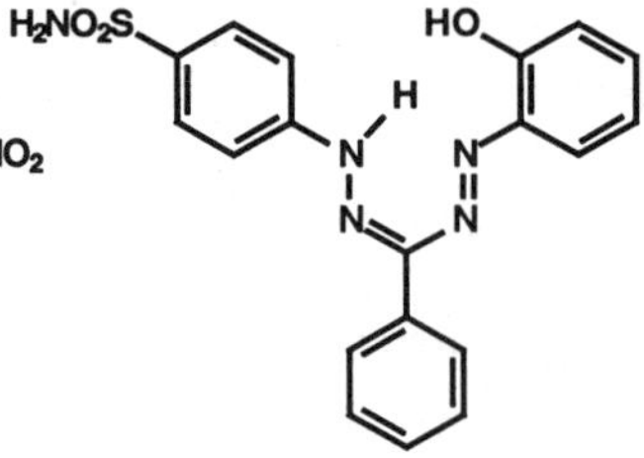

Figure 2. Formazan dye ligands used most frequently in this investigation.

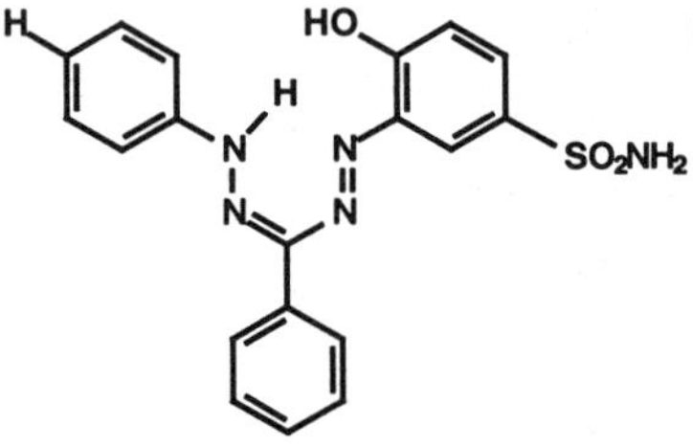

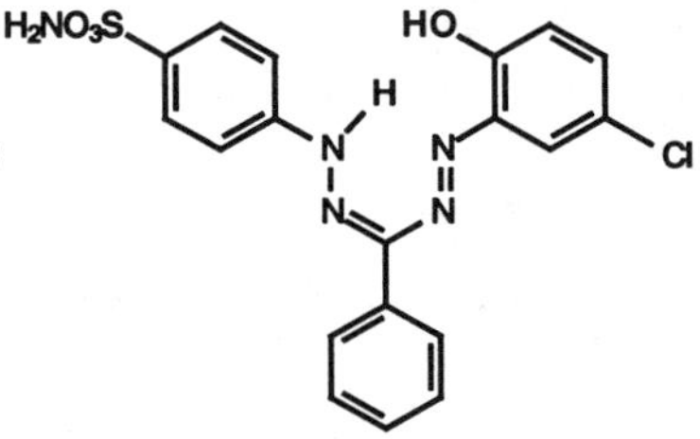

When used alone, dyes **22-25** produced black (**26**), bluish-violet (**27**), reddish-violet (**28**), and blue (**29**) 1:2 Fe-complexes, respectively. Note that blue dye **29** is the direct analog of prototype **13**. Thus, the formazan system provided the opportunity to generate red – blue colors with Fe used in place of Co and black dyes in the absence of Cr, a set of interesting achievements. Heretofore, it was generally believed that Fe-complexes were limited to muddy brown – brownish-black colors.

Like the Fe-complexed azo dyes described above, formazans **26-29** exhibited very good washfastness on wool and nylon. The ratings were 4.5–5.0 on a scale of 1.0–5.0, meaning that extremely little if any dye was removed during the laundering test. However, lightfastness varied widely in this group. Dyes **26-27** possessed very good lightfastness, while **28** was rated 'Fair' and **29** 'Good', in a standard test for lightfastness (*19*). It appears that presence of an SO_2NH_2 group in position-X of precursor **21** is important for lightfastness. This group is missing in dye **28**. Placing a nitro group in position-Y gave the best lightfastness; but this substitution broadened the absorption spectrum, resulting in the loss of colors in the red – blue region.

26
(Black Fe-complex from **22**)

27
(Blue-Violet Fe-complex from **23**)

28
(Red-Violet Fe-complex from **24**)

29
(Blue Fe-complex from **25**)

It should be noted that only dye **26** gave high lightfastness under the conditions employed in testing dyes for automotive use. In this case the evaluation of the black shades on nylon and wool was conducted at a 6% depth of shade. These test results constituted a significant step forward in our study, as the lightfast black shades produced with Fe-complexes of **19** and **20** were possible only on wool.

In addition to synthesizing symmetrical 1:2 Fe complexes of dyes **21**, we prepared unsymmetrical dyes such as **30-32**. In these examples, **23** and **25** were combined with **24** (cf. **30-31**) and dye **22** was combined with **25** (cf. **32**). Dyes **30** (violet) and **31** (bluish-violet) possessed low lightfastness, suggesting that formazan **24** should not be used alone or in combination with another formazan ligand when lightfastness in the Fe-complex is important. On the other hand, unsymmetrical dye **32** is a further example of a very lightfast black formazan for wool and nylon that arises from the use of dye ligand **25**.

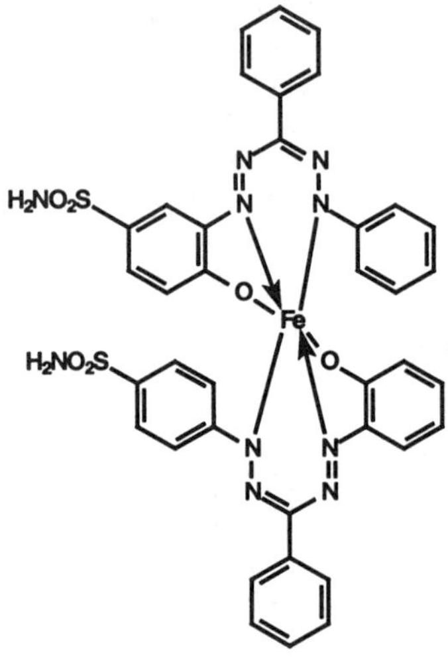

30 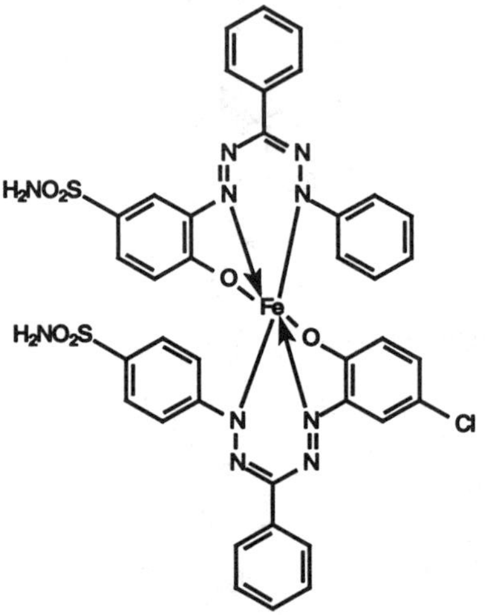   **31**

It should be pointed out that the unsymmetrical dyes, also known as "mixed" complexes, are statistical mixtures of the structure shown and the corresponding symmetrical complexes. For instance, the synthesis of dye **30** also produces lesser amounts of **27** and **28**.

Structure-Mutagenicity Relationships

Having demonstrated that Fe-complexed dyes that possess good technical properties could be prepared, the genotoxicity of these analogs was assessed. In this aspect of the study, genotoxicity was assessed in terms of mutagenicity. The test employed uses strains of *Salmonella* that cannot grow in the absence of histidine. Chemicals that alter these strains in a manner that causes them to grow (revert) are designated as mutagens. In the present experiments, a compound was deemed mutagenic if the number of revertant bacteria colonies produced was at least twice the number produced in the absence of the test chemical. Representative data from these studies are provided in Figure 3, which show that the Fe-complexed azo dyes possessed varying levels of mutagenicity. In the graphs that are labeled 3A-C, data are given for the unmetallized dye, the commercial dye, and the iron complex, successively.

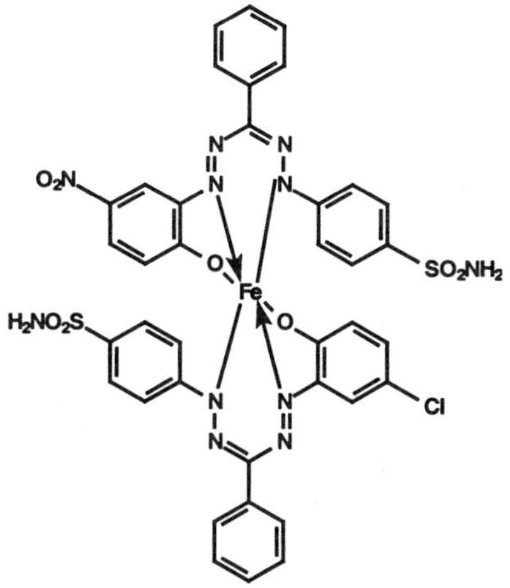

32

Results indicated that the yellow azo ligand (**17**) and the prototype 1:2 Co-complex (**8**) were nonmutagenic; however, the Fe-analog of **8** was a frame shift mutagen (i.e. active in TA98) following metabolic activation by S9 enzymes (cf. Figure 3A). On the other hand, in the orange dye series, the Cr-complex (**9**) was determined to be a frame shift mutagen. In this case, the unmetallized form (**18**) and the Fe-complex were nonmutagenic (cf. Figure 3B). Similarly, the Fe-complexed analogs of **10** and **11** were nonmutagenic. In the case of azo ligand **14**, the unmetallized dye, the prototype (**12**), and its 1:2 Fe analog were mutagenic (cf. Figure 3C). In this case, mutagenicity was attributed to the presence of an aromatic nitro group. These results were confirmed in the Prival assay.

Formazan ligands **21** and Fe-complexed dyes derived therefrom were nonmutagenic unless the structure contained an aromatic nitro group. In the latter case, the source of the mutagenicity was ligand **22**, an essential component in the formation of lightfast black Fe-complexes. Since the mutagenicity level was quite weak, we chose to retain **22** for further studies. It remains to be determined whether this dye ligand is also carcinogenic. This could be achieved *in vivo* using at least two animal models (e.g. mice and rats).

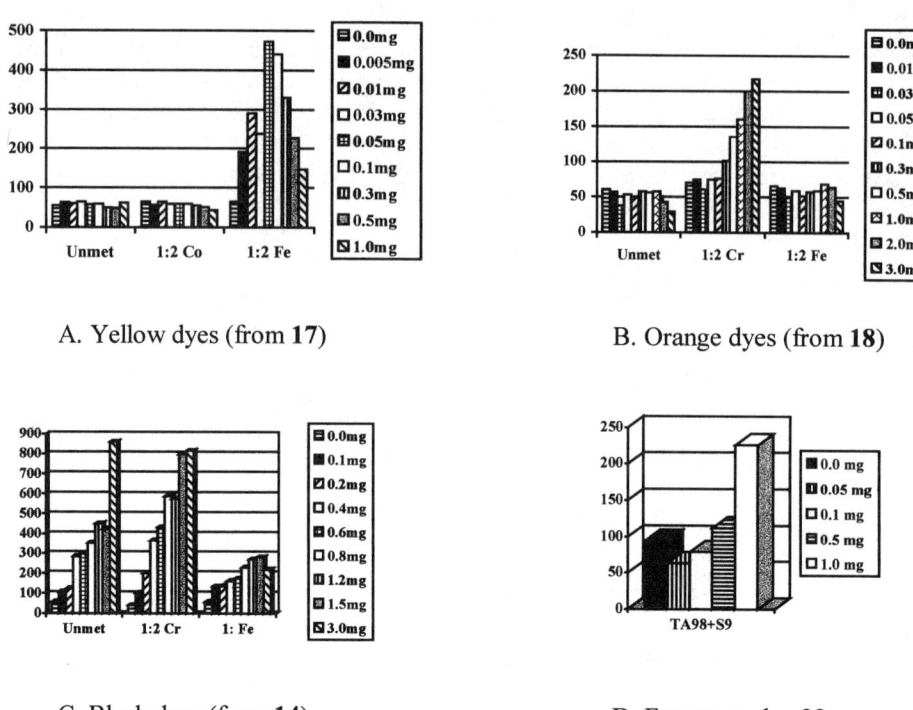

Figure 3. Representative mutagenicity data (number of revertants versus dye concentration), using TA98 with metabolic (S9) activation.

Structure-Aquatic Toxicity Relationships

The aquatic toxicity data generated in the *lemna minor* protocol was based on a seven-day test. After day three and day six the plants were placed in fresh dye baths, which simulated the periodic release of industrial wastewater into a local body of water. The dye and *lemna minor* plants were placed together at dye concentrations ranging from 10 to 1,000mg/L. Aquatic toxicity was assessed based on the survival rate of the initial plant leaves (fronds) and plant reproduction. The starting frond count was 22 to 24. In these experiments frond count was plotted against dye concentration on days 3, 6, and 7. A lack of aquatic toxicity was concluded if there were no adverse effects on frond count at 300mg/L concentration.

When the metal salts employed in dye complexation were tested, all except Fe(II) gave an EC_{50} below 300mg/L. The results are summarized in Figure 4. The EC_{50} values for $FeSO_4$, $Cr_2(SO_4)_3$, and $CoCl_2$, were 1,000mg/L, 36mg/l, and 14mg/L,

Fe(II) sulfate

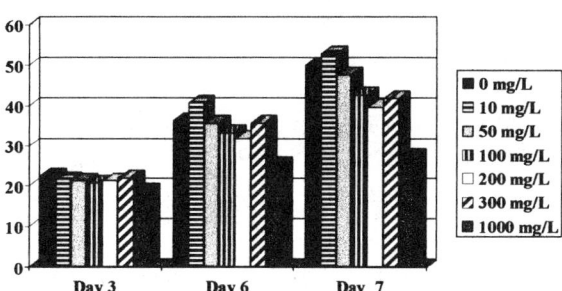

Co(II) chloride

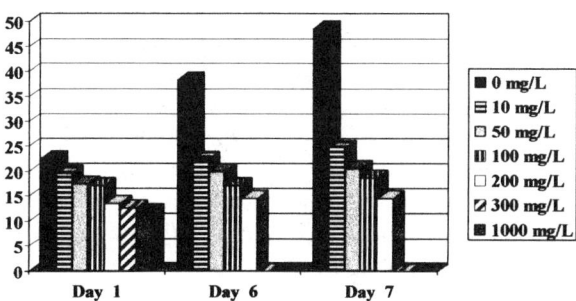

Cr(III) sulfate

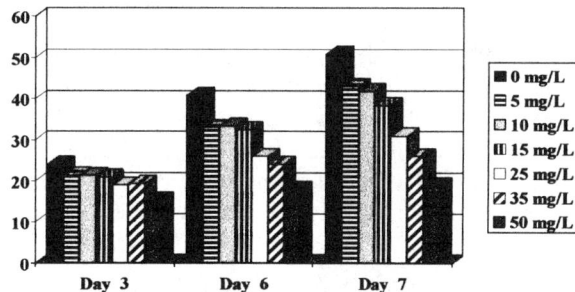

Figure 4. Plots of frond count versus concentration for metal salts employed in this investigation.

respectively. These data support the idea of considering Fe as an alternative to Cr and Co in dye-metal complex formation.

When the aquatic toxicity of the dyes was evaluated, in most cases the unmetallized dyes were sparingly soluble in water. Therefore, it was not surprising to find that the EC_{50} values were >1000mg/L (cf. **17**, Figure 5). An exception was dye **14**, which had EC_{50} = 122mg/L. Forming the Cr or Fe complex of dye **14** did not reduce toxicity. Substituting Fe for Co or Cr in the metallization of **17**, **19**, and **20** had a beneficial effect on aquatic toxicity, giving EC_{50} values of >1,000mg/L versus 100mg/L for **17**, 255mg/L versus 214mg/L for **19**, and >1,000mg/L versus 580mg/L for **20**. In the case of **18** no improvement was observed, but the Fe and Co complexes were both non-toxic.

In the formazan series, only the Fe complexes were examined. EC_{50} values of >400mg/L for **26**, 97mg/L for **27**, >300mg/L for **28**, >400mg/L for **29**, and >300mg/L for **32** were measured. These results indicated that the nitro group of dye ligand **22** does not adversely impact aquatic toxicity. This is important because of the very desirable color (black) and photostability of **26** and **32**. Similarly, the red/blue dyes were generally non-toxic. The exceptions (e.g. **27**) lack a substituent *para* to the phenolic oxygen atom.

It was also found that metals present in dye complexes used in this study did not contribute to the toxicity of wastewater solutions following treatment with ozone to decolorize the dye bath. This was surprising, until it was determined that the metals tended to form insoluble oxides following ozone treatment. This led to data of the type shown in Figure 6, where it is clear that the aquatic toxicity of the degraded dye (**9**) is significantly lower than what would be expected for the release of free Cr(III) ions. In this case the EC_{50} was 92mg/L. It was found that the pH of the decolorized solution was 3.15 and that the azo dyes gave toxic ozone-treated solutions whenever the pH was ≤3.5.

Conclusions

The results of this work indicate that it is possible to make iron complexed azo dyes for polyamide and protein substrates that are black, nontoxic and nonmutagenic, making the new dyes potential candidates for replacing commercial chromium-based black dyes. It is also believed that useful black dyes exist in the formazan series, though in this case the genotoxicity of a key dye ligand needs to be firmly established. Perhaps to the surprise of many, it has been shown that Fe can be used to produce non-traditional colors (red, violet, and blue) of reasonable brightness. There is now the need to enhance the photostability of these interesting new colors.

Acknowledgments

The authors are grateful to the contributions of Dr. Yolanta Sokolowska and Mr. Abraham Reife, and for funding from the National Textile Center. The aquatic toxicity testing was carried out in the textile engineering laboratories at Auburn University.

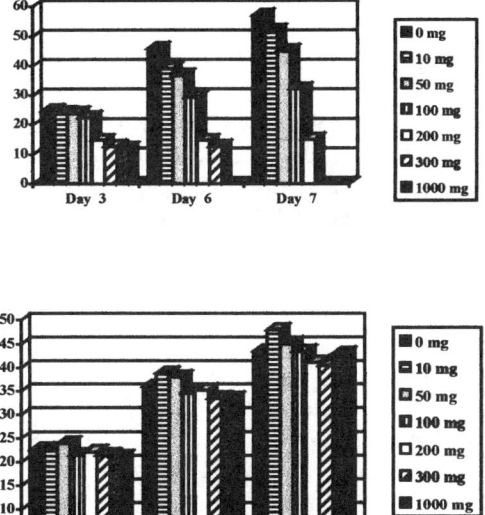

Figure 5. Aquatic toxicity data (frond count versus concentration) for dyes 14 (top) and 17 (bottom).

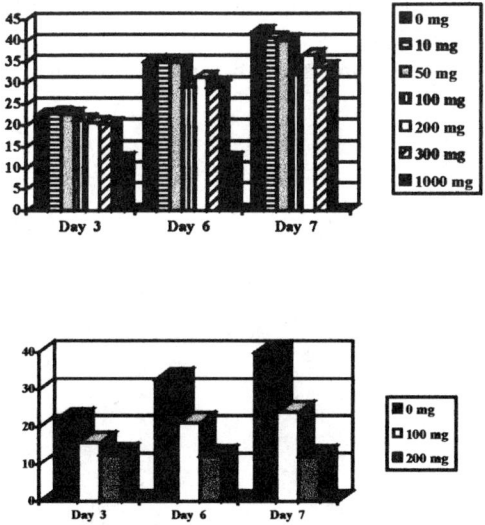

Figure 6. Aquatic toxicity data (frond count versus concentration) for Cr complex 9 before (left) and after (right) ozone decolorization.

Literature Cited

1. E.R. Trotman, *Dyeing and Chemical Technology of Textile Fibers*, 5th ed., Charles Griffin Publishers, 1975.
2. H. Zollinger, *Colour Chemistry*, 2nd ed., VCH Publishers, 1991.
3. A.C. Welham, *J. Soc. Dyers Colour.*, **102**, 126 (1986).
4. F. Beffa and G. Back, *Rev. Prog. Color. Relat Top.*, **14**, 33 (1984).
5. B.J. Berenguer, H.A. Holliger, and S.J. Rocus, Ger. Pat. 4133167 (1991).
6. R.E. Smith, *Kirk-Othmer Encyclopedia of Chemical Technology*, 4th ed., Vol. 8, 809 (1993).
7. *NIOSH Pocket Guide to Chemical Hazards*, NIOSH Publication No. 90-117 (1990).
8. B.N. Ames, J. McCann, and E. Yamasaki, *Mutat. Res.*, **31**, 347 (1975).
9. D.L. Maron and B.L. Ames, *Mutat. Res.*, **113**, 173 (1983).
10. M.J. Prival and V.D. Mitchell, *Mutat. Res.*, **143**, 75 (1979).
11. M.A. Brown and S.C. Devito, *Crit. Rev. Environ. Sci. and Technol.*, **23**, 249 (1993).
12. In *Federal Register*, 50(188), 39331 (1985).
13. I.R. Hardin, P.V. Wingler, P.J. Lasier, and M.S. Brewer, *AATCC Book of Papers – Int. Conf. & Exhibit.*, p.347 (1996).
14. H.S. Freeman, J. Sokolowska-Gajda, and A. Reife, U.S. Pat. 5,376,151 (1994).
15. H.S. Freeman, J. Sokolowska-Gajda, and A. Reife, U.S. Pat. 5,677,434 (1997).
16. A. Reife, E. Weber, and H.S. Freeman, *Chemtech*, **27**(10), 17 (1997).
17. J.E. Bullock and D.L. Garrett, *J. Ind. Fabrics*, **4**(2), 23 (1985).
18. Technical Manual, American Association of Textile Chemists and Colorists, TM107-1991, Vol. 67, 157 (1992).
19. Technical Manual, American Association of Textile Chemists and Colorists, TM16E-1990, Vol. 67, 33 (1992).

Chapter 4

In Vivo Synthesis in Yeast of Insect Sex Pheromone Precursors: An Alternative Synthetic Route for the Production of Environmentally Benign Insect Control Agents

Douglas C. Knipple, W. L. Roelofs, C.-L. Rosenfield,
P. Marsella-Herrick, and S. J. Miller

Department of Entomology, New York State Agricultural Experiment Station, Cornell University, Geneva, NY 14456

Elucidation of the chemical structures that constitute the sex pheromones of hundreds of species of moths has enabled the development of a highly selective, nontoxic pest control approach based on mating disruption. The widespread adoption of this approach is currently limited by the high cost of synthesizing pheromones due to the high chemical and stereospecific purity they require for biological activity. Toward the goal of enhancing the commercial prospects of this very promising, environmentally benign insect control technology, we have developed an alternative methodology to produce pheromones based on a partial enzymatic synthesis *in vivo* using a recombinant system. In this paper, we describe our rationale for developing this approach and the results of our initial efforts to produce useful unsaturated fatty acid precursors from recombinant yeast strains for the low-cost synthesis of pheromone products.

Need for Less Toxic Insect Pest Control Strategies

Environmental Aspects of Insect Pest Control

Insect pests affect virtually every major crop throughout the world and cause substantial losses to human food and fiber supplies. At present, the principal tools for controlling the insect pest species responsible for these losses are broad-spectrum neurotoxic insecticides. Despite their dominant position in the worldwide market for insect pest control products, these insecticides have a number of serious drawbacks, including their toxicity to humans and non-target organisms as well as their persistence in the environment. Another major problem associated with their use is the tendency of many pest species to develop genetically based resistance to the limited number of compounds that can be used against them, in some cases leading to catastrophic control failures. Thus, the development of novel pest control products that are specific, nontoxic, and cost-effective is both desirable and necessary to ensure both the continued productivity of worldwide agriculture and a safe environment.

Because the environmental and human health problems associated with insecticide manufacture and use are serious and widespread, government and public research universities have expended considerable effort to develop and implement integrated pest management (IPM) approaches to reduce the reliance on toxic chemicals for insect control (1). Such IPM strategies typically incorporate several pest control techniques, such as the use of pest-resistant host plants, more intensive cultural practices, and biological agents that kill insect pests. The latter include not only classical "biological control agents" such as predatory or parasitic insects, but also pathogenic bacteria and viruses that infect and kill their insect hosts. Interfering with insect reproduction using techniques such as sterile male release and mating disruption mediated by insect pheromones can also suppress insect pest populations below economic thresholds.

Economic Limitations to the Use of Pheromones for Insect Control

Many insect species have evolved a mate recognition system based on chemical communication (2-4). For example, in the lepidopteran species (moths), which includes many major agricultural pests, sexually mature females release a volatile sex pheromone from a specialized organ known as the pheromone gland. Male moths with highly specialized receptors on their antennae recognize the unique chemical signature of the pheromone released by a female of their own species, and fly upwind in an effort to locate the "calling" female and copulate with her.

Knowledge of the specific structures of insect pheromones has permitted their chemical synthesis, allowing their commercial use for the suppression of pest populations by mating disruption (5) (Figure 1). The practical use of pheromone-based pest control products typically takes one of two forms. Either the pheromones are used as discrete baits in traps so that males are drawn into the traps and removed

Figure 1. A male moth, flies upwind in a flight tunnel toward a sex pheromone lure (left panel). Upon landing on the lure, the male attempts to copulate with it (right panel).

from the population prior to mating or the pheromones are released over a large area, preventing mate location and mating by overwhelming the males' sensory system.

Pheromone-based pest control has several environmental advantages over the use of insecticides (5). First and foremost, each pheromone product targets a specific insect and is virtually nontoxic to vertebrates. Because of their low toxicity and high specificity, pheromone are extremely compatible with other IPM practices, in contrast to insecticides, whose effects are often at odds with IPM. From the standpoint of government regulation, many pheromone products are already registered for use as pesticides, and the expense and time required to register new pheromones are minimal compared to conventional insecticides. Finally, pheromones provide effective control at milligrams per acre and break down readily in the environment, whereas insecticides typically require kilograms per acre and often persist in the environment.

Despite their environmentally desirable qualities, pheromones account for less than 1% of the $6 billion worldwide market for insect control products (5). The two major factors that relegate pheromones to only a few minor market niches are, perversely, their high target specificity, which makes them relatively unattractive in crop systems requiring the control of multiple pest species, and their high cost of synthesis compared to insecticides. The latter factor represents a significant market barrier even in the best of circumstances where only a single pest species predominates.

The principal reason that pheromones are so expensive to synthesize is that they require very high chemical and stereospecific purity in order to elicit a biological response. Furthermore, each species' pheromone is comprised of unique chemical structures, which require the development of many discrete syntheses. Characterization of the pheromones of hundreds of moth species has revealed a remarkable diversity of unique chemical structures, the vast majority of which are aliphatic compounds with a specific chain length, one or more double bonds with specific location(s) and geometry (Z or E, alternatively designated *cis* or *trans*, respectively), and a terminal functional group such as an aldehyde, alcohol, or acetate (*2-4*). In general terms, pheromones are synthesized via coupling reactions that use moisture- and oxygen-sensitive organometallic reagents to establish the correct position(s) and geometry of one or more double bonds (*5,6*). These coupling reactions use especially pure feedstocks and solvents, require elaborate manipulations, and generate large amounts of organic wastes that must be treated. Thus, the coupling reactions typically drive the cost of the process. Moreover, because pheromones are effective in such small quantities, the cost of their synthesis via standard techniques is unlikely to decrease significantly if high volumes are produced.

Development of a Biologically-based Synthesis Route for Pheromone Production

Pheromone Chemical Structures and Biosynthetic Pathways

Because the high costs of synthetic pheromones limit their use, we were led to investigate the feasibility of cheaper, biologically-based synthetic processes. We chose the cabbage looper moth, *Trichoplusia ni*, for use as a model system to develop the technical precedents for this effort (*7*). This species is easy to rear in the laboratory, and the adult females have a relatively large pheromone gland that is a discrete eversible sack at the tip of the abdomen, which can be easily removed for biochemical and molecular biological studies.

The pheromone biosynthetic pathway of *T. ni* is well defined and is relatively simple compared to those of many other moth species (*8*) (Figure 2). The initial substrate is the common 16-carbon saturated fatty acid thioester of Coenzyme A (palmitoyl:CoA), which is derived from the combined actions of acetyl-CoA carboxylase and fatty acid synthase. An acyl-CoA Δ11 desaturase acts upon palmitoyl:CoA to produce a Z double bond between carbon atoms 11 and 12 (Z11-16:CoA). The multienzyme β-oxidation complex subsequently acts on this compound in two successive rounds of β-oxidation to produce Z9-14:CoA followed by Z7-12:CoA. The active component of the *T. ni* pheromone, Z7-12:OAc, results from the sequential action on Z7-12:CoA of a reductase and an acetyltransferase. A minor pheromone component, Z5-12:OAc is produced by the same enzymatic steps as

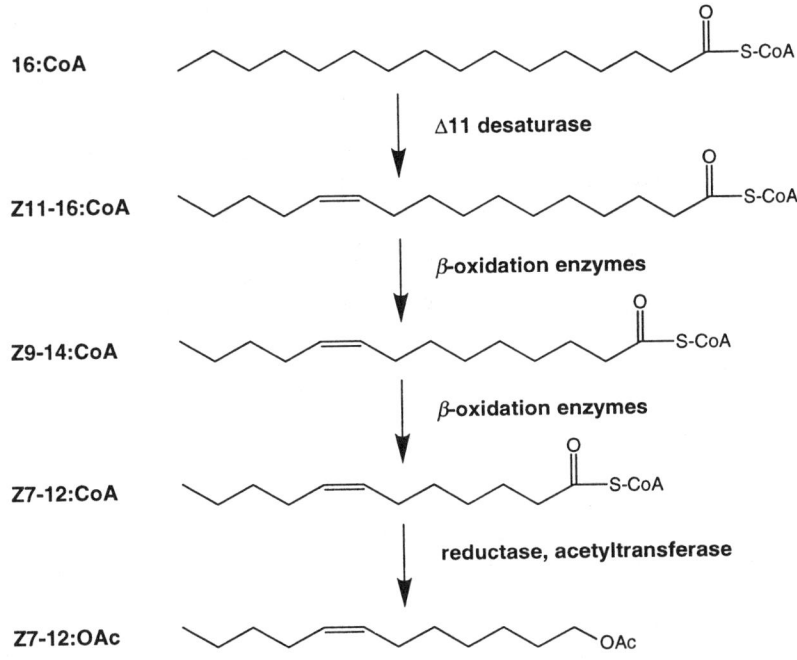

Figure 2. Biosynthetic pathway leading to the major active component of the T. ni sex pheromone.

shown in Figure 2, but beginning instead with the saturated 18-carbon fatty acid thioester of Coenzyme A (stearoyl-CoA) and having three rounds of β-oxidation after the desaturation reaction (not shown). The characterization of pheromone active components and their precursors from many other lepidopteran species has revealed a great diversity of chemical structures, reflecting evolutionary changes in pheromone biosynthetic pathways and in the functional properties of the individual enzymes comprising them.

Pheromone Desaturase Evolution and Homology to Metabolic Acyl-CoA Desaturases

Analysis of many lepidopteran pheromone biosynthetic pathways suggested that the evolution of novel catalytic mechanisms among pheromone gland acyl-CoA desaturases has played a major role in the generation of structural diversity of moth sex pheromone components. Diverse regio- and stereo-selective desaturation mechanisms have been found among the pheromone biosynthetic pathways, including those using Z9, E9, Z10, Z11, E11, Z12, E12, Z14, and E14 mechanisms (*2-4, 9,10*).

Despite this diversity of functional properties of insect pheromone desaturases, biochemical studies (*8,11*) indicate that these enzymes have many properties in common with the functionally conserved (Z9) acyl-CoA desaturases of yeast (*12,13*) and vertebrates (*14,15*). These biochemical similarities include a requirement for the electron transport proteins, cytochrome b5 and cytochrome b5 reductase, NADH, and molecular oxygen for catalytic function; an extremely high content of hydrophobic residues; and biochemical localization in the cell membrane fraction containing the endoplasmic reticulum. On the basis of these findings, we postulated that the genes encoding the commonly occurring metabolic acyl-CoA desaturase and the family of pheromone biosynthetic desaturases were descended from a common ancestral sequence, which predicts the existence of some structural conservation between them.

We exploited the existence of histidine-rich amino acid sequence motifs, conserved in both yeast and rat acyl-CoA desaturases and shown to be essential for catalytic function (*16*), to isolate a cDNA homolog (designated PDesat-TnΔ^{11}Z) from the cabbage looper moth pheromone gland (*17*). In brief, RNA was extracted from pheromone glands of sexually mature adult female moths and used to make cDNA by reverse transcription with an RNA-dependent DNA polymerase. Next, oligonucleotide mixtures possessing all the possible DNA sequences that encode two specific conserved amino acid sequence motifs (so-called degenerate primers) were used in the polymerase chain reaction (PCR) to replicate desaturase-homologous sequences present in the cDNA. The specific cDNA fragments that were replicated by this procedure were then hybridized to pheromone gland cDNA libraries in order to identify probe-positive clones containing full-length cDNAs. DNA sequencing of several isolated cDNA clones revealed a single large open reading frame encoding a 349-amino acid protein that has 52% sequence identity and 70% similarity (identical amino acids plus conservative substitutions) compared to the rat desaturase, and 30% identity and 50% similarity compared to the yeast desaturase. Conservation is somewhat higher in the core domain of the protein (delimited by the target sequences of the degenerate primers used in the PCR), which has 11 conserved histidine residues. The encoded protein is extremely hydrophobic, and its Kyte-Doolittle (*18*) hydrophobicity profile is consistent with transmembrane topology that is very similar to those of the yeast and rat desaturases.

Functional Expression of Cloned Pheromone Desaturase cDNAs

Having demonstrated a significant degree of structural conservation between the acyl-CoA desaturases of rat and yeast and the protein encoded by PDesat-TnΔ^{11}Z, we next tested the latter in a functional assay, both to establish its identity (as a desaturase possessing the Z11 catalytic specificity) and to establish a technical precedent for developing an alternative pheromone synthesis route. Two previously published methods were available to us for this purpose. One involves reconstituting a functional desaturase complex *in vitro* in an artificial membrane by combining the recombinant desaturase isolated from *E. coli* with phospholipids, biochemical fractions from bovine liver containing the electron transport proteins cytochrome *b5* and cytochrome *b5* reductase, NADH, and saturated fatty acid substrates (*15*). Although it is adequate for establishing the functional identity of desaturases encoded by cloned cDNAs, this procedure is labor intensive, requires a high degree of technical skill, and is relatively expensive. It is, thus, poorly suited for the practical goal of developing an alternative biologically-based synthesis for pheromone production.

The second method available to us for a desaturase functional assay involves the expression of the desaturase cDNA in a desaturase-deficient yeast strain *ole1* (*12,13*). This mutant has an absolute nutritional requirement for unsaturated fatty acids, which can be met by supplementing the growth medium with the natural Z9-16 and Z9-18 acids (palmitoleic and oleic acids, respectively) as well as with the unusual Z11-16 and Z11-18 acids (*12,13*). The *ole1* strain's unsaturated fatty acid auxotrophy can also be relieved by transformation with a rat stearoyl-CoA Δ9 desaturase cDNA (*13*). On the basis of these findings, we reasoned that an acyl-CoA Δ11 desaturase cDNA would also complement the *ole1* defect.

We tested the above prediction by constructing a yeast plasmid vector designed to express the PDesat-TnΔ^{11}Z cDNA under the control of the *OLE1* promotor (*17*) (Figure 3), similar to the previously described plasmid that expresses the rat cDNA

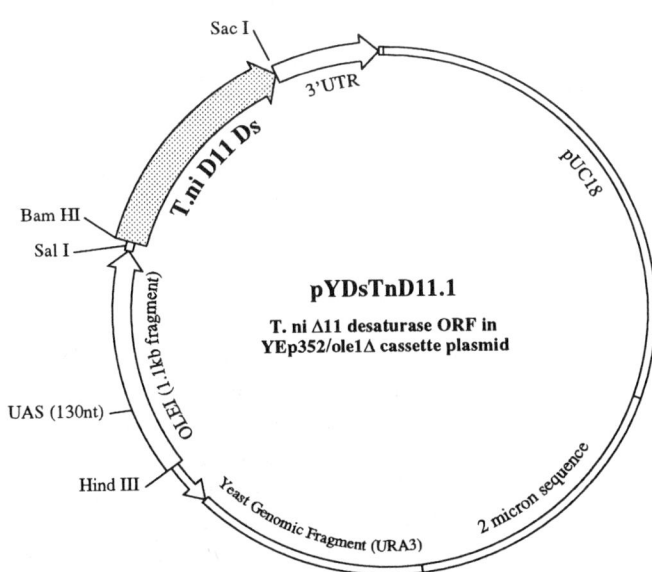

Figure 3. Structure of the yeast expression plasmid pYDsTnD11.1.

(*13*). Consistent with our prediction, *ole1* cells that were genetically transformed with the pYDsTnD11.1 expression plasmid were able to grow on medium in the absence of supplemental unsaturated fatty acids (*17*). The unsaturated fatty acids present in the total lipid extract of the transformed strain were characterized by GC/MS analysis of their fatty acid methyl ester derivatives (Figure 4). The major unsaturated products

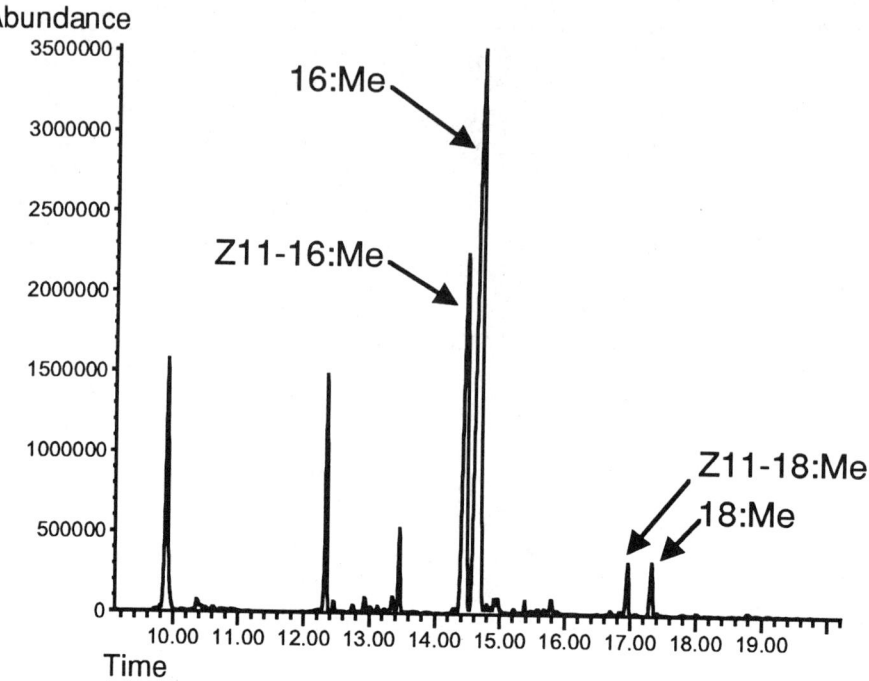

Figure 4. GC/MS total ion spectrum of fatty acid methyl ester derivatives of the lipid extract of pYDsTnD11.1-transformed ole1 yeast cells. (Reproduced with permission from reference 17. Copyright 1998 by The National Academy of Sciences.)

detected were (Z)11-hexadecenoic acid and (Z)11-octadecenoic acid in about a 10-to-1 ratio. The double bond positions of both products were confirmed by making dimethyl disulfide adducts and identifying the diagnostic degradation products in their mass spectra. There was no detectable (Z)9-hexadecenoic acid or (Z)9-octadecenoic acid present in the total lipid extract of the strain. This analysis demonstrated unequivocally that the PDesat-TnΔ^{11}Z encodes an acyl-CoA Δ11 desaturase Quantitation of the unsaturated fatty acid products (data not shown) showed that a liter of cells grown to about mid-log phase yields about 100 µg of (Z)11-hexadecenoic acid.

Conclusions and Future Directions

Our research has focused on developing basic knowledge and materials to reduce the cost of pheromone synthesis and, thus, increase the acceptance of environmentally benign pheromone products for insect control. Our results to date establish important technical precedents upon which an alternative commercial process can be developed to produce insect pheromone products (*17*). Specifically, we have established general procedures that enable us to isolate cDNAs encoding moth pheromone desaturases with novel catalytic properties and to demonstrate their functional expression in a desaturase-deficient yeast strain. The unique unsaturated fatty acid products from this system can be used directly as substrates in alternative pheromone syntheses, offering the potential of simplicity and reduced costs compared to conventional chemical syntheses. For example, the major unsaturated product of the the PDesat-TnΔ^{11}Z-expressing strain, (Z)11-hexadecenoic acid, can be reduced to (Z)11-hexadecenal, which is a major active component in several economically important pest species; and at least four other discrete pheromone products can easily be derived from this substrate. A preliminary economic analysis of an alternative commercial synthesis afforded by this system (*19*) indicates that these products can be produced for about a third the cost of conventional chemical syntheses.

Before commercializing this technology, it will be necessary to develop bioengineering process steps to isolate appropriate quantities of unsaturated intermediates. The yeast *S. cerevisiae* has a number of major advantages for this purpose. First, it is amenable to bioengineering process development and scale-up. Second, its fatty acids are relatively simple compared to those of other organisms, being predominantly 16 and 18 carbon atoms in length. The latter feature lends itself to the development of scalable purification procedures for the isolation of desired intermediates. Third, *S. cerevisiae* possesses well-defined metabolic pathways and powerful genetics, which makes it possible to make over-producing strains, for example, by modifying promotors and manipulating the expression of genetic loci encoding other metabolic enzymes. Accomplishment of these technical objectives has the potential to greatly enhance the commercial prospects of pheromone products, significantly reducing the release of toxic insecticides into the environment.

Acknowledgements

We acknowledge the financial support of our sponsors at the US EPA/NSF Environmental Technologies Initiative (BES-9728367), the USDA/ARS Competitive Grants Initiative (97-353024345), the NSF (IBN-9004979) and Verdant Brands, Inc. (formerly Consep, Inc.). We also wish to thank the following individuals for their helpful contributions to this work: Dr. Richard Pederson (Chemica Technologies) for his encouragement, helpful discussions and preliminary economic analysis; Dr. Timothy Dennehy (University of Arizona) for his encouragement and helpful entomological discussions; Dr. Charles Martin (Rutgers University) for his gift of yeast strains; Drs. Robert Staten (USDA APHIS, Phoenix AZ), Peter Landolt (USDA APHIS, Yakima WA) and Daryl Hathaway (USDA APHIS, Yakima WA) for their assistance in obtaining insects; and, Ms. Kathy Poole (Cornell University) for her technical support in insect rearing.

Literature Cited

1. Metcalf, R. L., *American Entom.* **1996**, *42*, 216.
2. *Pheromone Biochemistry*; Prestwich, G. D., Blomquist, G. J., Eds.; Academic Press, New York, 1987, p. 77.
3. Roelofs, W. L. & Wolf, W. A., *J. Chem. Ecology* **1988**, *14*, 2019-2031.
4. *Biocatalysis in Agricultural Biotechnology*, Whitaker, J. R., Sonnet, P. E., Eds.; American Chemical Society, Washington, D.C. 1989, p. 323.
5. *Behavior-Modifying Chemicals for Insect Management, Vol. 9* **1990**, Silverstein, R. M., Insco, M., Eds.; Marcel Dekker, Inc., New York, p. 1.
6. *The Total Synthesis of Natural Products, Vol. 9*; Simon, J. A. Ed.; Wiley-Interscience Publication, J. Wiley and Sons, New York, p. 1.
7. Tang, J., Wolf, W. A., Roelofs, W. L., Knipple, D. C., *Insect Biochem.* **1991**, *21*, 573.
8. Wolf, W. A., Roelofs, W. L., *Arch. Insect Biochem. Physiol.* **1991**, *3*, 45.
9. Foster, S. P., Roelofs, W. L., *Arch. Insect Biochem. Physiol.* **1988**, *8*, 1.
10. Zhao, C., Löftsedt, C., Wang, X., *Arch. Insect Biochem. Physiol.* **1990**, *15*, 57.
11. Rodriguez, F., Hallahan, D. L., Pickett, J. A., Camps, F. *Insect Biochem. Molec. Biol.* **1992**, *22*, 143.
12. Stukey, J. E., McDonough, V. M., Martin, C. E., *J. Biol. Chem.* **1989**, *264*, 16537.
13. Stukey, J. E., McDonough, V. M., & Martin, C. E., *J. Biol. Chem.* **1990**, *265*, 20144.
14. Thiede, M. A., Ozols, J., Strittmatter, P., *J. Biol. Chem.* **1986**, *261*, 13230.
15. Strittmatter, P., Thiede, M. A., Hackett, C. S., Ozols, J., *J. Biol. Chem.* **1988**, *263*, 2532.
16. Shanklin, J., Whittle, E., Fox, B. G., *Biochem.* **1994**, *33*, 12787.
17. Knipple, D. C.; Rosenfield, C.-L.; Miller, S. J.; Liu, W; Tang; Ma, P. W. K., Roelofs, W. L. *Proc. Natl. Acad. Sci. USA* **1998**, *95*,15287.
18. Kyte, J. & Doolittle, R. F. (1982) *J. Mol. Biol.***1982**, *157*, 105.
19. Bend Research, Inc., personal communication.

Chapter 5

Chrome-Free Single-Step In-Situ Phosphatizing Coatings

Tao Yu, Mary C. Whitten, Carmen L. Muñoz, and Chhiu-Tsu Lin[1]

Department of Chemistry and Biochemistry, Northern Illinois University, DeKalb, IL 60115-2862

The current process of organic coating on metals involves multiple steps and considerable energy, labor, and process control. It also generates toxic wastes such as chlorinated solvents, cyanide, cadmium, lead, and carcinogenic chromates. The green chemistry technology of *in-situ* phosphatizing coatings (ISPCs) developed in our laboratory is a one-step self-phosphating process in which the formation of a metal phosphate layer on the substrate surface and the curing of polymer paint film take place independently, but simultaneously. The generation of a metal phosphate layer *in-situ* essentially eliminates the metal surface pre-treatment step which employs a phosphating line/bath. The use of chemical bonds linked to the paint polymers to seal the pores of the metal phosphate layer *in-situ* enhances coating adhesion and suppresses metal corrosion without post-treatment final rinses containing chromium (Cr^{6+}). The successful application of ISPCs in three types of low-VOC commercial paints on bare and pre-treated cold-rolled steel and 2024 T3 aluminum coupons are presented in this chapter. The protective performance of ISPCs is shown to be superior to that of the current multi-step coating practice.

Introduction

The current practice of applying state-of-the-art organic coatings to metal substrates is a multi-step process. Normally, the metal surface is cleaned, phosphated or chromated, possibly sealed (with hot water or carcinogenic chromates) (*1,2*), dried, and finally painted. The surface pre-treatment process is error prone and costly, but it is necessary in the metal finishing industry (*3*) in order to enhance paint film adhesion.

[1]Corresponding author.

Unfortunately, multi-step coating technologies produce wastes including organic solvents, heavy metals, and other toxic and deleterious materials (4).

Recently, a novel surface conversion coating technique, namely "*in-situ* phosphatizing coatings (ISPCs)" has been developed in our laboratory (5-12). Using the ISPC process, phosphate conversion coatings and polymer films can be formed independently and simultaneously in a single-step. The simplicity of applying a single-coat phosphate/paint over the present multi-step coating practice makes the ISPC system attractive to many manufacturers of metal products. It provides increased quality without the capital and operating expense of a separate phosphating line/bath. In addition, the disposal of toxic wastes produced by the phosphating bath and waste treatment is avoided.

In this study, ISPC formulations of three low-VOC (volatile organic compound) commercial paints were made: a high-solids polyester baking enamel, a water-reducible alkyd paint, and a VOC-free thermoset acrylic latex system. The ISPC formulation's room-temperature shelf-storage stability was studied by monitoring its rheological behavior change upon aging. Thermal analysis was used to ensure the coating's successful crosslinking and film formation. The enhanced coating adhesion of ISPCs on cold-rolled steel and aluminum substrates was verified by cathodic delamination measurements and water immersion tests in a 3% NaCl solution. Both electrochemical impedance spectroscopy (EIS) and ASTM corrosion test standards (e.g., salt spray (fog) test) were used to evaluate the coating's corrosion resistance performance. The results are discussed in light of the environmental, economical, and technical advantages of the ISPCs in comparison to current multi-step coating practices.

Experimental

Three commercial baking enamels obtained from the Sherwin-Williams Company—a gloss ivory high-solids polyester-melamine paint (PERMACLAD® 2500) with a solids content of 82.4%, a water-reducible alkyd paint (KEM AQUA® 1400) with a solids content of 30-33%, and a solvent-free acrylic latex paint (KEM AQUA® 1800T) with a solids content of 50 ± 2% by volume—were used as the control paint formulas. The dry paint film of the commercial baking enamels was achieved by thermal curing at 163°C (325°F) for 15 minutes. An arylphosphonic acid (or arylphosphoric acid) together with a proper pH-adjusting-agent (e.g., an organic amine) were used as *in-situ* phosphatizing reagents (ISPRs) (7) to formulate the corresponding ISPC systems. The metal coupons tested for the control paints and ISPC systems were bare cold-rolled steel (CRS), iron phosphated (Bonderite 1000, BD), and iron phosphated plus Parcolene 60 chromated (BD+P60) mild steel panels (Q-PANEL Co., Cleveland, OH and ACT Laboratories, Inc., Hillsdale, MI), and bare and chromated 2024 T3 aluminum coupons.

The ISPC formulation stability was monitored by rheological measurements. Thermal analysis of cured paint films for the control paints and ISPC systems was conducted by differential scanning calorimetry (DSC) measurements. Electrochemical impedance spectroscopy (EIS) was used. This technique reduced the time needed to evaluate the corrosion protective performance of the organic coatings, and provides insight into the chemical nature of the failure mechanisms. The experimental details and instruments used for rheological, DSC, and EIS measurements have been described elsewhere (*10-12*). All measurements and tests reported here were conducted in duplicate, and the experimental reproducibility was verified.

In order to study the formation of interfacial metal phosphate layers, the untreated bare CRS panels were mechanically polished to a mirror finish and used as coating substrates. The ISPC was applied on mirror-finished CRS and cured at 163°C for 15 min. The polymer layer on each panel was removed without damaging the interface by soaking the panel in tetrahydrofuran solvent. The interfacial metal phosphate layer was rinsed with deionized water, dried, and characterized by a Bruker Fourier transform infrared (FTIR) spectrophotometer model Vector 22 equipped with a Spectra Tech FT-80 grazing angle accessory.

The disbonding resistance of the cured paints on metal substrates was studied by salt water (3% NaCl) immersion testing and also by cathodic delamination using the same apparatus as for the EIS measurements. In cathodic delamination, a 20.0 cm^2 area of cured paint film on the metal coupon was assembled in a delamination cell and used as the working electrode. The paint film was exposed to a 3% NaCl solution and polarized to -1.1 V *versus* a saturated calomel electrode (SCE) throughout all testing periods. Initially, a holiday (a hole of *ca*. 1.0 mm diameter) was drilled through the polymer film to the substrate to observe and measure the coating delamination expanding around the holiday. Dynamic control of the applied potential and data acquisition of the delamination area (or the delamination current at a fixed potential) as a function of time were obtained by using EG&G model 342C corrosion measurement software.

Results and Discussion

Rheology and Storage Stability of ISPCs

The rheological profile (not shown) of the solventless acrylic latex enamel (KEM AQUA® 1800T) indicated that no significant change in rheology from the original latex control was observed for the latex-ISPC formula. The viscosities of the latex-ISPC formula after various storage times at room temperature were measured as 82 ± 1 cp (freshly prepared), 110 ± 1 cp (two weeks storage), 175 ± 2 cp (one month storage), and 325 ± 3 cp (two months storage), whereas the corresponding values for the control latex formula were 108 ± 1 cp, 125 ± 1 cp, 194 ± 2 cp, and 368 ± 4 cp, respectively. Only a very small decrease in the viscosity was seen in the latex-ISPC formula, that may have resulted from relatively smaller latex particles. The reduction of particle radius in the latex-ISPC formula has been known to reduce "effective" hydration by latex particles, leading to a decrease in the apparent viscosity of the latex-ISPCs. However, the "up-and-down" shear rate cycle in the rheological profile was an appreciably smooth rheological response, indicating good paint stability for both latex-ISPC and control latex formulas.

The rheological profile of water-reducible alkyd ISPC (KEM AQUA® 1400) formula and its corresponding control coating sample was also recorded. The ISPC alkyd paint demonstrated a flow that was close to "Newtonian." There was a small shear thinning effect in the alkyd control formula, suggested by the splitting observed at the low shear end of the "up and down" shear rate cycle. The decreased viscosity value was considered to be a "memory" effect of its high shear history. Both rheological profiles remained smooth throughout the shear rate range tested, indicating that the alkyd control and ISPC-alkyd baking enamels displayed a homogeneous coating with no significant pigment flocculation (*13*).

When an ISPR-amine weak complex was dispersed into the KEM AQUA® 1400 control sample, the resultant water-reducible ISPC formula showed an apparent viscosity increase along with a clear shear thinning effect, while the shear stress-shear rate curve deviated from the straight line of "Newtonian" flow. However, no change in the viscosity of the water-reducible ISPC formula was observed during a long-term shelf-storage. Apparently, the initial viscosity change in the ISPC formula was not related to a chemical reaction of the ISPRs with the backbone polymers or oligomers, but rather was probably due to a physical modification of ionic strength in the water-based coating system by the incorporation of an ISPR-amine complex. The system's ionic strength in water-based paint normally affects the hydration status of the "hydrophilic" sites or segments of the alkyd resin, altering its molecular conformation, and in turn increasing the apparent viscosity of the paint formula.

Incidentally, the ISPC formula high-solids polyester-melamine paint (PERMACLAD® 2500) showed an initial increase in absolute viscosity at room temperature, owing most probably to the acid catalyzed polymer chain extension of polyester molecules in the formula. This is a classic problem with melamine baking enamels, and usually requires the addition of a volatile base to temporarily neutralize the acid (*14*).

Polymer Chemistry and Paint Film Quality

Crosslinking is one of the most important aspects of polymer film formation in coatings (*15*). It influences many paint film properties such as chemical stability, solvent resistance, network morphology, and mechanical properties (*15,16*). The degree of crosslinking can be monitored by DSC measurements, wherein the glass transition temperature (T_g), T_g span (T_f (glass transition end point) - T_i (glass transition start point)), and specific heat capacity jump (ΔC_p in mJ/mg·°C) are obtained. A higher T_g value and a lower ΔC_p during the glass transition are indicative of higher crosslinking density in a paint film (*17*).

The thermal properties obtained from DSC analysis of cured paint films, including the polyester control, ISPC polyester, alkyd control, ISPC alkyd, latex control, and ISPC latex baking enamels are listed in Table I. The T_g values are slightly higher for the ISPC formulas as compared to the control samples, indicating a relatively higher average polymer crosslinking density for the ISPC paint films of polyester and alkyd systems. The small increase in T_g value is consistent with a slightly lower heat capacity change (ΔC_p) associated with the glass transition. For latex systems, on the other hand, the observed thermal properties may be again attributed to the slight ionic strength increase in the latex-ISPC system following the addition of ISPR-amine weak complexes. The effect of ionic strength could promote a reduction in the packing distance of the latex particles in the latex-ISPC system, leading to a minor increase in latex particle coalescence and molecular entanglement. In all cases, the cured coating films of ISPC and control baking enamels had similar T_g values, and therefore the control and modified coatings were considered comparable.

Table I. Thermal Chemical Data for the Cured Paint Films of ISPC and Control Enamels

Enamel	T_g(onset) (°C)	T_g(offset) (°C)	T_g span (°C)	T_g (°C)	ΔCp (mJ/mg·°C)
Polyester Control	17.3	40.6	23.3	28.9	0.088
ISPC Polyester	20.6	44	23.4	32.5	0.085
Alkyd Control	18.2	54.0	35.8	36.1	0.173
ISPC Alkyd	17.6	60.8	43.2	39.2	0.163
Latex Control	14.0	32.7	18.7	23.9	0.070
ISPC Latex	16.4	33.1	16.7	24.9	0.069

In-Situ Phosphatization of Metal Surface

The dispersion of phosphatizing reagents in ISPCs was designed especially for synthesizing a metal phosphate thin layer *in-situ* at the interface of the metal substrate and polymer coating. The nature of metal phosphate bonds in ISPCs was found to have an acid-base type interaction, P-O⁻ -- M^{n+}, rather than an induced dipole interaction of the P=O/metal complex type (9). The metal phosphate products synthesized by thermal-cured solvent-borne polyester ISPC and acrylic latex ISPC on polished bare CRS coupons were investigated by means of reflectance FTIR spectra (not shown). The formation of a metal phosphate product was confirmed by peaks at 1073 cm^{-1} and 574 cm^{-1} for the solvent-borne system and at 1063 cm^{-1} and 561 cm^{-1} for the water-borne system, corresponding respectively to the υ_3 (stretching type) and υ_4 (deformation type) vibration modes of the phosphate group being distorted by the crystal field. For the water-borne system, the band intensity of the υ_4 mode was higher than that of the υ_3 vibration which is in contrast to the band intensity observed in the spectra of the solvent-based system. It is not known at this time why the metal phosphate layer synthesized by water-borne ISPCs has a different chemical nature from that produced by solvent-borne ISPCs. In principle, the higher spectral band intensity in the reflective mode of FTIR spectrum should correspond to a larger projection of electric dipole transition perpendicular to the substrate surface. The ISPRs in latex-ISPCs are situated in the hydrophilic phase of the latex emulsion system, whereas those in solvent-borne ISPCs are incorporated into a slightly polar solvent/polymer environment. Thus, the acid-base nature of ISPRs' interaction with the metal surface in latex-ISPCs is different from that in solvent-borne ISPCs. This may cause the differing bond dipoles of the metal phosphate layer seen in the FTIR spectra.

Besides the metal surface phosphatization, the pre-dispersed ISPRs have the function of reacting with backbone polymers in the coating to form covalent linkages with the polymer. This was confirmed spectroscopically by the FTIR peak around 944 cm^{-1}, corresponding to the P-O bond distortion caused by formation of the P-O-C linkage. In addition, by comparing the FTIR spectra of the metal-phosphate layer produced on a polished CRS substrate to those on the polished 2024 Al substrate, the P=O absorption band on the Al substrate was higher in frequency by about 50 cm^{-1}. This spectral shift to higher frequency was attributed to a higher bond order of the P-O bond due to the larger ionicity of the Al-O bond compared to the Fe-O bond (*18*).

Coating Protective Performance

Electrochemical impedance spectroscopy (EIS) has proven to be a powerful tool for the determination of coating performance and underfilm metallic corrosion (*19,20*). To show the effect of substrate pre-treatments on the coating protective performance, the control formula of solvent-free acrylic latex enamel was applied on three different types of cold-rolled steel panels: (i) untreated bare CRS panel, (ii) iron phosphated (B-1000) panel, and (iii) iron phosphated plus Parcolene 60 chromated (BD+P60) panel. A dry film thickness of about 1.1 mil cured at 163°C for 30 min was prepared. The coated panels were soaked in a 3% NaCl solution for 72 hours before the EIS measurements. Figure 1 shows the Bode-magnitude plots for latex control coating on bare CRS (curve 1a), B-1000 (curve 1b) and BD+P60 (curve 1c) panels. In the high frequency region, all three curves 1a, 1b, and 1c, were practically unified, as the impedance in this region is dominated by the pure dielectric properties of the non-conductive organic paint film. This indicated that the control latex dry film coated on three different types of substrates had similar dielectric properties.

In the low frequency region of Figure 1, the three Bode-magnitude curves split, as the impedance becomes more and more affected by the degree of pre-treatment at the coating/substrate interface while responding to the AC signal. In general, all curves deviated from the linear relationship of log |Z| vs. log f (f = frequency in hertz), and became more frequency-independent. A frequency-independent horizontal line in the Bode-magnitude diagram is characteristic of a pure resistor. At $f = 1.0 \times 10^{-2}$ Hz, the |Z| values were measured as 7.2×10^7 $\Omega \cdot cm^2$, 2.6×10^8 $\Omega \cdot cm^2$, and 5.3×10^8 $\Omega \cdot cm^2$ for curves 1a, 1b, and 1c, respectively. The results followed nicely the expectation that pre-treatment of iron phosphate on CRS tripled the |Z| value, and that the additional chromate rinse of B-1000 substrate further doubled the impedance value of the control

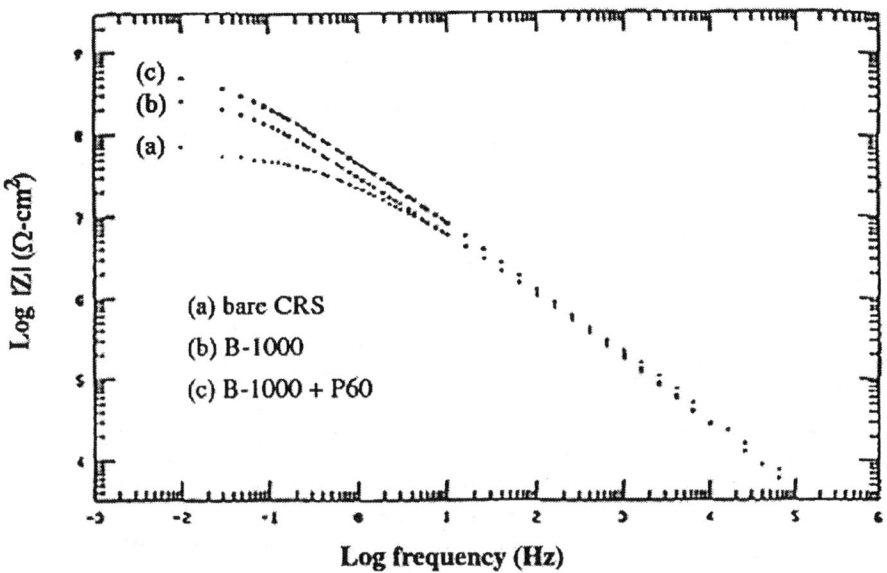

Figure 1. Bode-magnitude plots: Control formula of solvent-free acrylic latex baking enamel on (a) bare CRS, (b) iron phosphated, B-1000, and (c) iron phosphated and chromated, B-1000 + P60.

latex coating. It is commonly expected when measuring paint resistance (21) that, in most cases, a measured resistance of 1×10^8 $\Omega \cdot cm^2$ predicts good corrosion protection properties, a resistance of 1×10^6 $\Omega \cdot cm^2$ is poor, and paints with resistance between 1×10^6 and 1×10^8 $\Omega \cdot cm^2$ are borderline. In short, the latex control coated on pretreated substrates (curves 1b and 1c) had good protective properties, whereas that on the untreated panel (curve 1a) was borderline.

Upon application of the single-step latex-ISPC system to a metal substrate, the pre-dispersed phosphatizing reagent reacted effectively with the metal substrate. Figure 2 shows the Bode-magnitude plots of latex-ISPCs coated on bare CRS (curve 2a), on iron phosphated B-1000 (curve 2b) and on iron phosphated and chromated BD+P60 (curve 2c) coupons. All three curves appeared to overlap throughout the experimental frequency region, and all behaved as pure capacitors (i.e., the curve of log |Z| versus log f gives a straight line with a slope of -1). At $f = 1.0 \times 10^{-2}$ Hz, the |Z| value was measured as 1.0×10^{10} $\Omega \cdot cm^2$, 1.2×10^{10} $\Omega \cdot cm^2$, and 1.4×10^{10} $\Omega \cdot cm^2$ for curves 2a, 2b, and 2c, respectively. This observation clearly indicated that the cured latex-ISPC film on metal substrates (both untreated and pre-treated) could be classified as excellent in terms of corrosion protective performance. An increase in coating resistance (|Z|) of more than two orders of magnitude was observed for the latex-ISPC (curve 2a) as compared to the control latex formula (curve 1a) on bare CRS. This supported the conclusion that the *in-situ* metal phosphate layer was successfully generated at the metal/paint interface by the ISPC coating. Moreover, the corrosion inhibition of latex-ISPC on bare CRS (curve 2a, |Z| = 1.0×10^{10} $\Omega \cdot cm^2$) is equivalent to or better than that of the control latex formula on BD+P60 panel (curve 1c, |Z| = 5.3×10^8 $\Omega \cdot cm^2$). Similar results were also observed for the ISPC polyester and ISPC alkyd systems. The results provided strong evidence that a successful latex-ISPC could effectively replace the traditional phosphate/chromate bath/line.

The metal phosphate layer produced on 2024 T3 Al substrate had a higher bond order of the P-O bond than that on the CRS substrate. It is expected that the coating protection of ISPCs on 2024 T3 Al panel would be equal to or better than that on the CRS coupon. The Bode-magnitude plots for a cured paint film (about 1.1 mil dry film thickness) of polyester-melamine white paint (AKZO resin 26-1612) on bare and chromated 2024 T3 Al substrates recorded after soaking in a 3% NaCl solution for 72 hours and 2500 hours, are shown in Figures 3 and 4, respectively. The ISPCs on both bare and chromated 2024 T3 Al substrates displayed a pure capacitive behavior in Figure 3 and their polymer coatings remained intact even after 2500 hours of soaking in 3% NaCl solution, as evidenced in Figure 4. The polyester ISPCs painted on both the bare and the chromated aluminum panels displayed very high impedance values of $>10^{10}$ $\Omega \cdot cm^2$ at $f = 1.0 \times 10^{-2}$ Hz. Under the same processing conditions, the |Z| values obtained for the ISPC-polyester system were at least 1000 times higher than those of the polyester-control system. In Figure 3, the polyester control painted on bare Al gave a maximum impedance of less than 10^6 $\Omega \cdot cm^2$, which is considered indicative of a poor protective barrier (21). The polyester control painted on chromated Al substrate increased the impedance at low frequencies by only an order of magnitude.

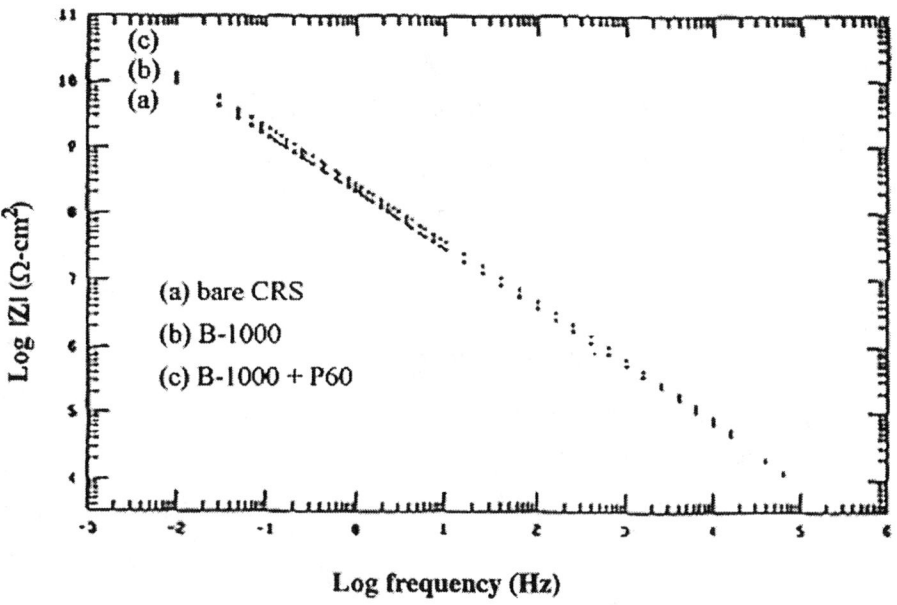

Figure 2. Bode-magnitude plots: ISPC formula of solvent-free acrylic latex baking enamel on different CRS coupons.

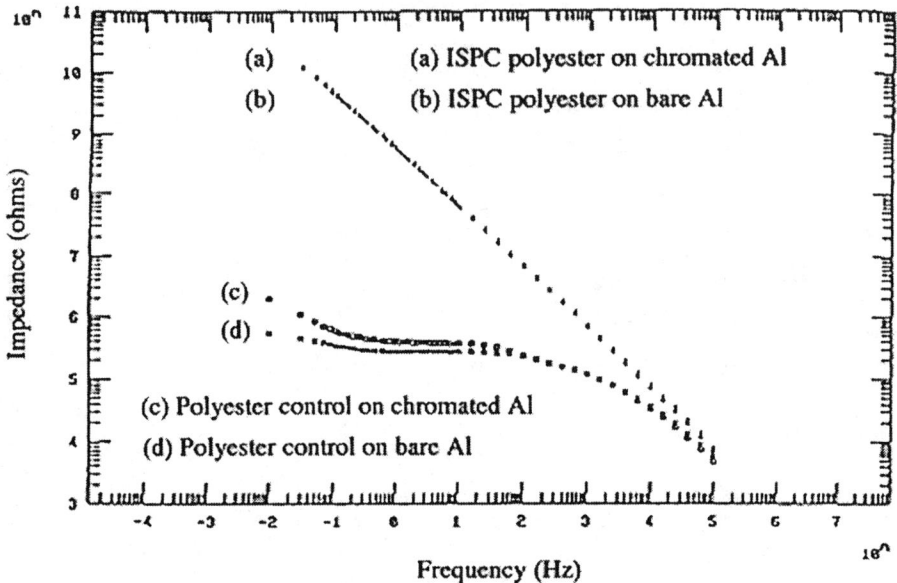

Figure 3. Bode-magnitude plots: Control and ISPC formulas of polyester-melamine white paint on 2024 T3 Al coupons after soaking in a 3% NaCl solution for 72 hours.

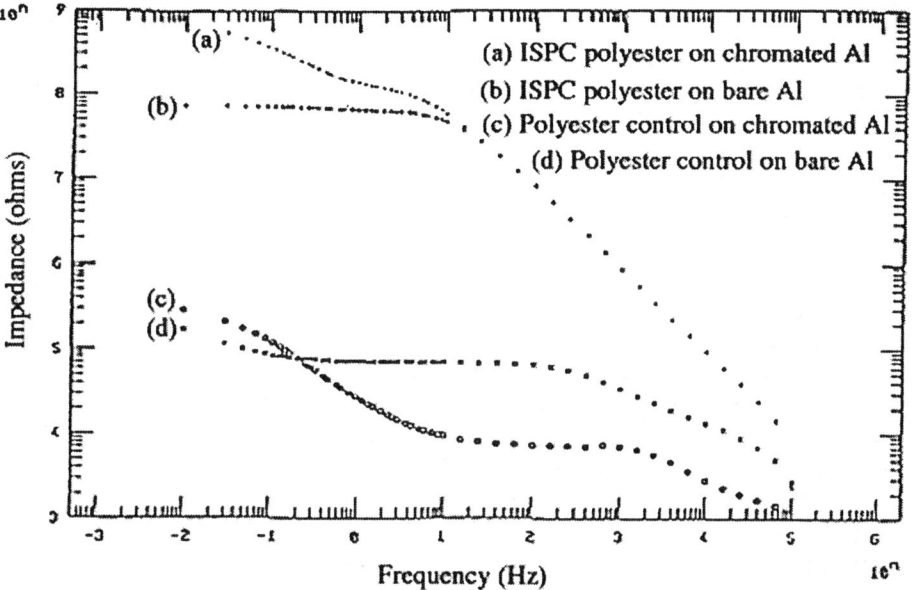

Figure 4. Bode-magnitude plots: Control and ISPC formulas of polyester-melamine white paint on 2024 T3 Al coupons after soaking in a 3% NaCl solution for 2500 hours.

Cathodic Delamination

In cathodic delamination, the delamination rate of an organic coating under a cathodic potential depends upon the applied potential, electrolyte solution, and metal substrate (22,23). Cathodic delamination tests were conducted for all three commercial paints. As an example, Figure 5 shows the cathodic delamination plots of control alkyd enamel and water-reducible ISPC coated on bare CRS coupons. Curves 5a and 5b represent delamination area for the control alkyd enamel and the water-reducible ISPC formulation, respectively; curves 5c and 5d are the respective plots of delamination current for the two formulations. The delamination test was conducted in a 3% NaCl solution; the alkyd coated CRS coupon served as a cathode and was polarized at -1.1V *versus* a saturated calomel electrode. A significantly slower delamination rate was obtained for the ISPC formula (curve 5b) as compared to the control alkyd enamel (curve 5a). At 44 hours of delamination time, the entire painted area of the working electrode (almost 20 cm × 20 cm) for the control alkyd paint had been delaminated, whereas the delamination area of the water-reducible ISPC was only as little as 1 cm^2, indicating a remarkable coating adhesion improvement for the alkyd ISPC painted on bare CRS coupon.

Studies of cathodic delamination processes indicate that the pH beneath the organic coating where the cathodic reaction occurs is highly alkaline (24-26). The *in-situ* self-phosphating layer on the metal surface produced in alkyd ISPCs may have made the substrate/alkyd film interface more acidic and thus more resistant to alkaline attack. The insulating effect of a metal phosphate layer generated on the substrate surface by ISPR greatly suppressed the delamination current, reduced the surface activity from the cathodic reaction, and thus prevented coating delamination. The right-hand portion of Figure 5 illustrates a remarkably lowered delamination current as recorded during the entire experimental process for ISPC as compared to the control alkyd formula. The enhanced cathodic delamination resistance of the alkyd ISPC paint film (a similar result was also obtained for the ISPC-polyester and ISPC-latex systems) on CRS panel represents a good coating/substrate adhesion and offers a superior corrosion protective performance as illustrated by the EIS measurements.

Water Disbonding Resistance and ASTM Salt Spray [Fog] Tests

The salt water immersion test (SWI) and the standard salt spray (fog) test (SS) were conducted for both ISPCs and control baking enamels applied on bare CRS, iron phosphated B-1000, and BD+P60 panels. The painted coupons of about 1.0 mil dry film thickness cured at 163°C for 15 minutes were X-cut through the film to the substrate and then either immersed in a 3% NaCl solution (SWI test) or subjected to a continuous salt-solution spray in a test chamber (SS test). After a specified duration of testing, the specimens were removed from the salt solution, and the coated surface was immediately dried. A DUCK® brand tape (Manco, Inc., Westlake, OH) was applied over the X-cut and then removed, and the protective performance was

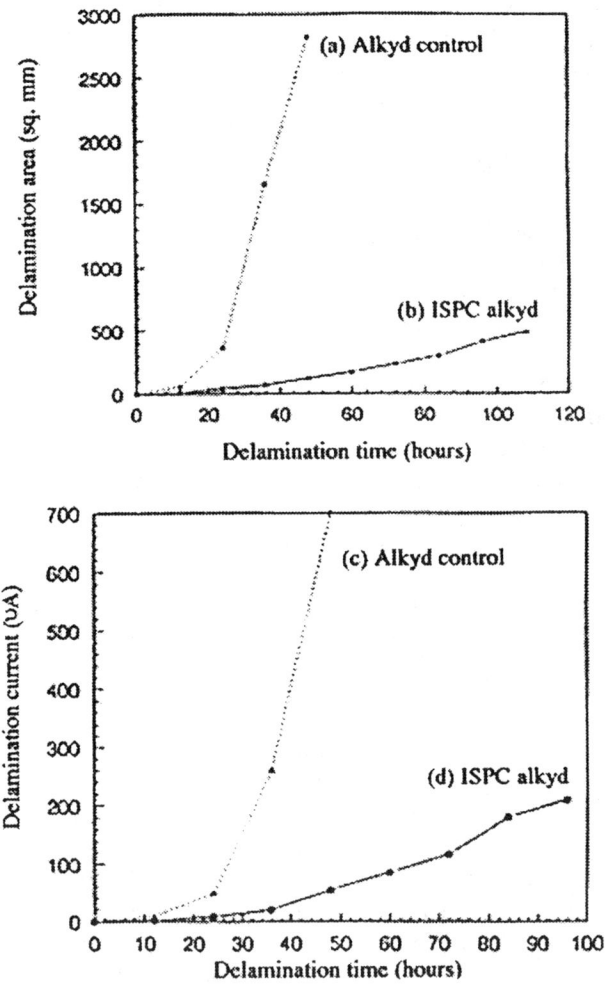

Figure 5. Cathodic delamination tests for the control and ISPC formulas of water-reducible alkyd baking enamel on CRS panel. Left: delamination area vs. delamination time. Right: delamination current vs. delamination time.

assessed quantitatively by measuring the distance of the paint disbondment (SWI test) and substrate corrosion (SS test) across the X-scribe.

Figure 6 shows the results of the salt spray (fog) test of corrosion resistance of the alkyd ISPC as compared to that of the control alkyd formula. On all three types of substrates, the control alkyd coating provided much less corrosion protection, whereas less paint deterioration was shown for the alkyd ISPC formula. The water-reducible ISPC applied on bare CRS panel (i.e., a single-step *in-situ* phosphatizing coating, Figure 6b) out-performed the control alkyd paint on pre-phosphated and chromated substrate (i.e., a state-of-the-art multi-step coating process, Figure 6e), indicating another advantage of the single-step ISPC technique over the traditional multi-step surface pre-treatment/coating process.

Measurements of blistering width across the "X" scribe (in mm) for SWI and SS tests are summarized in Table II. The enhancement of paint disbonding resistance (SWI test) and corrosion inhibition (SS test) in all three ISPCs is evident. A common phenomena observed for the B-1000 + P60 panels in Table II is that the chromate rinse (i.e., on the iron phosphated and chromated coupon) increased coating protective performance, justifying the market value of chromate rinses despite the costs of regulatory compliance. Moreover, the superior performance of the ISPC on this substrate, as shown in Table II and Figure 6f, indicates that the chemical bonds generated in ISPCs are capable of further sealing the pores of iron phosphated and chromated panel, thus providing additional coating adhesion enhancement and substrate corrosion inhibition.

Table II. Blistering Width Measured Across the "X" Scribe (mm)

Painted Coupons	High-Solids Polyester (144 h)		Water-Reducible Alkyd (100 h)		VOC-Free Acrylic Latex (200 h)
	SS	SWI	SS	SWI	SWI
Control on bare CRS	22	failure	failure	22–33	7.0–8.0
ISPC on bare CRS	5.0	28–40	4.0–7.0	3.0–7.0	3.0–5.0
Control on B-1000	15	35	26	29–32	7.0–8.0
ISPC on B-1000	3.0	8.0	3.0–4.0	3.0–5.0	3.0–5.0
Control on B-1000+P60	1.0	5.0	14	2.0–7.0	3.0
ISPC on B-1000+P60	< 0.5	2.0	<0.1	< 0.5	0.5–1.5

Conclusions

The innovative technique of *in-situ* phosphatizing coating (ISPCs) has been successfully applied to three commercial paints: (i) solvent-borne high-solids polyester, (ii) water-reducible alkyd, and (iii) VOC-free acrylic latex baking enamels. The technology of ISPCs is **a smarter chemistry** because the "simultaneous" chemical reactions of *in-situ* phosphatizing reagents (ISPRs) with metal substrate and polymer resin form a defect-free superior surface coating, without contamination of

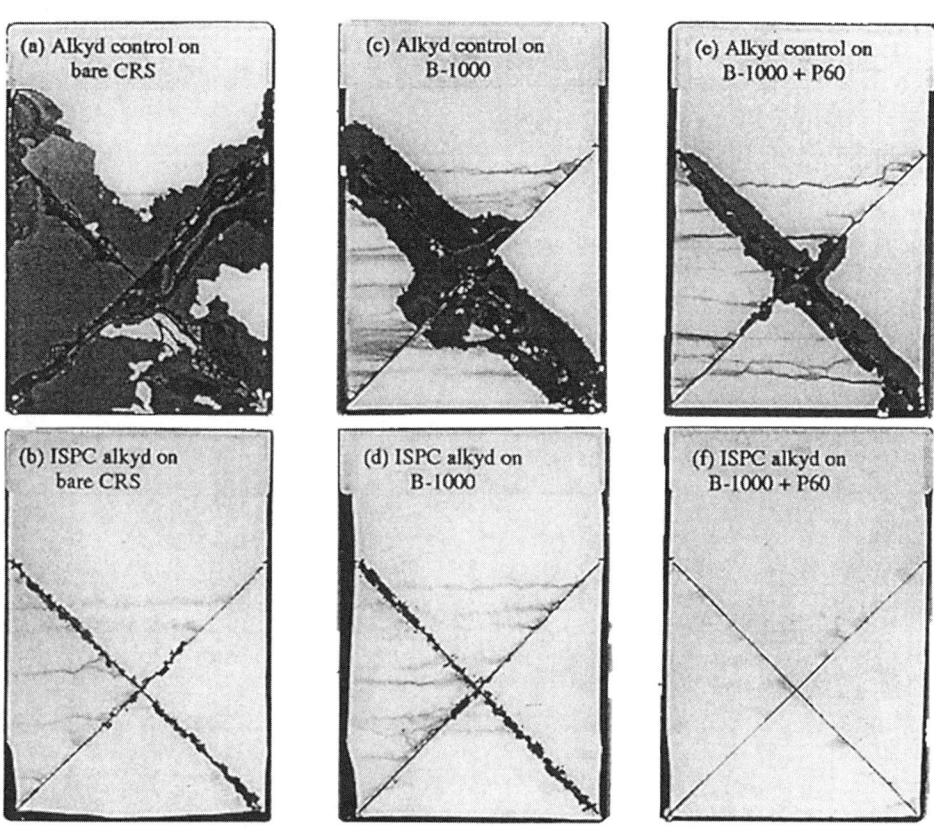

Figure 6. Salt spray (fog) tests (100 hours) for the control and ISPC formulas of water-reducible baking enamel on bare CRS, iron phosphated B-1000, and iron phosphated plus chromated, B-1000 + P60 coupons.

the pre-treatment by parts carrying impurities from bath to bath. The technology of ISPCs is **a cheaper chemistry** because the elimination of the phosphating/chromating line/bath saves time, energy, materials, and labor. The technology of ISPCs is **a cleaner chemistry** because the toxic wastes produced from the phosphating/chromating bath/line are avoided. With its use, landfills and treatment centers would see a reduction in hazardous waste, and transportation of such waste would be avoided.

Acknowledgments

Financial support is acknowledged from the National Science Foundation, Grant CTS-9312875; Illinois Department of Commerce and Community Affairs; Illinois Hazardous Waste Research and Information Center, ENR contract No. RRT 19; Cargill Corporation; Finishes Unlimited, Inc.; Rust-Oleum Corporation; and Sherwin-Williams Company.

Literature Cited

1. Freeman, D. B. *Phosphating and Metal Pretreatment;* Industrial Press, Inc.: New York, 1986.
2. Spadafora, S. J.; Hegedus, C. R.; Hirst, D. J.; Eng, A. T. *Modern Paints and Coatings* September 1990, pp. 36-48.
3. Rausch, W. *The Phosphating of Metal;* ASM International: Metal Park and Finishing Publications, Ltd.: Teddington, Middlesex, England, 1990.
4. "Paint Waste Reduction and Disposal Options," Hazardous Waste Research and Information Center, Champaign, IL, June 1992 (Vol. 1, HWRIC RR-060) and February 1993 (Vol. II, HWRIC TR-008).
5. Lin, C. T.; Lin, P.; Hsiao, M. W.; Meldrum, D. A.; Martin, F. *Ind. Eng. Chem. Res.* **1992**, *31*, 424.
6. Lin, C. T.; Lin, P.; Quitian-Puello, F. *Ind. Eng. Chem. Res.* **1993**, *32*, 818.
7. Lin, C. T. U.S. patent 5,322,870, 1994.
8. Lin, C. T.; Li, L. *Ind. Eng. Chem. Res.* **1994**, *32*, 3241.
9. Yu, T.; Li, L.; Lin, C. T. *J. Phys. Chem.* **1995**, *99*, 7613.
10. Yu, T.; Lin, C. T. *J.Coatings Tech.* **1999**, *71(892)*, 69.
11. Yu, T.; Lin, C. T. *J. Coatings Tech.*, **1999**, *71(895)*, 87.
12. Muñoz, C. L.; Yu, T.; Lin, C. T. In *Proceedings of the 25th Annual International Symposium on Waterborne, Higher Solids, and Powder Coatings;* Storey, R. F.; Thames, S. F., Eds.; University of Southern Mississippi: Hattiesburg, MS, 1998; pp. 332-343.
13. Patton, T. C. *Paint Flow and Pigment Dispersion: A Rheological Approach to Coating and Ink Technology*, 2nd ed; Wiley-Interscience Inc.: New York, 1979.

14. Wicks, Z.R. Jr.; Jones, F.N.; Pappas, P.S. *Organic Coatings: Science and Technology*; Wiley-Interscience, Inc.: New York, 1992; , Vol. I, Film Formation, Components, and Appearance.
15. Wicks, Z. W.; Jones, F. N.; Pappas, S. P. *Organic Coatings: Science & Technology*; John Wiley & Sons: New York, 1992; Vol. 1, Chapter 3.
16. Mijovic, J.; Fishbain, A.; Wijaya, J. *Macromolecules* **1992**, *25*, 979.
17. Montserrat, S. *Polymer* **1995**, *36*, 435.
18. Akiyoshi, O.; Takahashi, K.; Ikeda, M. *J. Mater. Sci. Lett,* **1984**, *3*, 36.
19. Scantlebury, J. D.; Ho, K. N.; Eden, D. A. In *Electrochemical Corrosion Testing*; Mansfeld, F.; Bertocci, U., Eds.; Amer. Soc. Test. Mater. (ASTM): Philadelphia, 1981; STP 727, p.187.
20. Atrivens, T. A.; Taylor, C. C. *Mater. Chem.* **1982**, *7*, 199.
21. Leidheiser, Jr., H.*Corrosion* **1982**, *38*, 374.
22. Leidheiser, Jr., H. *J. Adhesion Sci. Technol.* **1987**, *1*, 79.
23. Sugana, T.; Carciello, N. R. *J. Appl. Polymer Sci.* **1993**, *50*, 1701.
24. Castle, J. E.; Watis, J. F. In *Proc. Natl. Assoc. Corrosion Engrs.;* Leidheisr, Jr., H., Ed.; National Association of Corrosion Engineers: Houston, TX, 1981; Vol. 28.
25. Ritter, J. J.; Kruger, J. In *Proc. Natl. Assoc. Corrosion Engrs.;* Leidheiser, Jr., H., Ed.; National Association of Corrosion Engineers: Houston, TX, 1981; Vol. 28.
26. Dickie, R. A.; Hammond, J. S.; Holubka, J. W. *Ind. Eng. Chem. Prod. Res. Dev.* **1981**, *20*, 339.

Green Chemical Synthesis

Chapter 6

Water as Solvent for Organic and Material Synthesis

Chao-Jun Li

Department of Chemistry, Tulane University, New Orleans, LA 70118

Water as a potentially benign solvent for reactions and syntheses is discussed. While most laboratory and industrial organic reactions have been carried out in organic solvents, organic reactions in water have attracted more and more attention within the last decade. This resurgence is partly due to the drive for the development of more environmentally safe reaction processes. The use of water for organic and material syntheses, among many other benefits, reduces the volume of organic solvents used in various processes which will result in pollution-prevention.

Introduction

When talking about organic and chemical synthesis in water, probably the first question that many people might ask is "why water?" Perhaps a better question is: "why not water?" This question was asked by Professor Breslow in the early 1980's during his study of Diels-Alder reactions and is an area of considerable research interest in the chemical community today. After all, aqueous chemistry is one of the oldest forces of change in the solar system (1), and water has been the solvent for all the biological transformations in nature.

Although it is possible not to use solvents in certain reactions for carrying out syntheses the use of solvents is necessary in most occasions for mixing purposes and to avoid by-product formations. When selecting a solvent, water is a good choice for many reasons (2). The most obvious one is cost. Water is the most abundant solvent on earth; therefore, it is much cheaper to use a water solvent for both chemical synthesis and material synthesis. Second, water is not flammable, toxic, or explosive. Third, it is related to synthetic efficiency: when organic synthesis is carried out in water, it may be possible to avoid having to protect functionalities such as hydroxy groups that would potentially increase the overall efficiency of a synthetic scheme. Also, consider operational convenience: if a large scale synthesis is carried out in water in industry, the organic product (which is often insoluble in water) can, in principle, be simply isolated by phase separation. Finally, using water as a solvent has the potential to reduce organic emissions to the environment. Despite these advantages, the most abundant substance on earth is probably the least explored solvent for organic synthesis.

Reactions That Can Be Carried Out in Water

Another question people may ask is: what kind of reactions can be carried out in water? To put it simply, we know that most major reaction types that have historically been carried out in anhydrous organic solvent in organic chemistry, have their counterparts in water. Among these reactions types, the more notable ones are pericyclic reactions, organometallic reactions, transition metal catalyzed reactions, and lewis acid catalyzed reactions.

Pericyclic Reactions

In 1980, Breslow (3) made the dramatic observation that the reaction of cyclopentadiene with butenone in water was more than 700 times faster than the same reaction in isooctane. The reaction rate in methanol is comparable to that in a hydrocarbon solvent. Such an unusual acceleration of the Diels-Alder reaction by water was attributed to the "hydrophobic effect" (4) in which the hydrophobic interactions brought together the two nonpolar groups in the transition state. The use of β-cyclodextrin, which simultaneously forms an inclusion complex with the diene and dienophile, and the use of 4.86 M LiCl aqueous solution as solvent, which salts out nonpolar materials dissolved in water (5), further enhanced the rate of aqueous Diels-Alder reactions. The second-order rate constant of the reaction between hydroxymethylanthracene and N-ethylmaleimide in water at 45°C was over 200 times larger than in acetonitrile (eq. 1).

In addition, Grieco has contributed extensively to the studies of aqueous Diels-Alder reaction toward the syntheses of a variety of complex natural products. When the Diels-Alder reaction in Scheme 1 was carried out in water, a higher reaction rate and reversal of the selectivity were observed, compared to the same reaction in a hydrocarbon solvent (6). It should be noted that, for the aqueous reaction, the sodium salt of the diene was used. Water soluble co-solvents caused a rapid reduction in rate. The best result was obtained when the reaction was conducted with a five-fold excess of the sodium salt of diene carboxylate.

In addition to the study of Diels-Alder reaction in aqueous media, Grieco and co-workers also studied the hetero Diels-Alder reactions (7) with nitrogen-containing dienophiles and [3,3]-sigmatropic rearrangement extensively. For example, the rearrangement of an allyl vinyl ether in water generates the aldehyde efficiently in water (eq. 2) (8). The corresponding methyl ester underwent the facile rearrangement similarly. A solvent polarity study on the rearrangement rate of the allyl vinyl ether was conducted in solvent systems ranging from pure methanol to water at 60°C. The first order rate constant for the rearrangement of the allyl vinyl ether in water was 18×10^{-5} s^{-1}, compared to 0.79×10^{-5} s^{-1} in pure methanol.

These pioneer works generated the early academic interest of using water as solvent for synthesis. Theoretical and synthetic studies by many renown researchers have greatly advanced the pericyclic reactions in aqueous media (9).

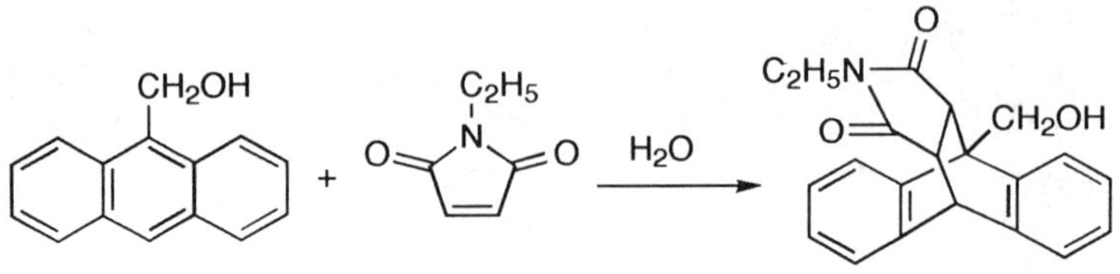

(1)

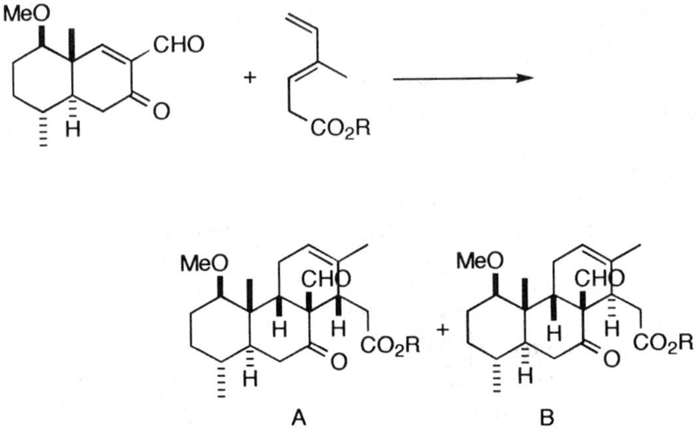

ratio A/B = 3 : 1 (in water) R= Na
1 : 0.85 (in benzene) R= Et

Scheme 1

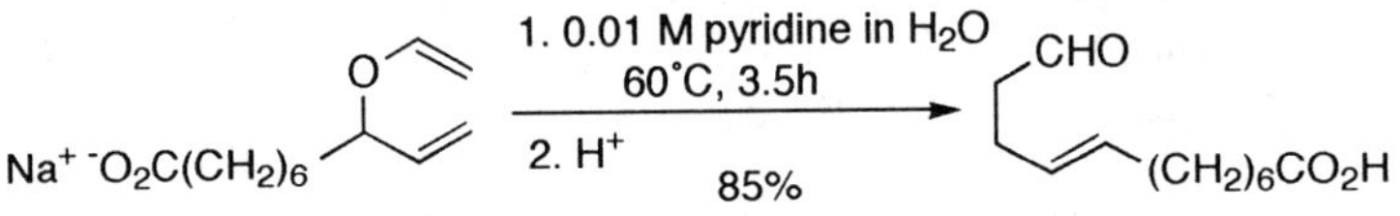

(2)

Organometallic Reactions

Metal-mediated carbon-carbon bond formations, using organometallic reagents such as Grignard reagents, are among the most fundamental and important reactions in organic chemistry (*10*). Such reactions are notorious for their sensitivity toward moisture. Very often, their generation and subsequent reactions have to be handled in a dry-box of inert environment, with the utmost exclusion of water. Thus, it would seem surprising that people would consider carrying out these reactions in aqueous media at all. Such organometallic-type reactions, if successful, could have profound implications for organic chemistry. Yet in the last decade or so, studies have shown that such reactions are in fact very possible in water. Our work and that of others have shown that various metals including tin (*11*), zinc (*12*), indium (*13*), bismuth (*14*), manganese (*15*), and even magnesium (*16*) can be used for such reactions. The most successful study is on reaction of carbonyl compounds with allyl halides (eq. 3) and propargyl halides. The most effective metal is indium which was discovered by Li and Chan (*12*), and further developed by Whitesides (*17*), Paquette (*18*), Schmidt (*19*), and others. Progress, although slow, will be made to extend to alkylation of other than allyl and propargyl halides.

Transition Metal Catalyzed Reactions

The use of transition metal catalysts for effecting organic transformations has gained increased role in synthesis due to the fact that these catalyzed reactions are usually more selective, the conditions are milder, and these processes are more atom economical (*20*). They also generate less waste than the use of stoichiometric amounts of reactants. Traditionally, transition metal catalysis has been carried out in organic solvent. Recent studies have shown that many transition metal catalyzed reactions can be carried out efficiently in water (*21*). For example, palladium catalyzed carbon-carbon bond formations were shown by several groups, e.g. Sinou (*22*) Genet (*23*) Beletskaya (*24*). Casalnuovo and Calabrese reported that by using the water-soluble palladium(0) catalyst $Pd(PPh_2(m-C_6H_4SO_3M))_3$ (M= Na^+, K^+) various aryl bromides and iodides reacted with aryl and vinyl boronic acids, terminal alkynes and dialkyl phosphites to give the cross-coupling products in high yields in water (eq. 4) (*25*).

Research from our own group in this area resulted in a novel functional group reshuffling reaction by using ruthenium catalyst (eq. 5) and and a palladium catalyzed simple/practical bis(aryl)acetylene synthesis with acetylene gas (eq. 6) (*26*).

Lewis Acid Catalyzed Reactions

Lewis acids were generally considered water and moisture sensitive. However, recent studies predominantly by Kobayashi and co-workers have shown that many Lewis acids tolerate the presence of a large amount of water in their reactions. For example, a catalytic amount of lanthanide triflate (a stronger Lewis acid) greatly

$$\underset{R}{\overset{O}{\overset{\|}{C}}}\text{H} + \diagup\!\!\!\diagdown\text{Br} \xrightarrow[H_2O]{M} \underset{R}{\overset{OH}{\underset{|}{C}}}\diagup\!\!\!\diagdown \qquad (3)$$

$$R\text{-}X + R'\text{-}Y \xrightarrow[47\text{-}100\%]{[Pd]} R\text{-}R' \qquad (4)$$

$$\text{XPh}\overset{OH}{\underset{|}{\diagup\!\!\!\diagdown}}\diagup\!\!\!\diagdown \xrightarrow[\substack{H_2O,\ \text{air atmosphere}\\49\text{-}75\%}]{2\text{-}4\ \text{mol\%}\ RuCl_2(PPh_3)_3} \text{XPh}\diagup\!\!\!=\!\!\!\diagdown\overset{OH}{\underset{|}{C}}\text{CH}_3 \qquad (5)$$

$$2\ \text{X-C}_6\text{H}_4\text{-I} + \text{H}\!\!\equiv\!\!\text{H} \xrightarrow[CH_3CN/H_2O,\ R.T.]{[Pd]/Et_3N} \text{X-C}_6\text{H}_4\text{-C}\!\!\equiv\!\!\text{C-C}_6\text{H}_4\text{-X} \qquad (6)$$

improved the rate and the yield of the following Mukaiyama Aldol reaction (eq. 7) (27). Subsequently, various reactions have been found to be catalyzed with water-tolerant Lewis acids by several groups (28).

Synthesis in Water

Chemical Synthesis

The use of water as solvent for chemical synthesis could, in principle, lead to huge environmental and economic benefits if designed properly. The best example in this regard is the Ruhrchemie/Rhône-Poulenc's hydroformylation process (29). The hydroformylation process is one of the most successful applications of aqueous medium catalysis in industrial manufacture. It started in 1982 with a series of patents including the synthesis of aldehydes (29), the recovery of rhodium catalyst (30), and the preparation of water-soluble sulfonated phosphane ligands (31). The Ruhrchemie/Rhône-Poulenc's process went into operation two years later. The present annual production of butyraldehyde alone exceeds a quarter of million tons by this process. The product is separated from the catalyst solution by a simple phase separation, and the catalyst solution is recharged to the reactor for further reaction. During the process, the loss of rhodium catalyst in the organic phase is negligible. The reaction provides predominantly n-butyraldehyde (eq. 8).

Organic Synthesis

Reactions in water developed by many groups have been elegantly applied in the synthesis of natural products (1). Although it is difficult to include all of the studies within this paper, a particular example is noted here. That is the sialic acid synthesis initially developed by Chan and Li (13), and advanced by Whitesides and co-workers (32). In this example, (+)-3-deoxy-D-*glycero*-D-*galacto*-nonulosonic acid (KDN, **1**) was synthesized by using indium mediated cross-coupling of a carbohydrate with an allyl bromide derivative followed by ozonolysis (Scheme 2). During this synthesis, no protection is required, the water soluble carbohydrate is used directly without the usual required protection-deprotection process involving carbohydrates. The synthesis was later further improved to two steps by Chan's group (33). Stepwise, these syntheses are equivalent to or better than natural syntheses. Such reactions have been applied to many other syntheses in our group (34) as well as in others.

Material Synthesis in Water

New Polymerization Processes

Many polymerization processes including radical, cationic, and anionic polymerizations can proceed in water (35). Recent studies have shown transition

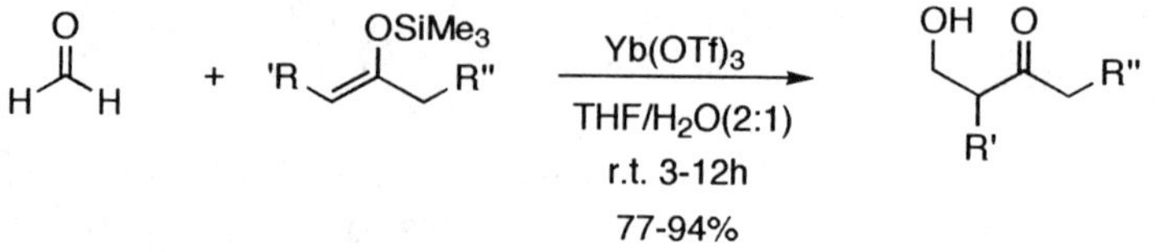

(7)

$$CH_3\text{-}CH{=}CH_2 + CO + H_2 \xrightarrow[H_2O]{\text{cat. } HRh(CO)(tppts)_3} CH_3CH_2CH_2CHO \qquad (8)$$

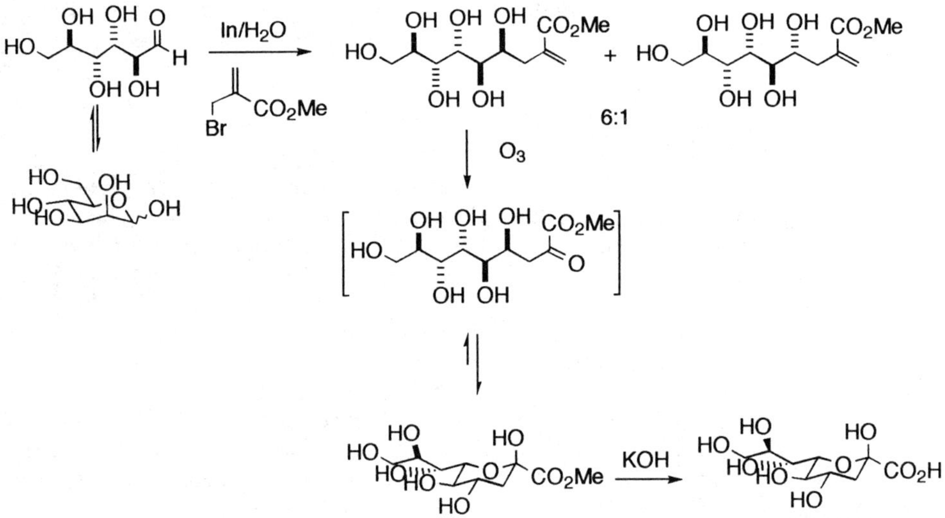

1 (+) KDN

Scheme 2

metal catalyzed polymerization reactions to proceed well in water. Such processes are advantageous not only from the standing point of being environmentally friendly, but also from the consideration of properties of material. The best known example regarding transition metal catalyzed polymer synthesis is olefin-metathesis *via* the use of the Grubbs' catalyst (eq. 9) (*36*). Compared with the same reaction carried out in organic solvent, the initiation time decreased from 22-24 h to 30-35 min. After the polymerization, the aqueous catalyst solution not only could be reused but became more active in subsequent polymerizations and the initiation period dropped to only 10-12 seconds. Solutions containing these aqueous catalysts have been recycled for 14 successive polymerizations without any detectable loss of activity. By using the Ru catalyzed ring-opening-polymerization in water, Kiessling synthesized neoglycopolymers for intercellular recognition studies (eq. 10) (*37*). Water-soluble palladium compounds have been used by Sen for the alternating copolymerization of olefins with carbon monoxide in aqueous medium (*38*). More recently, polymerization of *p*-methoxystyrene catalyzed by a water-stable Lewis acid [$Yb(OTf)_3$] (*39*) and an alternating co- and terpolymerization of carbon monoxide and olefins catalyzed by a water-soluble palladium catalyst (*40*) was successfully carried out in water.

New Materials

A hydrocarbon non-metallic conducting polymer with a rigid rod of benzene rings was synthesized recently from two biphenyl compounds *via* the Suzuki reaction in water (eq. 11) (*41*). The reaction was catalyzed by a water soluble palladium catalyst. In our own study, by using the palladium catalyzed coupling between aryl halides with acetylene gas, a variety of poly(arene ethynylene)s were prepared from aryl diiodides (eq. 12) (*42*). Such polymers have important potential as nonlinear optical materials, light-emitting materials, organic conduction materials, and sensors.

Previously the synthesis of such polymers include several steps starting from the dialdobenzene. The diiodobenzene was then treated with for example, TMS acetylene, followed by deprotection and then finally polymerization between the diacetylene system with the dialdobenzene to give the corresponding product (*43*). The new synthesis used a one-step catalytic process with water as solvent. More recently, we have shown that palladium catalyzed copolymerization of 3,5-diiodobenzoic acid with acetylene gas in a basic aqueous medium provides a high molecular weight (~60,000), zig-zag phenylethynylene polymer (eq. 13) (*44*). The polymer has a high thermostability and is soluble in basic solutions. Solubility and processibility is a common problem associated with all polyaromatic materials. However, in the present case, our polymer is reversibly switchable from soluble to hydrogel states in water by changing the pH of the solvent. In addition to overcoming the solubility and processing problem of using the water-based synthesis, such hydrogel properties have potential biological and environmental applications such as developing artificial skin and performing drug delivery.

Conclusion

In conclusion, the use of water as solvent for organic and material synthesis not only is environmentally benign and economically advantageous, but provides

(9)

(10)

$$\text{Br}\underset{\text{HOOC}}{\overset{\text{COOH}}{\text{C}_6\text{H}_3\text{-C}_6\text{H}_3}}\text{Br} + \text{(glycol boronate)}-\text{C}_6\text{H}_4-\text{C}_6\text{H}_4-\text{(glycol boronate)} \tag{11}$$

$$\xrightarrow[\text{L=P(C}_6\text{H}_5)_2(m\text{-C}_6\text{H}_4\text{SO}_3\text{Na})]{\text{NaHCO}_3,\ \text{PdL}_3} \left[\underset{\text{HOOC}}{\overset{\text{COOH}}{\text{-C}_6\text{H}_3\text{-C}_6\text{H}_3\text{-C}_6\text{H}_4\text{-C}_6\text{H}_4\text{-}}}\right]_n$$

$$\text{I-Ar-I} + \text{H}\equiv\equiv\text{H} \xrightarrow[\text{CH}_3\text{CN/H}_2\text{O, R.T.}]{[\text{Pd}]/\text{Et}_3\text{N}} \left(\equiv\!\!\equiv\!-\text{Ar}\right)_n \tag{12}$$

$$\underset{\text{I}\quad\quad\text{I}}{\overset{\text{CO}_2\text{H}}{\text{C}_6\text{H}_3}} + \text{H}\equiv\equiv\text{H} \xrightarrow[\text{NaOH/H}_2\text{O, R.T.}]{[\text{Pd}]/\text{Et}_3\text{N}} \left(\text{3,5-(CO}_2\text{H)-C}_6\text{H}_3\text{-C}\equiv\text{C-}\right)_n \tag{13}$$

opportunities to explore new reactions, new chemistry, and new materials. In addition, both water-soluble and water-insoluble reactants can be used for most reactions carried out in aqueous medium.

Acknowledgments

We would like to acknowledge the colleagues and co-workers who carried out the research described in this paper (whose names appeared in the references). We would like to acknowledge the support especially by Tulane University, by the American Chemical Society, the Louisiana Board of Regents, and the NSF/EPA Technology for a Sustainable Environment Program.

Literature Cited

1. Endress, M.; Bischoff, A.; Zinner, E. *Nature* **1996**, *379*, 701.
2. For general reviews on organic reactions in water, see: Li, C. J.; Chan, T. H. *Organic Reactions in Aqueous Media*, John Wiley & Sons, New York 1997. See also: C. J. Li *Chem. Rev.* **1993**, *93*, 2023; C. J. Li *Tetrahedron* **1996**, *52*, 5643; Lubineau, A.; Auge, J.; Queneau, Y. Lubineau, A.; Auge, J.; Queneau, Y. *Synthesis* **1994**, 741; *Organic Synthesis in Water*, Ed. Grieco, P. A. Blackie Academic & Professional, Glasgow, 1998.
3. Rideout, D. C.; Breslow, R. *J. Am. Chem. Soc.* **1980**, *102*, 7816.
4. Ben-Naim, A. *Hydrophobic Interactions*, Plenum Press, New York, 1980; Tanford, C. *The Hydrophobic Effect*, 2nd ed., John Wiley, New York, 1980.
5. von Hippel, P. H.; Schleich, T. *Acc. Chem. Res.* **1969**, *2*, 257.
6. Grieco, P. A.; Garner, P.; He, Z. M. *Tetrahedron Lett.* **1983**, *24*, 1897;
7. Grieco, P. A.; Larsen, S. D. *J. Am. Chem. Soc.* **1985**, *107*, 1768.
8. Brandes, E. ; Grieco, P. A.; Gajewski, J. J. *J. Org. Chem.* **1989**, *54*, 515.
9. For reviews, see: Breslow, R. *Acc. Chem. Res.* **1991**, *24*, 159; Grieco, P. A. *Aldrich. Acta* **1991**, *24*, 59..
10. Wakefiled, B. J. *Organomagnesium Methods in Organic Chemistry*, Academic Press, 1995.
11. Nokami, J.; Otera, J.; Sudo, T.; Okawara, R. *Organometallics* **1983**, *2*, 191.
12. Petrier, C.; Luche, J. L. *J. Org. Chem.* **1985**, *50*, 910.
13. Li, C. J.; Chan, T. H. *Tetrahedron Lett.* **1991**, *32*, 7017; Chan, T. H.; Li, C. J. *J. Chem. Soc., Chem. Commun.* **1992**, 747.
14. Wada, M.; Ohki, H.; Akiba, K. Y. *J. Chem. Soc., Chem. Commun.* **1987**, 708; Katritzky, A. R.; Shobana, N.; Harris, P. A. *Organometallics* **1992**, *11*, 1381.
15. Li, C. J.; Meng, Y.; Yi, X. H.; Ma, J. H.; Chan, T. H. *J. Org. Chem.*, **1998**, *63*, 7498; Li, C. J.; Meng, Y.; Yi, X. H.; Ma, J. H.; Chan, T. H. *J. Org. Chem.*, **1997**, *62*, 8632.
16. Li, C. J.; Zhang, W. C. *J. Am. Chem. Soc.* **1998**, *120*, 9102.
17. Kim, E.; Gordon, D. M.; Schmid, W.; Whitesides, G. M. *J. Org. Chem.* **1993**, *58*, 5500.

18. For a review, see: Paquette, L. A. in *Green Chemistry, frontiers in benign Chemical Syntheses and Processes*, Eds. Anastas, P. and Williamson, T. C. Oxford University Press, New York 1998.
19. Prenner, R. H.; Binder, W. H.; Schmid, W. *Libigs Ann. Chem.* **1994**, 73.
20. Trost, B. M. *Science* **1991**, *254*, 1471.
21. For reviews, see: Joo, F.; Toth, Z. *J. Mol. Catal.* **1980**, *8*, 369; Sinou, D. *Bull. Soc. Chim. Fr.* **1987**, 480; Kuntz, E. G. *Chemtech* **1987**, *17*, 570; Kalck, P.; Monteil, F. *Adv. Organomet. Chem.* **1992**, *34*, 219; Hermann, W. A.; Kohlpainter, C. W. *Angew. Chem., Int. Ed. Engl.* **1993**, *32*, 1524; Roundhill, D. M. *Adv. Organomet. Chem.* **1995**, *34*, 155.
22. Safi, M.; Sinou, D. *Tetrahedron Lett.* **1991**, *32*, 2025; Genet, J. P.; Blart, E.; Savignac, M. *Synlett* **1992**, 715.
23. Genet, J. P.; Blart, E.; Savignac, M. *Synlett* **1992**, 715.
24. Bumagin, N. A.; More, P. G.; Beletskaya, I. P. *J. Organomet. Chem.* **1989**, *371*, 397.
25. Casalnuovo, A. L.; Calabrese, J. C. *J. Am. Chem. Soc.* **1990**, *112*, 4324.
26. Li, C. J., Wang, D. and Chen, D. L. *J. Am. Chem. Soc.* **1995**, *117*, 12867; C. J. Li, D. L. Chen, C. W. Costello, *Org. Res. Process. Develop.* **1997**, *1*, 315.
27. Kobayashi, S.; Hachiya, I. *J. Org. Chem.* **1994**, *59*, 3590; Kobayashi, S. *Synlett.* **1994**, 689.
28. Kobayashi, S. in *Organic Synthesis in Water*, ed. Grieco, P. A. Blackie Academic & Professional, Glasgow, 1998.
29. Cornils, B.; Hibbel, J.; Konkol, W.; Lieder, B.; Much, J.; Schimd, V.; Wiebus, E.(Ruhrchemie AG), DE-B 3234701, **1982**; for more information in this area, see: Aqueous-Phase Organometallic Catalysis: Concepts and Applications, eds. Cornils, B.; Herrmann, W. A., WILEY-VCH, Weinheim 1998.
30. Gärter, R.; Cornils, B.; Bexten, L.; Kupies, D (Ruhrchemie AG), DE-B 3235029, **1982**.
31. Bexten, L.; Cornils, B.; Kupies, D. (Ruhrchemie AG), DE-B 3431643, **1984**.
32. Gordon, D. M.; Whitesides, G. M. *J. Org. Chem.* **1993**, *58*, 7937.
33. Chan, T. H.; Lee, M. C. *J. Org. Chem.* **1995**, *60*, 4228.
34. For a review, see: Li, C. J.; Chan, T. H. *Tetrahedron* **1999**, *55*, 11149.
35. Wang, X. S.; Lascelles, S. F.; Armes, S. P. *Chem. Commun.* **1999**, 1817; and references cited therein.
36. Novak, B. M.; Grubbs, R. H. *J. Am. Chem. Soc.* **1988**, *110*, 7542.
37. Mortell, K. H.; Gingras, M.; Kiessling, L. L. *J. Am. Chem. Soc.* **1994**, *116*, 12053; Mortell, K. H.; Weatherman, R. V.; Kiessling, L. L. *J. Am. Chem. Soc.* **1996**, *118*, 2297.
38. Jiang, Z.; Sen, A. *Macromolecules* **1994**, *27*, 7215.
39. Satoh, K.; Kamigaito, M.; Sawamoto, M. *Macromolecules* **1999**, 32, 3827.
40. Bianchini, C.; Lee, H. M.; Meli, A.; Moneti, S.; Patinec, V.; Petrucci, G.; Vizza, F. *Macromolecules* **1999**, 32, 3859.
41. Wallow, T. I.; Novak, B. M. *J. Am. Chem. Soc.* **1991**, *113*, 7411.
42. Li, C. J.; Slaven, W. T, IV.; John, V. T.; Banerjee, S. *Chem. Commun.* **1997**, 1569.
43. Giesa, R.; *J. Macromol. Sci.-Rev. Macromol. Chem. Phys.* **1996**, *C36(4)*, 631.
44. Li, C. J.; Slaven, W. T., IV.; Chen, Y. P.; John, V. T.; Rachakonda, S. H. *Chem. Commun.* **1998**, 1351.

Chapter 7

The Greening of a Fundamental Reaction: Metal-Mediated Reactions in Water

Chao-Jun Li

Department of Chemistry, Tulane University, New Orleans, LA 70118

The increasing environmental consciousness of the chemical community has led to the search for alternative, non-polluting media and processes for chemical and organic synthesis (*1*). Due to the natural abundance of water as well as the inherent advantages of using water as a solvent, interests have been growing in studying organic reactions in water. Many reactions that are traditionally carried out in organic solvent can be carried out in water with additional interesting features (*2*). Since water is clean and non-toxic, the study of water as a reaction solvent contributes to the advancement of chemical technologies that are intrinsically "green". The focus of this paper is on one particular class of reactions whose properties in water are currently being researched. These are organometallic reactions.

Introduction

If a carbonyl compound is reacted with an organic halide in the presence of metal, it is possible to create a carbon-carbon bond---an essential step in chemical and organic synthesis. This is one of the fundamental reactions of organic chemistry. Such reactions proceed in either one (Barbier-type) or two steps (Grignard-type) (eq. 1). These reactions have earned the reputation of being highly water-sensitive. For example, we know that every time an organometallic reaction is run the reaction system must be dried completely. Thus, it seems surprising that people would consider carrying out these reactions in aqueous media at all. Such organometallic-type reactions, if successful, could have profound implications in organic chemistry. There is an ancient philosophy that says, "things will develop in the opposite direction when they become extreme". This philosophy suggests that it is possible that those reactions requiring such extremes of dryness, might actually occur in water. Since 1980's, it has been discovered that metal-mediated reactions could in fact be carried out in water. Nakomi initially developed the tin chemistry (*3*); Luche (*4*) developed the zinc chemistry; Li and Chan developed the indium chemistry (*5*) which

© 2000 American Chemical Society

was further developed by Whitesides (6), Paquette (7), Schmidt (8), and others; and Wada and Katrizky developed the bismuth chemistry (9). Such reactions have been extended to manganese and even magnesium by Li *et al.* (10). This paper will focuses on selected recent work from our research group at Tulane University.

Features of Organometallic Reactions in Water

New Chemoselectivity

One of the major concerns when carrying out organic synthesis is the selective transformation of certain functionalities in the presence of others (chemoselectivity). From a "green chemistry" point of view, a more chemoselective reaction would generate cleaner products and thus decrease the amount of by-products and chemical waste that are potentially harmful to the environment. In addition, the generation of by-products also decreases profit and increases operation costs. Recent studies in our group as well as in others have shown that the aqueous organometallic reactions exhibit some unusual chemoselectivity that is absent from equivalent reactions in organic solvent. For example, an aldehyde usually can be alkylated selectively in the presence of a ketone. As shown by Yamamoto and co-workers (11), the selectivity is higher than 99%. Even a cyclohexanone can be selectively allylated in the presence of cyclopentanone (eq. 2). Other functional groups, such as hydroxyl, esters, carboxylic acids, amides, phthalimides, nitriles, phosphonate esters and acetals are not reactive under the aqueous metal-mediated reaction conditions. These functionalities can be present either in the substrates or as part of the allylic halide.

Another interesting chemoselectivity was reported by Li *et al.* (9). It was shown that by using manganese as the metal mediator together with a catalytic amount of copper (metal or salts), aromatic aldehydes are exclusively alkylated in the presence of an aliphatic aldehyde both intramolecularly and intermolecularly; whereas classical methods exhibit little or no chemoselectivity in such reactions (eq. 3). Interestingly enough, no reaction was observed with either manganese or copper alone. In addition, by using manganese together with a catalytic amount of acid or ammonium chloride, exclusive pinacol coupling of aromatic aldehydes in the presence of aliphatic aldehydes is also achieved (eq. 4). The use of metal mediators has a very different effect regarding the reactivity of various functionalities in aqueous reactions. Thus, it can be expected that new and unusual chemoselectivity related to aqueous organometallic reactions will be continuously discovered with the choice of different metal mediators.

Stereoselectivity

The study by several groups, most extensively by Paquette and co-workers have shown that the aqueous metal reaction also has a high diastereoselectitvity (7). The selectivity is comparable yet sometimes different from the corresponding reactions in organic solvent. Chelation-control plays an important role in the diastereoselectivity of these reactions. Even long range stereoselectivity is also possible in water (12).

$$\underset{R}{\overset{O}{\|}}\underset{H}{\|} + R'X \xrightarrow{M} \underset{R}{\overset{OH}{\|}}\underset{R'}{\|} \quad (1)$$

cyclohexanone + cyclopentanone (1:8) → tetraallyltin / 2N HCl/THF → 1-allylcyclohexanol + 1-allylcyclopentanol 99 : 1 (2)

OHC-(CH$_2$)$_3$-O-CH$_2$-C$_6$H$_4$-CHO → allyl chloride/Mn/Cu (3:3:1) → OHC-(CH$_2$)$_3$-O-CH$_2$-C$_6$H$_4$-CH(OH)-CH$_2$-CH=CH$_2$ (only allylation product)   (3)

$$2\ \underset{R}{\overset{O}{\|}}\underset{H}{\|} \xrightarrow[H_2O]{Mn/HOAc} \underset{R}{\overset{HO\ \ OH}{\|}}\underset{R}{\|} \quad (4)$$

Another interesting stereoselectivity was observed in our group druing the manganese mediated pinacol coupling in water. An intramolecular pinacol coupling produced a single *trans* diastereomer was observed in this reaction (eq. 5).

New Reactivity

For metal-mediated carbonyl alkylations in organic solvent, an essential requirement is the character of the carbonyl group. Because of this, alkylations on carbonyls that are readily enolizable and carbonyls that are easily equilibrated with other type of functionalities will give poor results or will not work. Our research, as well as several groups in this field, has shown that such problems encountered in organic solvent are not an issue for the aqueous medium organometalllic reactions. For example, a simple, efficient, and general approach to the carbonyl allylation of the enolizable 1,3-dicarbonyl compounds can be carried out based on the Barbier-type reaction using water as the solvent (13). Both indium and tin give high yields of the products for this transformation. The use of zinc is less effective. Various 1,3-dicarbonyl compounds were allylated in this way. Even though there is a high tendency of enolization for these compounds, the allylation only occurs on the dicarbonyl form which drives the reaction into completion (eq. 6).

Similarly, carbohydrates often exist in an equilibrium between a cyclic hemiacetal form and a carbonyl form with the latter in an extremely low quantity (14). Therefore, for alkylation of carbohydrate derivatives in organic solvent, extensive functional group transformation and derivatization is required to generate the desired carbonyl. However, in the aqueous reaction, the hemiacetal-form and the carbonyl form is in a constant equilibrium. And the carbonyl form will react, which drives the equilibrium until all the starting material has reacted (eq. 7).

Similarly, aldehydes that form hydrates easily or that are only available as aqueous solutions can be used directly (15).

Magnesium Reaction

The introduction of magnesium for carbon-carbon bond formations by Barbier and Grignard about a century ago (16), through the addition of an organometallic reagent to a carbonyl group, was an important step in the history of organic chemistry. The study of magnesium-based reactions since then has sparked the development of new reagents based on electronically more negative and more positive metals; as well as semi-metallic elements to tailor reactivities and selectivities (chemo, regio and stereo). For carbonyl additions based on organomagnesium reagents, it is generally accepted that strict anhydrous reaction conditions are required for a smooth reaction. Grignard reagent itself reacts violently with water. Surprisingly, we found that the allylation of benzaldehyde with allyl bromide and iodide mediated by magnesium proceeds effectively in 0.1 N HCl or 0.1 N NH$_4$Cl aqueous solution (eq. 8) (10). It is possible that the classical Barbier-Grignard reaction in organic solvent may one day be replaced by the aqueous reaction completely.

In the absence of allyl halide, a simple and effective pinacol-coupling was carried out with magnesium (17). The pinacol coupling is another fundamental reaction in

(5)

(6)

(7)

(8)

organic chemistry. The reaction was highly effective in water in the presence of a catalytic amount of ammonium chloride. Under these conditions, various aromatic aldehydes and ketones underwent carbonyl coupling, generating 1,2-diols in good yields (eq. 9). The effectiveness of the reaction was strongly influenced by steric environment surrounding the carbonyl group. Aliphatic aldehydes appeared inert under the reaction conditions.

Palladium Catalyzed Carbon-Carbon Bond Formation in Air

Palladium-catalyzed reductive couplings are among the most important carbon-carbon bond forming reactions in modern synthetic organic chemistry (18). These reactions are generally air sensitive. Recently we found that, in water, palladium catalysts are air-stable. Under such conditions several important palladium catalyzed reactions are more reactive and proceed smoothly at room temperature. For example, a zinc mediated Ullmann-type coupling provides high yield of the products catalyzed by palladium in water and under air-atmosphere (eq. 10) (*19*).

Synthetic Applications

Cyclizations

Cyclopentanoids constitute one of the most common structural features of many natural and synthetic products. Many research efforts have been devoted to their formation. An annulation methodology was developed in our group based on the indium-mediated allylation reaction in water (eq. 11) (*20*). Various compounds have been transformed into bicyclic, spiral, or simple cyclopentanoids.

Ring Expansions

The importance of medium size (8, 9, 10) rings in organic chemistry is exemplified by their presence as the structural core of a large number of biologically important natural products, such as albolic acid, ceroplastol II, and ophiobolic C. We have been interested in the development of a simple two-atom ring expansion method toward the synthesis of such compounds. By using the indium-mediated Barbier-type reaction in water, five-, six-, seven-, eight-, and twelve-membered rings were enlarged by two carbon atoms into seven-, eight-, nine-, ten-, and fourteen-membered ring derivatives, respectively (*21*). The use of water as a solvent was found to be critical for the success of the reaction. Attempts to effect a similar ring expansion in various organic solvents were not successful. In a more complicated system, the ring expansion was equally successful (eq. 12). This study provides a basis for the synthesis of the target natural products. One carbon ring expansion (*22*) and heterocyclic ring expansion (*23*) have also been successful by the method.

$$\underset{R}{\overset{O}{\|}}\underset{H}{\overset{}{C}} \xrightarrow[\text{H}_2\text{O/cat. NH}_4\text{Cl}]{\text{magnesium}} \underset{R}{\overset{HO}{\underset{}{C}}}\underset{R}{\overset{OH}{\underset{}{C}}} \qquad (9)$$

$$2\ \text{R–C}_6\text{H}_4\text{–X} \xrightarrow[\text{Zn, H}_2\text{O/acetone, air}]{\text{Pd(0) cat.}} \text{R–C}_6\text{H}_4\text{–C}_6\text{H}_4\text{–R} \qquad (10)$$

(11) 1,3-diketone + 3-chloro-2-(chloromethyl)-1-propene / base, then In/H$_2$O → 2-acyl-1-hydroxy-4-methylenecyclopentane

(12) Bicyclic ketoester with allyl bromide side chain → In/H$_2$O, DBU/THF → two diastereomeric bicyclic products

63% overall
d.e = 2.5/1

Polyhydroxylated Natural Products

The most important application of the aqueous organometallic reaction in natural product synthesis probably is in the field of carbohydrate synthesis (24). Among them, the sialic acid synthesis, initially developed by Chan and Li (25), and further advanced by Whitesides and co-workers (26), provides an excellent example of the greening factor related to organometallic reactions in water. For example, (+)-3-deoxy-D-*glycero*-D-*galacto*-nonulosonic acid (KDN, **1**) was synthesized by Li and Chan using indium mediated cross-coupling of a carbohydrate with an allyl bromide derivative followed by ozonolysis (Scheme 1). During this synthesis, no protection was required. The water soluble carbohydrate was used directly without the usually required protection-deprotection process involving carbohydrates. The synthesis was further improved to two steps by Chan (27). In fact, step for step, these synthesis are equivalent to or better than synthesis by nature. The elimination of protection and deprotection steps for similar synthesis under classical conditions saves manpower as well solvent and reagents for reactions and isolations. It also reduces considerably the chemical waste generated.

Recently, we have also been engaged in the synthesis of several polyhydroxylated natural products. Among them are styryl carbohydrate derivatives (Figure 1, **2-8**). Many compounds of such structures have shown anti-tumor activities. To design synthetic approaches to these compounds, we and others studied the metal-mediated reaction of propargyl bromide with aldehydes in water and found it to give either the propargylation or the allenylation product (or a mixture of propargylation and allenylation products). The reaction mediated by indium at room temperature was found to be highly regioselective (28, 6). The product formation was found to be dictated by both electronic and steric environments of the propargyl bromide moiety. The effects of a variety of metals on the product distribution were examined using an aliphatic as well as an aromatic aldehyde and propargyl bromide as a standard (eq. 13) (29).

In the initial study we completed the total synthesis of (+)goniofufurone (**3**) (eq. 14) (30), the key component of styryl lactones which showed anti-tumor activities. As a first generation synthesis, we felt that even though it uses water as a solvent, it was still too long. Presently, we are developing a second generation synthesis in which only three steps are involved (Scheme 2). With this method, a variety of structurally related natural products, such as papulacandin D (**9**) and chactiacandin (**1 0**), can be synthesized readily (Scheme 3) (*31*).

Besides the gioniofufrone family, the synthesis of several natural products by the aqueous methodology is in progress in our laboratory. For example, (+)-bergenin (**1 1**) can be readily synthesized from arabinose in a few steps (Scheme 4) (*32*). We are currently also synthesizing anti-sense DNA and RNA analogs by using the aqueous metal-mediated reactions.

Conclusion

The take-home message from this paper is that the chemistry of tomorrow might not be the chemistry that is known today. There would be a fundamental changes in the way organic reactions are run and in how chemical synthesis or chemical

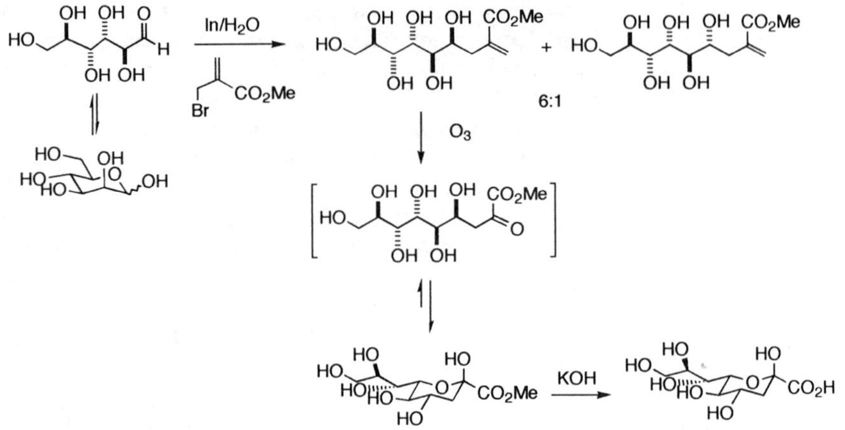

1 (+) KDN

Scheme 1

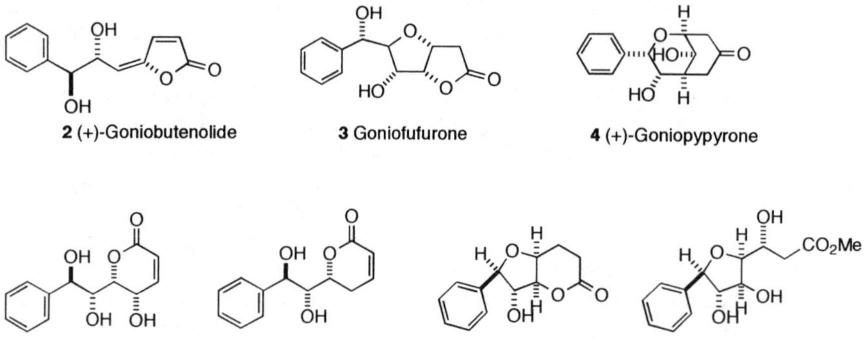

2 (+)-Goniobutenolide

3 Goniofufurone

4 (+)-Goniopypyrone

5 (+)-Gonioniotriol

6 (+)-Gonioniodiol

7 (+)-Altholactone

8 Gonioheptolide A

Figure 1

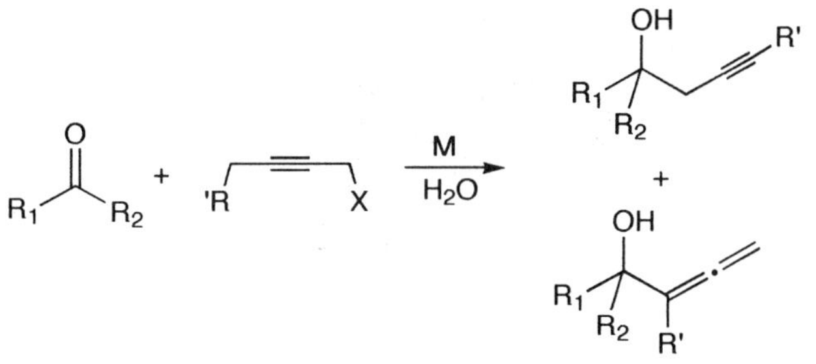

$$R_1\underset{R_2}{\overset{O}{\parallel}}\!\!\!\!\!\!\!\! + \ \ 'R\!\!-\!\!\equiv\!\!-\!\!X \ \ \xrightarrow[H_2O]{M} \ \ \begin{array}{c} R_1\underset{R_2}{\overset{OH}{|}}\!\!\!\!\!\!\!\!-\!\!\equiv\!\!-R' \\ + \\ R_1\underset{R_2}{\overset{OH}{|}}\!\!\!\!\!\!\!\!\underset{R'}{=}\!\!\!\!=\!\!\!\!= \end{array} \quad (13)$$

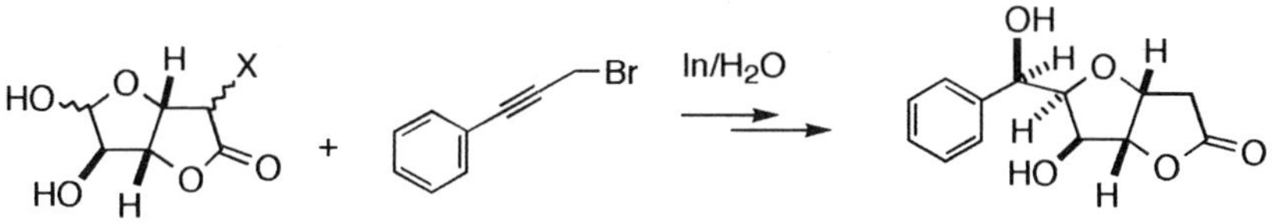

3 (+)-Goniofufurone

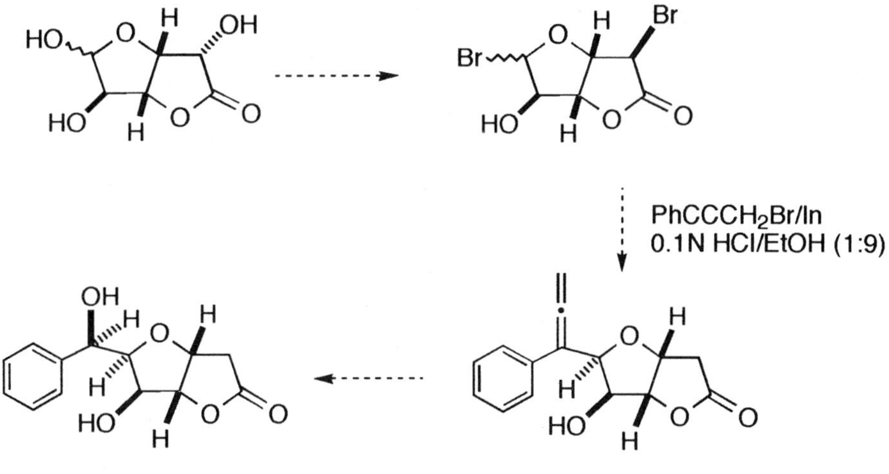

Scheme 2

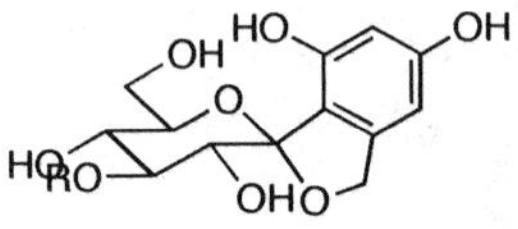

9 Papulacandin D 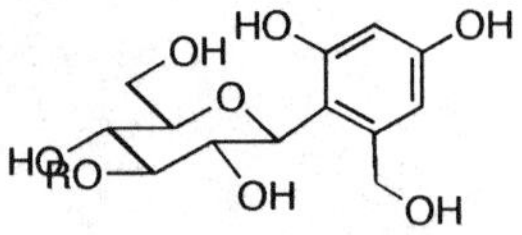   **10** Chactiacandin

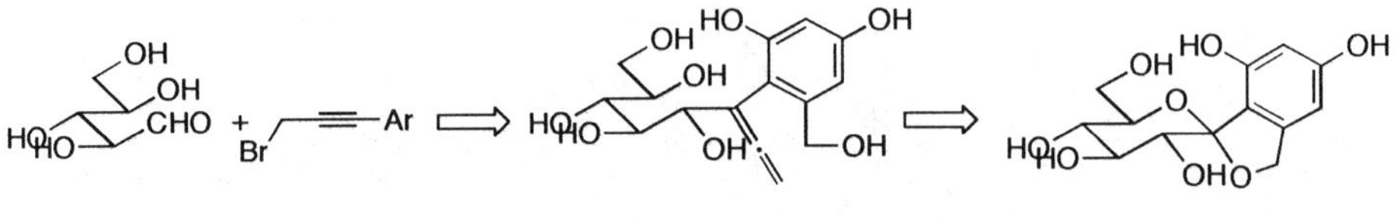

Scheme 3

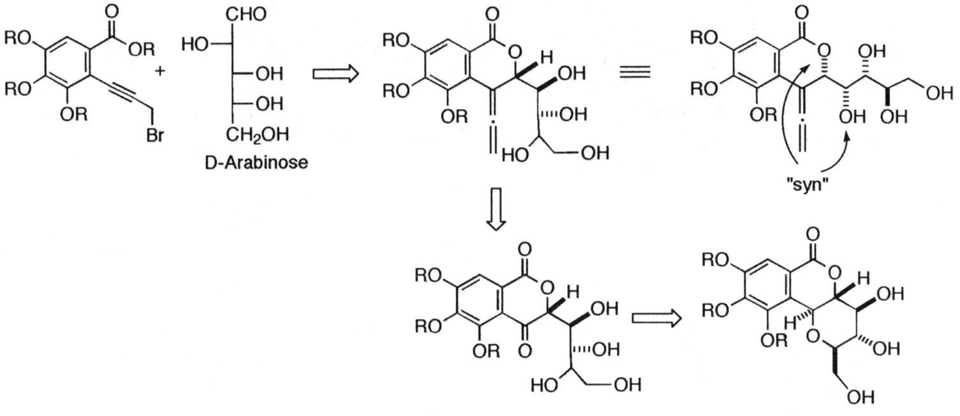

R=H, Me   **11** (+)-Bergenin

Scheme 4

production is performed. And, "environmental consciousness" will play a role. The use of water as a solvent has, among other advantages, the potential of greatly simplifying protection and deprotection steps for a synthesis which reduces environmental burden and results in pollution-prevention. The non-toxic feature and its natural abundance further enhance water as a viable benign media for organic reactions and synthesis. Thus metal-mediated carbon-carbon bond formations in water will be a fertile area of research in future.

Acknowledgments

We would like to acknowledge the colleagues and co-workers (whose names appear in the references) who has carried out research discussed in this paper. We also thank the support by Tulane University, by the American Chemical Society (the Petroleum Research Fund), the Louisiana Board of Regents, the NSF Career Award, and the NSF/EPA Technology for a Sustainable Environment Program.

Literature Cited

1. For general references on green chemistry, see: Anastas, P. T. and Williamson, T. C. (1996) Green Chemistry: an Overview. In *Green Chemistry: Designing Chemistry for the Environment*, American Chemical Society Symposium Series, No. 626, (ed. P. T. Anastas and T. C.Williamson). American Chemical Society; Anastas, P. T. and Farris, C. A. (1994).
2. For general reviews on organic reactions in water, see: Li, C. J.; Chan, T. H. *Organic Reactions in Aqueous Media*, John Wiley & Sons, New York 1997. See also: Li, C. J.*Chem. Rev.* **1993**, *93*, 2023; Li, C. J. *Tetrahedron* **1996**, *52*, 5643; Lubineau, A.; Auge, J.; Queneau, Y. Lubineau, A.; Auge, J.; Queneau, Y. *Synthesis* **1994**, 741; *Organic Synthesis in Water*, Ed. Grieco, P. A. Blackie Academic & Professional, Glasgow, 1998.
3. Nokami, J.; Otera, J.; Sudo, T.; Okawara, R. *Organometallics* **1983**, *2*, 191.
4. Petrier, C.; Luche, J. L. *J. Org. Chem.* **1985**, *50*, 910.
5. Li, C. J.; Chan, T. H. *Tetrahedron Lett.* **1991**, *32*, 7017;
6. Kim, E.; Gordon, D. M.; Schmid, W.; Whitesides, G. M. *J. Org. Chem.* **1993**, *58*, 5500.
7. For a detail review, see: Paquette, L. A. in *Green Chemistry, Frontiers in Benign Chemical Syntheses and Processes*, Eds. Anastas, P. and Williamson, T. C. Oxford University Press, New York 1998.
8. Prenner, R. H.; Binder, W. H.; Schmid, W. *Libigs Ann. Chem.* **1994**, 73.
9. Wada, M.; Ohki, H.; Akiba, K. Y.*J. Chem. Soc., Chem. Commun.* **1987**, 708; Katritzky, A. R.; Shobana, N.; Harris, P. A. *Organometallics* **1992**, *11*, 1381.
10. Li, C. J.; Meng, Y.; Yi, X. H.; Ma, J. H.; Chan, T. H. *J. Org. Chem.* **1998**, *63*, 7498; Li, C. J.; Meng, Y.; Yi, X. H.; Ma, J. H.; Chan, T. H. *J. Org. Chem.* **1997**, *62*, 8632; Li, C. J.; Zhang, W. C. *J. Am. Chem. Soc.* **1998**, *120*, 9102.
11. Yanagisawa, A.; Inoue, H.; Morodome, M.; Yamamoto, H. *J. Am. Chem. Soc.* **1993**, *115*, 10356.

12. Paquette, L. A., Bennett, G. D., Chhatriwalla, A., Isaac, M. B. *J. Org. Chem.* **1997**, *62*, 3320; Maguire, R. J.; Mulzer, J.; Bats, J. W. *J. Org. Chem.* **1996**, *61*, 6936.
13. Li, C. J.; Lu, Y. Q. *Tetrahedron Lett.* **1995**, *36*, 2721.
14. T. H. Chan; C. J. Li.; M. C. Lee; Z. Y. Wei *Can. J. Chem.* **1994**, *72*, 1181.
15. Loh, T. P.; Li, X. R. *J. Chem. Soc., Chem. Commun.* **1996**, 1929.
16. Barbier, P. *Compt. Rend.* **1898**, *128*, 110; Barbier, P. *J. Chem. Soc.* **1899**, *76*, Pt. 1, 323; Grignard, V. *Compt. Rend.* **1900**, *130*, 1322.
17. Zhang, W. C.; Li, C. J. *J. Chem. Soc. Perkin Trans. I*, **1998**, 3131.
18. Tsuji, J. *Organic Synthesis with Palladium Compounds*, Springer-Verlag, Berlin 1980; Heck, R. H. *Palladium Reagents for Organic Synthesis*, Academic Press, New York, 1985.
19. Venkatraman, S.; Li, C. J. *Org. Lett.* **1999**, *1*, 1133.
20. Li, C. J.; Lu, Y. Q.*Tetrahedron Lett.* **1996**, *37*, 471.
21. Li, C. J.; Chen, D. L.; Lu, Y. Q.; Haberman, J. X.; Mague, J. T. *J. Am. Chem. Soc.* **1996**, *118*, 4216. Li, C. J.; Chen, D. L.; Lu, Y. Q.; Haberman, J. X.; Mague, J. T. *Tetrahedron* **1998**, *54*, 2347.
22. Haberman, J.X.; Li, C. J.*Tetrahedron Lett.* **1997**, *38*, 4735.
23. Li, C. J.; Chen, D. L. *Synlett* **1999**, 735.
24. Schmid, W.; Whitesides, G. M. *J. Am. Chem. Soc.* **1991**, *113*, 6674.
25. Chan, T. H.; Li, C. J. *J. Chem. Soc., Chem. Commun.* **1992**, 747.
26. Gordon, D. M.; Whitesides, G. M. *J. Org. Chem.* **1993**, *58*, 7937.
27. Chan, T. H.; Lee, M. C. *J. Org. Chem.* **1995**, *60*, 4228.
28. Isaac, M. B.; Chan, T. H. *J. Chem. Soc., Chem. Commun.* **1995**, 1003.
29. Yi, X. H.; Meng, Y.; Hua, X. G.; Li, C. J. *J. Org. Chem.* **1998**, *63*, 7472.
30. Yi, X. H.; Meng, Y.; Li, C. J. *Chem. Commun.* **1998**, 449.
31. Ventrakaman, S.; Meng, Y.; Li, C. J. *unpublished results*.
32. For our preliminary study, see: Hua, X. G.; Mague, J. T.; Li, C. J. *Tetrahedron Lett.* **1998**, *39*, 6837; Mague, J. T.; Hua, X. G.; Li, C. J. *Acta Crystallogr., Sect. C: Cryst. Struct. Commun.* **1998**, *C54*, 1934.

Chapter 8
Dimethylcarbonate as a Green Reagent

Pietro Tundo[1], Maurizio Selva[1], and Sofia Memoli[2]

[1]Department of Environmental Science, Cà Foscari University,
Dorsoduro 2137, Venice 30123, Italy
[2]Interuniversity Consortium "Chemistry for the Environment",
Via della Libertá 5/12, 30175 Marghera, Venice, Italy

Dimethylcarbonate (DMC) is an environmentally friendly substitute for dimethylsulfate (DMS) and methyl halides in methylation reactions. It is also a very selective reagent. The reactions of DMC with methylene-active compounds produce monomethylated derivatives with a selectivity not previously observed. The batchwise monomethylation of arylacetonitriles, arylacetoesters, aroxyacetonitriles, methyl aroxyacetates, benzylarylsulfones and alkylarylsulfones with DMC achieve >99% selectivity at 180-220°C in the presence of K_2CO_3. Mono-N-methylation of primary aromatic amines at 120-150 °C in the presence of Y- and X-type zeolites, achieved selectivities up to 97%. At high temperature (200°C) and in the presence of potassium carbonate as the catalyst, DMC splits benzylic and aliphatic ketones into two methyl esters; in contrast, DMC converts ketone oximes bearing a methylene group to 3-methyl-4,5-disubstituted-4-oxazolin-2-ones. Dibenzylcarbonate (DBzlC) exhibits similar reactivity, selectively mono-benzylating methylene-active compounds.

Dimethylcarbonate (DMC) is a non-toxic, environmentally safe reagent that can be used in organic synthesis as a "green" substitute for toxic intermediates such as phosgene in carbonylation reactions, and dimethylsulphate (DMS) and methyl chloride in methylation reactions (*1*).

However, the limit for the use of DMC in industrial practice was in its preparation, far from eco-friendly, that involved the reaction of methanol with phosgene. Among the alternative phosgene-free routes to DMC considered in the last two decades, the most attractive is the metal ion-catalyzed oxidative carbonylation of methanol, set up by EniChem in 1983 (*2*). This technology is now currently used in the industry for the production of DMC.

The ever increasing numbers of industrial applications of DMC include its use as a solvent for removing asphalt and metals from the residue of crude oil distillation, as a lubricant, as a component of oxygenated gasoline and as an expanding system for polyurethane foams.

As far as its use in organic synthesis is concerned, the carbonyl group and the methyl group of DMC represent the two reactive centers at which a nucleophile may react. Under batch conditions and in the presence of a weak bases (e.g. an alkaline carbonate), DMC may act as a methylating agent (Eq. 1), in the place of DMS and methylchloride (*3*), or as a carbomethoxylating agent (Eq. 2) and a phosgene substitute (*3*).

$$Y^- + (CH_3O)_2C=O \longrightarrow YCH_3 + CH_3O^- + CO_2 \quad (1)$$

$$Y^- + (CH_3O)_2C=O \longrightarrow YCOOCH_3 + CH_3O^- \quad (2)$$

The reactivity of DMC can be influenced by experimental parameters. Methylation reactions (Eq. 1) occur at high temperatures (T>180 °C) when a nucleophilic anion attacks the methyl group (instead of the acyl carbon) of the organic carbonate. The leaving group (methoxycarbonate anion, CH_3OCOO^-) is not stable and decomposes rapidly into CO_2 and methoxide, which is converted into methanol by reaction with the substrate. In this way, catalytic amounts of the alkaline carbonate are sufficient to initiate the reaction. At lower temperatures, the attack of the nucleophile on the acyl group of DMC gives the transesterification product (Eq. 2).

The reactivity of DMC toward nucleophilic compounds is somewhat lower than that observed when analogous reactions are performed with phosgene and DMS. However, both carboxylation with phosgene and methylation with DMS generate stoichiometric quantities of inorganic salt as a byproduct because a base must be used as a reagent. The corresponding processes with DMC do not involve disposal problems since no salts are produced and the co-product methanol can be easily recycled in the DMC production plant (*4*).

DMC can be used profitably to carry out methylation reactions under both continuous-flow and batch conditions. When performed under Gas-Liquid Phase-

Transfer Catalysis (GL-PTC) conditions (1), the reactions of DMC with methylene-active compounds produce monomethylated derivatives, with a selectivity not previously observed. It is worth noting that industrial monomethylation reactions of methylene-active compounds are not a one-step process because the usual methylating agents produce a significant quantity of dimethyl derivatives.

Methylation Reactions

Selective Monomethylations of Arylacetonitriles and Arylacetoesters

The monomethylation reactions of arylacetonitriles and arylacetoesters start from readily available intermediates and produce 2-arylpropionic acid (antiinflamatory drugs, e.g. ketoprofen, naproxen, etc.). Using a 10-30 molar excess DMC, either under GL-PTC (5) or batch conditions (6), it is possible to synthesize 2-arylpropionic acid derivatives with >99% purity in monomethyl derivatives (Eq. 3).

$$ArCH_2X + DMC \xrightarrow{K_2CO_3} ArCH(CH_3)X + CH_3OH + CO_2 \quad (3)$$
$$X = CN, COOCH_3$$

Experimental evidence (detection of $ArCH(COOCH_3)X$ and $ArC(CH_3)(COOCH_3)X$ as reaction intermediates) strongly supports the hypothesis that the high monomethyl selectivity is not due to the S_N2 displacement of the nucleophile $ArCH(^-)X$ on DMC. Instead, DMC acts first as a carboxymethylating agent ($B_{Ac}2$ mechanism), which allows the protection of the methylene-active derivatives and permits nucleophilic displacement ($B_{Al}2$) to occur with another molecule of DMC. The proposed mechanism is reported in Scheme 1.

This pattern shows the peculiar action of the methoxycarbonyl group, which plays a two-fold role in 1) increasing the acidity of $ArCH(COOCH_3)X$, favoring the formation of the corresponding anion, and 2) acting as a protecting group, preventing further methylation. The key step is the attack of the anion $ArC(^-)(COOCH_3)X$ onto the DMC molecule. Kinetic studies (7) on the DMC-mediated methylation of phenylacetonitrile at 140°C showed that $k_5 > k_{-2}$ and, since reaction 5 is the only non-equilibrium reaction, the formation of $PhC(CH_3)(COOCH_3)CN$ is the driving force of the process.

Reaction of $[K(^+)PhC(^-)(COOCH_3)CN]$, the potassium salt of 2-carboxymethylphenylacetonitrile, with DMC yields $PhC(CH_3)(COOCH_3)CN$ as the sole product. For this reaction, the activation energy evaluated using the Arrhenius equation was found to be 23.4 kcal mol^{-1}; this is higher, as expected, than the value observed performing the reaction with other usual methylating agents.

$$PhCH_2CN + B \underset{k_{-1}}{\overset{k_1}{\rightleftharpoons}} PhC^{(-)}HCN + BH^+ \qquad 1$$

$$PhC^{(-)}HCN + (CH_3O)_2CO \underset{k_{-2}}{\overset{k_2}{\rightleftharpoons}} PhCH(COOCH_3)CN + CH_3O^- \qquad 2$$

$$BH^+ + CH_3O^- \underset{k_{-3}}{\overset{k_3}{\rightleftharpoons}} B + CH_3OH \qquad 3$$

$$PhCH(COOCH_3)CN + B \underset{k_{-4}}{\overset{k_4}{\rightleftharpoons}} PhC^{(-)}(COOCH_3)CN + BH^+ \qquad 4$$

$$PhC^{(-)}(COOCH_3)CN + (CH_3O)_2CO \overset{k_5}{\rightarrow} Ph-\underset{CH_3}{\overset{COOCH_3}{\underset{|}{\overset{|}{C}}}}-CN + CO_2 + CH_3O^- \qquad 5$$

$$Ph-\underset{CH_3}{\overset{COOCH_3}{\underset{|}{\overset{|}{C}}}}-CN + CH_3O^- \underset{k_{-6}}{\overset{k_6}{\rightleftharpoons}} PhC^{(-)}(CH_3)CN + (CH_3O)_2CO \qquad 6$$

$$PhC^{(-)}(CH_3)CN + BH^+ \underset{k_{-7}}{\overset{k_7}{\rightleftharpoons}} PhCH(CH_3)CN + B \qquad 7$$

---

total reaction

$$PhCH_2CN + (CH_3O)_2CO \longrightarrow PhCH(CH_3)CN + CO_2 + CH_3OH$$

Scheme 1. Proposed mechanism for the reaction of DMC with phenylacetonitrile.

Selective Monomethylation of Aroxyacetonitriles and Methyl Aroxyacetates

Similar to the mechanism reported in Scheme 1, the methylation o aroxyacetonitriles and methyl aroxyacetates proceeds with a selectivity up to 99%. The monomethyl derivatives (*8*) (2-aroxy propionitriles and methyl 2-aroxypropionates) are the corresponding products (Eq. 4).

$$ArOCH_2X + CH_3OCOOCH_3 \xrightarrow{\text{base}} ArOCH(CH_3)X + CH_3OH + CO_2 \quad (4)$$

The reaction proceeds under batch conditions, using K_2CO_3 or t-BuOK, and is maintained at 180-200 °C in a stainless steel autoclave. Although DMC acts as both the alkylating agent and the solvent (30 molar excess with respect to the substrates), no dialkylated byproducts are formed.

The results of the methylation of different methyl aryloxyacetates and aryloxyacetonitriles are reported in Table I.

Table I. Reactions of Aryloxyacetates and Aryloxyacetonitriles with DMC

	Substrate	T (°C)	React. time (h)	Conv. (%)	Product (%)
1	$PhOCH_2COOMe$	190	40	99	$PhOCH(Me)COOMe$ (94)
2	$p\text{-}MeC_6H_4OCH_2COOMe$	190	70	100	$p\text{-}MeC_6H_4OCH_2COOMe$ (92)
3	$m\text{-}ClC_6H_4OCH_2COOMe$	190	26	100	$m\text{-}ClC_6H_4OCH(Me)COOMe$ (91)
4	$PhOCH_2CN$	190	32	100	$PhOCH(Me)CN$ (69)
5	$p\text{-}MeC_6H_4OCH_2CN$	180	40	100	$p\text{-}MeC_6H_4OCH(Me)CN$ (51)
6	$m\text{-}ClC_6H_4OCH_2CN$	180	24	100	$m\text{-}ClC_6H_4OCH(Me)CN$ (79)
7	$PhOCH_2COOH$	200	48	100	$PhOCH(Me)COOH$ (96)

Reproduced from Ref.(8) with permission of Elsevier Science Publisher, Amsterdam.

In general, the reaction occurs faster with nitriles than with esters, which require higher temperatures and longer reaction times for complete conversion (compare entries 1 and 4, 2 and 5, 3 and 6, respectively). This behaviour parallels the trend already observed with the methylation of arylacetonitriles and alkyl arylacetates by DMC (6). The nitriles may be more reactive because it is easier to form the corresponding carbanions $ArOCH^{(-)}CN$ under basic conditions.

The unusually high selectivity observed in the monomethylation of methyl aryloxyacetates and aryloxyacetonitriles may be explained by the mechanism previously described for the reaction of DMC with methyl arylacetates and arylacetonitriles (6) in which $ArCH(COOCH_3)X$ and $ArC(CH_3)(COOCH_3)X$ are the key intermediates. It is likely that this reaction proceeds through the formation of methyl-carboxymethyl intermediates, $ArOC(CH_3)(COOCH_3)X$, as these derivatives were actually detected by GC/MS during the course of the reaction. In this case, the attack on the acyl carbon, which produce the possible intermediate $PhC(COOCH_3)_2X$, does not affect selectivity, because it is an equilibrium reaction.

Selective Mono-C-Methylation of Alkylarylsulfones

Sulfones bearing α-methylene groups (benzylaryl- and alkylaryl-sulfones: $ArCH_2SO_2Ar'$ and RCH_2SO_2Ar') can be effectively mono-C-methylated (selectivity >99%) by DMC, even when a mild base (K_2CO_3) is used (9) (Eq. 5).

$$RCH_2SO_2R' + CH_3OCOOCH_3 \xrightarrow{K_2CO_3} RCH(CH_3)SO_2R' + CH_3OH + CO_2 \quad (5)$$

At 180-210 °C, batch mono-C-methylations of benzylaryl- and alkylaryl-sulfones (Table II) proceed with >99% selectivity, at conversions of 95-99%, and good to high yields (77-92%) of isolated products (Eq. 5).

Table II. Reactions of Benzylaryl- and Alkylarylsulfones with DMC

	R	R'	T (°C)	Product Yield (%)
1	Ph	Ph	180	78
2	p-ClC$_6$H$_5$	Ph	"	76
3	p-CH$_3$C$_6$H$_5$	Ph	"	92
4	Ph	p-ClC$_6$H$_5$	"	80
5	p-ClC$_6$H$_5$	p-ClC$_6$H$_5$	"	81
6	Ph	CH$_3$	200	85
7	p-ClC$_6$H$_5$	CH$_3$	"	77
8	p-CH$_3$C$_6$H$_5$	CH$_3$	210	76

Reprinted from Ref. (9). By courtesy of the Royal Society of Chemistry.

The various aryl and alkyl groups directly bound to the methylene reacting group exert a major influence on reactivity. Benzylaryl sulfones (entries 1-5) are efficiently monomethylated at 180 °C. Alkylaryl sulfones (entries 6-8) require a higher temperature (200-210 °C) for the reaction to be completed. This behaviour seems to be clearly related to the stabilization, by resonance with the adjacent Ar' group, of arylsulfonyl carbanions, ArSO$_2$CH$^{(-)}$Ar', formed during the reactions. In all likelihood, the methylation of sulfones 1-5 in Table II follows the mechanistic pattern reported for aryl- and aroxy-acetic acid derivatives.

Accordingly, the monomethyl selectivity is explained by two consecutive nucleophilic displacements, 1) methoxycarbonylation of the initial sulfonyl carbanion [ArSO$_2$CH$^{(-)}$R] (B$_{Ac}$2 mechanism) followed by 2) methylation of the resulting intermediate [ArSO$_2$CH(COOCH$_3$)R] to yield the methyl derivative [ArSO$_2$C(CH$_3$)(COOCH$_3$)R, B$_{Al}$2 mechanism], that undergoes de-methoxycarbonylation to yield the final product, [ArSO$_2$CH(CH$_3$)R]. Both intermediates ArSO$_2$CH(COOCH$_3$)R and ArSO$_2$C(CH$_3$)(COOCH$_3$)R are detected during the reaction of DMC with the sulfones in entries 1-5 of Table II.

Methylarylsulfones (ArSO$_2$CH$_3$) also react with DMC. Methoxycarbonylated compounds (ArSO$_2$CH$_2$COOCH$_3$) are formed as intermediates, allowing the homologation of the methyl group to an i-propyl group. Thus, PhSO$_2$CH$_3$ yields PhSO$_2$CH(CH$_3$)$_2$ and PhSO$_2$C(COOCH$_3$)(CH$_3$)$_2$ (12 and 81%, respectively; 14 h at 180 °C; conversion 93%). Likewise, PhCH$_2$SO$_2$CH$_3$ yields PhCH(CH$_3$)SO$_2$CH(CH$_3$)$_2$ and PhCH(CH$_3$)SO$_2$C(COOCH$_3$)(CH$_3$)$_2$ (14 and 50%, respectively; 21.5 h at 180 °C; conversion 98%). No t-butyl derivatives are produced.

Selective Mono-N-Methylation of Primary Aromatic Amines

The reaction of DMC with different primary aromatic amines (Eq. 6) has been investigated under batch conditions in the presence of Y- and X-type zeolites (10).

$$ArNH_2 + CH_3COOCH_3 \xrightarrow{\text{zeolite}} ArNHCH_3 + CH_3OH + CO_2 \quad (6)$$

Reactions of anilines have been performed at 120-150 °C, in the presence of Na^+- and K^+-exchanged Y faujasites as the catalyst (Table III). Highly selective mono-N-methylations are observed even when anilines are deactivated by electron-withdrawing and/or sterically hindered groups.

Table III. Reactions of Anilines with DMC

	Substrate	Catalyst	T (°C)	Reaction time (min)	Catalyst /substr. (w/w)	Conv. (%)	Selectivity (mono/di)
1	Ar: p-NO_2-C_6H_4	NaY	130	840	3.3	57	96.4
2	Ar: p-NO_2-C_6H_4	NaY	150	420	1.2	72	94.3
3	Ar: p-NO_2-C_6H_4	NaY	150	270	6.6	78	92.9
4	Ar: p-NO_2-C_6H_4	KY	150	600	3.3	90	92.9
5	Ar: p-CN-C_6H_4	NaY	130	300	1.2	76	90.7
6	Ar: p-CN-C_6H_4	NaY	150	120	3.3	89	91.9
7	Ar: p-CN-C_6H_4	KY	150	270	3.3	90	97.6
8	Ar: o-COOMe-C_6H_4	NaY	150	720	1.2	91	94.2
9	Ar: o-COOMe-C_6H_4	KY	150	240	2.0	88	92.8
10	Ar: 2,6-Me-C_6H_3	NaY	150	300	3.3	84	93.8

Reproduced from Ref. (*10*). By courtesy of the Royal Society of Chemistry.

In all cases, the mono-N-methylated anilines ($ArNHCH_3$) are obtained with a high selectivity (92-97%) at conversions ranging from 72 to 93%. However, deactivated substrates have much longer reaction times than does aniline.

Although the data in Table III do not suggest a general difference between the activities of NaY and KY, they clearly show that the reactions depend on the weight ratio of zeolite to substrate. As previously reported for the reaction of DMC with arylacetonitriles and methyl arylacetates (*6*), the reaction of DMC with aromatic amines over Y- and X-zeolites also provides evidence for the formation of the corresponding methoxycarbonylated compounds (carbamates). Small amounts (1-5%) of methyl arylcarbamates [$ArNH(COOCH_3)$] are observed for all the anilines investigated and, in the case of aniline itself, the corresponding N-methyl methyl phenylcarbamate [$PhN(CH_3)(COOCH_3)$] is also detected.

In the light of this evidence, the same mechanism reported for the DMC monomethylation of methylene-active compounds might be considered to operate for primary aromatic amines. In this case, the zeolite cages can assist the mono-N-methylation (*11*). Synergistic effect between the reactivity of DMC (acting both as a methylating and as a reversible methoxycarbonylating agent) and the dual acid-basic properties of zeolites (*12*) is considered to be responsible for the unusually high selectivity observed.

Reactions of Dimethylcarbonate with Ketones

Unlike the CH-acidic compounds so far examined, both benzylic and aliphatic ketones (13) are split by reaction with DMC into two methyl esters (Eq. 7).

$$RCH_2COR' + CH_3OCOOCH_3 \xrightarrow{base} RCH_2COOCH_3 + R'COOCH_3 \quad (7)$$
$$R, R' = Alkyl, Aryl$$

The reaction takes place at high temperature (about 200 °C) and in the presence of K_2CO_3 as the catalyst. Under these conditions, the β-ketoester arising from the condensation between DMC and the aryl/alkyl ketone is split in a reverse Claisen condensation, as outlined in the following scheme (Scheme 2).

$$RC(-)HCOR' + CH_3OCOOCH_3 \rightleftharpoons RCH(COOCH_3)COR' \quad 1$$

$$RCH(COOCH_3)COR' + CH_3O^- \rightleftharpoons RCH(COOCH_3)-\underset{O^-}{\overset{OCH_3}{C}}-R' \quad 2$$

$$RCH(COOCH_3)-\underset{O^-}{\overset{OCH_3}{C}}-R' \rightleftharpoons RCH(-)COOCH_3 + R'COOCH_3 \quad 3$$

$$RCH(-)COOCH_3 + BH^+ \rightleftharpoons RCH_2COOCH_3 + B \quad 4$$

Scheme 2. Proposed mechanism for the cleavage of the C-CO bond and the formation of two esters.

The overall reaction results in the dismutation of ketones with DMC without oxidative conditions and yields methyl (ethyl)esters as the final products. Some examples are listed in Table IV.

Alicyclic ketones produce esters of α, ω-dicarboxylic acids (entries 5, 6). An example is cyclohexanone, which gives dimethyl pimelate. These compounds, used for the production of polyesters and polyamides, have traditionally been synthetized in industry by oxidation of alicyclic ketones (14).

Table IV. Reactions of Benzylic and Aliphatic Ketones with DMC

	Substrate	T (°C)	Reaction time (h)	Conv. (%)	Products (%^{a})
1	$PhCH_2COCH_3$	195	6.4	88	$PhCH_2COOCH_3$ (71)
2	$PhCH_2COPh$	200	6.5	71	$PhCOOCH_3$ (99)
					$PhCH_2COOCH_3$ (66)
3	$p\text{-}CH_3PhCOCH_2Ph$	200	5.8	78	$p\text{-}CH_3PhCOOCH_3$ (99)
					$PhCH_2COOCH_3$ (65)
4	$(CH_3CH_2CH_2)_2CO$	200	6.7	99	$CH_3CH_2CH_2COOCH_3$ (34)
5	Cyclopentanone	200	4.0	80	$(CH_2)_4(COOCH_3)_2$ (14)
6	Cyclohexanone	200	11.4	88	$(CH_2)_5(COOCH_3)_2$ (11)
7	$PhCH(CH_3)COPh$	200	6.2	98	$PhC(CH_3)_2COPh$ (70)
8	$PhCH(C_4H_9)COCH_3$	200	7.6	38	$PhC(CH_3)(C_4H_9)COCH_3$ (32)

^{a} refers to converted ketone. Reproduced from Ref. (*13*) with permission of Wiley-VCH Publishers.

Reaction of Dimethylcarbonate with Oximes

When ketone oximes bearing a methylene group react with DMC at 190°C in the presence of K_2CO_3, the main products are N-methyl-4,5-disubstituted-4-oxazolin-2-ones (*15*) (Eq. 9).

O-carboxymethylation and N-methylation products are observed in relatively small amounts.

$$RCH_2\overset{NOH}{\underset{}{C}}R' \xrightarrow[DMC]{base} \begin{array}{c}CH_3\\ R'\text{-}N\\ R\text{-}\diagdown O\end{array}=O \quad (8)$$

The reaction seems to be general and can be applied to both aliphatic and aromatic ketone oximes. Moreover, when the oxime has two methylene groups near the C=N bond and one of them is benzylic, the reaction takes place regioselectively on the benzylic moiety (Table V).

The mechanism proposed for the reaction of ketone oximes with DMC is reported in scheme 3, for cyclohexanone oxime.

The key step is the initial N-methylation of the O-carbonate derivative of the oxime, which produces the enamine intermediate.

Table V. Reactions of Oximes with DMC

	Substrate	T (°C)	Reaction time (min)	Conv. (%)	Product	Yield (%)
1	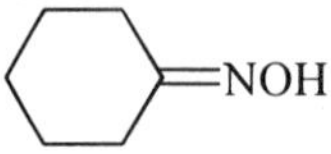 cyclohexanone oxime (=NOH)	190	435	97	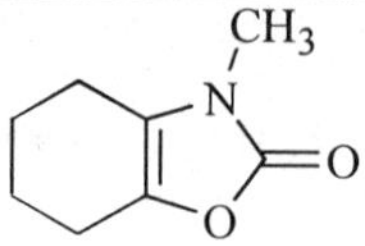 N-methyl hexahydrobenzoxazol-2-one	48
2	$(CH_3CH_2CH_2)_2CNOH$	180	300	100	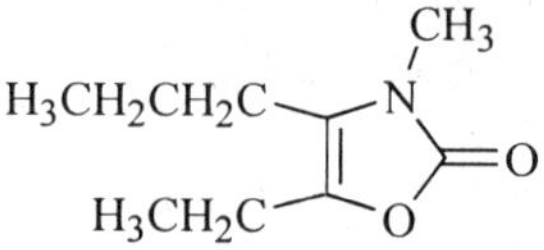 3-methyl-4,5-dipropyl oxazol-2-one ($H_3CH_2CH_2C$-, H_3CH_2C-)	22
3	$PhH_2C(H_3CH_2C)C{=}NOH$	180	465	95	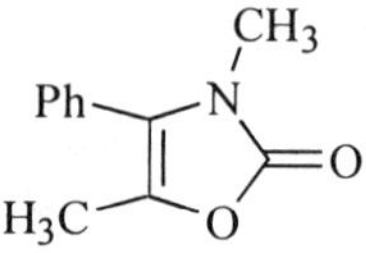 3-methyl-4-phenyl-5-methyl oxazol-2-one (Ph-, H_3C-)	37
4	$Ph(H_3CH_2C)C{=}NOH$	180	480	98	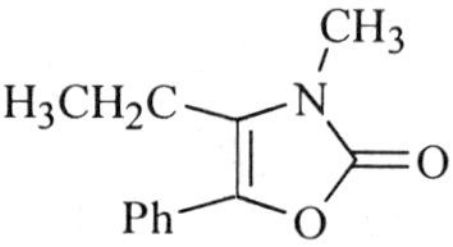 3-methyl-4-ethyl-5-phenyl oxazol-2-one (H_3CH_2C-, Ph-)	37

Reprinted with permission from Ref. (15). Copyright (1993) American Chemical Society.

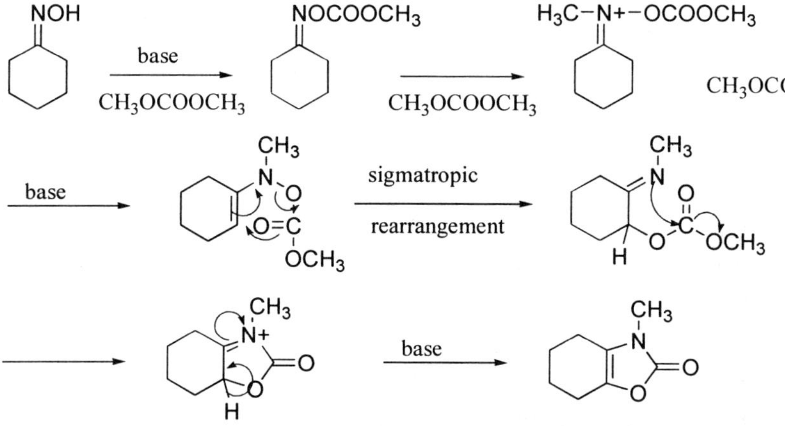

Scheme 3. Proposed reaction mechanism of cyclohexanone oxime with DMC. Reprinted with permission from Ref. (15). Copyright (1993) American Chemical Society.

This is followed by [3,3]sigmatropic rearrangement of the enamine, involving the C=O bond in the 5 position, which probably takes place by cleavage of the 3,4 N-O bond. This reaction represents a new one-step synthetic method to construct an oxazolinone ring, used as precursors of β-aminoalcohols in the synthesis of ephedrines (16).

Benzylation Reactions with Dibenzylcarbonate

At high temperatures (140-180 °C), dibenzylcarbonate (DBzlC) is an efficient benzylating agent of phenol and methylene acidic compounds, such as phenylacetonitrile and benzyl phenylacetate (17). At reflux temperature in N,N-dimethyl-formamide (DMF) solvent with K_2CO_3 as the catalyst, DBzlC react with phenol to yield the benzyl phenyl ether (Eq. 9). Phenylacetonitrile yields the monobenzylated compound (2,3-diphenyl-propionitrile) (Eq. 10).

$$\text{PhOH} + \text{PhCH}_2\text{OCOOPhCH}_2 \xrightarrow{K_2CO_3} \text{PhOCH}_2\text{Ph} + \text{PhCH}_2\text{OH} + CO_2 \quad (9)$$

$$\text{PhCH}_2\text{X} + \text{PhCH}_2\text{OCOOCH}_2\text{Ph} \xrightarrow{K_2CO_3} \text{PhCH}(CH_2\text{Ph})\text{X} + \text{PhCH}_2\text{OH} + CO_2 \quad (10)$$
X= CN, COOCH$_2$Ph

This behavior strongly resembles that of DMC; the main difference however is that methylation by DMC (at 170-220 °C) necessarily requires an autoclave system (DMC boils at 90°C) while the higher boiling point of DBzlC allows benzylation to occur at atmospheric pressure. Reactions with DBzlC require a solvent and a slight excess of DBzlC over the reagent substrate. The best solvents have proven to be DMF and diethylformamide (aprotic, polar solvents used at their reflux temperatures: 155 and 177 °C, respectively).

The effect of the reaction temperature and the molar ratio of substrate to alkylating agent has been investigated for the benzylation of phenylacetonitrile in DMF (Table VI).

Table VI. Reaction of Phenylacetonitrile with DBzlC

	React. time(h)	T (°C)	Alkylating agent	Sub/ Alkyl	Conv (%)	Selecti-vity (%)	PhCH(R)CN (%)	PhC(R$_2$)CN (%)
1	4.0	155	DBzlC	1:1.1	86.0	99.2	83.5	0.6
2	2.5	155	DBzlC	1:1.5	91.7	99.1	87.7	0.8
3	2.25	155	DBzlC	1:3.0	95.6	98.4	91.7	1.5
4	6.0	40	PhCH$_2$Cl	1:0.8	48.1	84.9	40.6	7.2

Reproduced from Ref. (17). By courtesy of Royal Society of Chemistry.

Table IV also shows a comparison between DBzlC and PhCH$_2$Cl as alkylating agents. High selectivity ($\geq$99%) in monobenzylation is always observed with DBzlC (entries 1-3), whereas PhCH$_2$Cl had lower selectivity (entry 4). This reaction probably uses the mechanism already described for DMC. As in the case for DMC, the driving force for the monobenzylation is the non-equilibrium reaction following the B$_{Al}$2 mechanism.

Conclusions

One of the primary methods for pollution prevention is the design and the development of chemical products with lower toxicity and of processes that reduce or eliminate the use of hazardous substances.

The use of DMC as a safe alternative to toxic intermediates such as phosgene, DMS and methyl chloride has manifold advantages as:
- non-toxic, non-polluting starting material is used, while phosgene, methyl chloride and DMS are extremely toxic, the latter being also carcinogen;
- no waste to be disposed of are formed, while both methylation with DMS or methyl chloride and carbonylation with phosgene produce stoichiometric amounts of inorganic salts;
- no solvent is needed, as DMC acts both as the reagent and the solvent;
- very high selectivity in monomethylation of anilines and methylene active compounds is achieved.

Acknowledgements

We gratefully acknowledge support for this work by the MURST (Ministero Università e Ricerca Scientifica e Tecnologica) and the Interuniversity Consortium "Chemistry for the Environment".

Literature Cited

1. Tundo, P. In *Continuous Flow Methods in Organic Synthesys*; Horwood, E. Pub.: Chichester (UK), 1991.
2. Rivetti, F., Romano, U.; Delledonne, D. In *"Green Chemistry: Designing Chemistry for the Environment"*; Anastas, P. T., Williamson, T.C. Eds, Acs Symposium Series 626, **1997** p.70.
3. Tundo, P., Selva, M. *ChemTech*. **1995**, *25(5)*, 31.
4. Delledonne, D. Rivetti, F., Romano, U., *J. Organomet. Chem.* **1995**, *448*, C15-C19.
5. a) Tundo, P.; Moraglio, G.; Trotta, F. *Ind. Eng. Chem. Res.* **1989**, 28, 881, b) Tundo, P.; Moraglio, G.; Trotta, F. *J. Chem. Soc., Perkin Trans. I* **1989**, 1070 .

6. a) Tundo, P., Selva, M, Marques, C.A. In *"Green Chemistry: Designing Chemistry for the Environment"*; Anastas, P. T., Williamson, T.C. Eds, Acs Symposium Series 626, 1997 p.81; b) M., Marques, C.A., Tundo, P. *J. Chem. Soc. Perkin Trans. I*, **1994**, 1323.
7. Franceschini, N.; *Thesis,* University Cà Foscari, Venice, Italy, 1998.
8. Bomben, A., Marques, C.A., Selva, M., Tundo, P. *Tetrahedron* **1995**, *51*, 11573.
9. Bomben, A., Selva, M., Tundo, P. *J. Chem. Res. (S)*, **1997**, 448.
10. Selva, M., Bomben, A., Tundo, P. *J. Chem. Soc., Perkin Trans. I*, **1997**,1041.
11. Hari, P. R., Massiani, P., Barthomeuf, D. *Catal. Lett.* **1995**, *31*, 115.
12. Fu, Z.-H., Ono, Y. *Catal. Lett.*, **1993**, *22*, 277.
13. Selva, M., Marques, C.A., Tundo, P. *Gazz. Chim. Ital.*, **1993**, *123*, 515.
14. Kirk-Othmer, *"Encyclopedia of Chemical Technology"*, Wiley, New York, 1981, Vol. 13, p. 924.
15. Marques, C.A., Selva, M., Tundo, P., Montanari, F. *J. Org. Chem.* **1993**, *58*, 5765.
16. Shono, T.; Matsumura, Y.; Kanazawa, T. *Tetrahedron Letts.*, **1983**, *24*, 4577.
17. Selva, M., Marques, C.A., Tundo, P. *J. Chem. Soc., Perkin Trans. I*, **1995**, 1889.

Chapter 9

Indium-Promoted Coupling Reactions in Water

Leo A. Paquette

Evans Chemical Laboratories, The Ohio State University, Columbus, OH 43210

Factors influencing the stereoselectivities observed for the 1,2-addition of various allylindium reagents to aldehydes and ketones in aqueous media have been investigated. Described herein are the mechanistic rationales based upon the adoption of either internally chelated (Cram-like) or Felkin-Anh transition states.

Introduction

The ideal chemical transformation involves the combination of two or more reagents in a preferably nonflammable solvent such that the single insoluble product that is formed can be isolated by simple filtration. The absence of by-products permits direct re-use of the reaction medium so recovered. The number of synthetic steps that satisfy these criteria is widely recognized to be minuscule. Nonetheless, efforts to develop technology leading to a reduction in risks to the environment posed by existing practices in chemical and pharmaceutical manufacturing have begun to bear fruit. Relevant to the present discussion is the discovery during the past decade that a number of carbon-carbon bond-forming reactions can indeed be promoted by select metals in water (1-4). For our part, we have opted to explore the feasibility of utilizing indium(0) for this purpose (5) in light of its distinctive physical and chemical properties (6), including particularly the ease with which it can be completely recovered from aqueous solutions of its salts by simple electrochemical means (7).

From the very outset of our studies, we recognized that the "benign by design" philosophy being espoused by the Environmental Protection Agency (EPA) was not a unique consideration. Pharmaceutical firms have concurrently been mandated by the Federal Drug Administration (FDA) to market a single enantiomer of all proprietary drugs whenever possible. This concern translates into a need to accomplish highly stereocontrolled coupling reactions in water if adoption of aqueous-based chemistry is ever to materialize at the large-scale production level. Our attention was therefore directed initially to investigating the level of diastereoselectivity attainable in aqueous solvent systems by means of intermolecular chelation. An understanding of the limits to which solvation forces provided by the water molecules might damp those factors that customarily control facial selectivity was specifically sought. More recent aspects of our research program have emphasized useful transformations to which the initially formed products may be subjected.

Diastereoselective Allylation of Chiral Aldehydes

The results obtained for the allylation of aldehydes possessing α-oxy substituents of widely differing basicities are exemplified by the data contained in Table I (8). If operational, chelate control should deliver the syn diastereomer. The 9.8:1 syn/anti

Table I. Stereoselection in C-Allylation of α-Oxygenated Aldehydes
($CH_2=CHCH_2Br$, In, solvent, 25 °C)

R	H_2O	H_2O-THF (1:1)	THF
–SiMe$_2$t-Bu	1 : 3.9	1 : 4.2	1 : 4.0
–CH$_2$OCH$_3$	2.1 : 1	1.7 : 1	1.6 : 1
–H	9.8 : 1	—	—

ratio observed for the α-hydroxy derivative provided the first indication that Lewis acid-base interactions may play an important role in directing indium-promoted reactions conducted in aqueous media. The dropoff in basicity of the α-oxygenated substituent was met with increased levels of anti product and a lowering of reactivity. Thus, the more selective substrates were also the most prone to react rapidly (Table II). One implication of these findings is that added protection-deprotection maneuvers would no longer be necessary and would actually be detrimental to reaction efficiency.

Table II. Competitive Indium-Promoted Allylations in Water at 25 °C.

first aldehyde	second aldehyde	reaction time, h	product ratio first ald : second ald	
Cy-CH(OMOM)-CHO	TBSO-epoxide-CHO	4.0	2.6	1
Cy-CH(OMOM)-CHO	Cy-CH(OTBS)-CHO	5.0	3.2	1
Cy-CH(OH)-CHO	Cy-CH(OMOM)-CHO	3.5	11.1	1

Although increases in the acidity of the aqueous solution were found not to impact on product stereoselectivity, salt effects can prove beneficial (8), presumably as a consequence of the increased internal pressure brought about in the system. The sense of asymmetric induction conforms to operation of the illustrated Cram-like transition state (Scheme 1). This working model is consistent with the nondirective effects brought on by the sterically bulky α-oxy (OBn, OTBS), α-thia (PhS, MeS), and α-amino (Bn$_2$N, isoindolyl) groups (9). Under the latter circumstances, chelation is not observed and π-facial discrimination is achieved instead via Felkin-Anh transition states under the steric control of the substituents. The dimethylamino

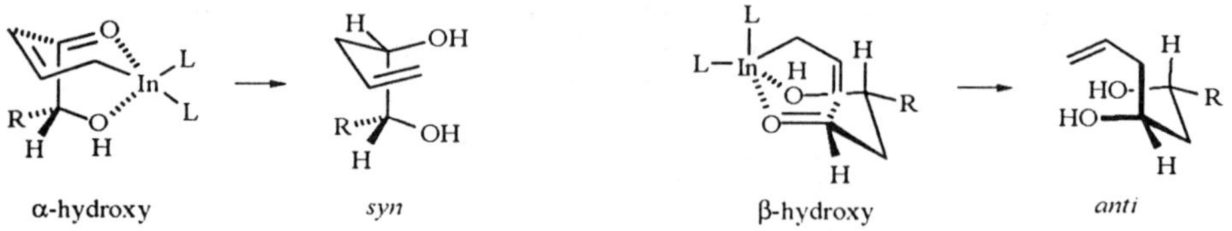

Scheme 1

functionality is responsive to chelation control and can lead to high levels of the syn diastereomer provided that the spatial demands of the second substituent are not excessive (Equation 1) (9).

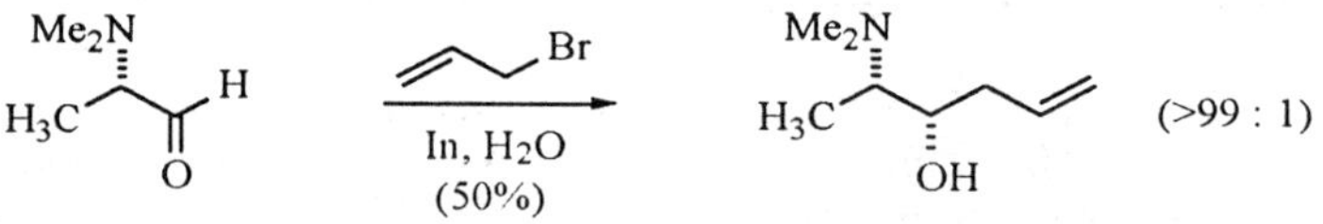

Equation 1

The diastereomeric ratios realized with several β-oxy aldehydes in three solvent systems including water are compiled in Table III. Adherence to a chelation-controlled pathway in this series translates into anti adduct formation. Good correlation with the α-alkoxy aldehydes is noted. For the β-hydroxy derivative, the pronounced facial selectivity suggests that structural rigidification once again occurs prior to nucleophilic attack (Scheme 1).

Competitive Intramolecular/Intermolecular Chelation Options

Despite the stereochemical outcomes and significant kinetic enhancements (10) detailed above, further proof that chelation can indeed operate in water with heightened efficiency was sought. The several reservations that had to be removed included any possibility that syn-1,2-diol production arises because the heteroatomic substituent constitutes the medium-sized group in the Felkin-Anh model (11,12), the inability of In(III) to form a chelation complex because of overriding solvation to water molecules, the alternative involvement of free-radical intermediates because of the

Table III. Indium-Promoted C-Allylations of β-Oxygenated Aldehydes in Various Solvents at 25 °C.

aldehyde	solvent	reaction time, h	product ratio syn	anti	yield,%
OH O (structure)	H₂O	2	1	8.5	77
	H₂O-THF (1:1)	2	1	8.2	74
	THF		No reaction		
OBn O (structure)	H₂O	2.5	1	1	80
	H₂O-THF (1:1)	2.7	1	1	84
	THF	10	1	1	72
	THF	8	1	1	82
TBSO O (structure)	H₂O	3.5	1	1	84
	H₂O-THF (1:1)	3.5	1.2	1	83
	THF	8.5	1.7	1	77
CH₃O O (structure)	H₂O	2.7	1	4	78
	H₂O-THF (1:1)	3	1	4	78
	THF	8.5	1	3.3	69
	THF	7.7	1	3.5	75

preferred operation of single electron transfer pathways, and the like (*13,14*)

For this purpose, recourse was made to allylic bromides that carry a stereocontrolling element of their own. The purpose was to elucidate the level and direction of the competitive involvement of intermolecular versus intramolecular modes of chelation when these options are allowed to vie for control of π-facial stereoselectivity (*15*). The contrasting response of methyl 4-bromocrotonate and methyl (Z)-2-(bromomethyl)-2-butenoate to 2- and 3-pyridinecarboxaldehyde and to glyoxylic acid (Table IV) constitutes a useful reference standard. In this instance, a

Table IV. Stereochemical Course of 1,2-Additions Involving Bromo Esters and Functionalized Aldehydes.

R	syn/anti	yield, %	R	syn/anti	yield, %
3-Py	22 : 78	62	3-Py	64 : 36	82
2-Py	94 : 6	81	2-Py	6 : 94	74
COOH	88 : 12	52	COOH	81 : 19	61

1:1 mixture of water and THF was used as the solvent in order to guarantee solubility on the part of all the reactants. It will be recognized that while the indium species derived from the bromo crotonate is not capable of internal coordination, its positional isomer is not similarly constrained. On this basis, the crossover in diastereoselectivity observed for 3-PyCHO, where extramolecular chelating events are not feasible, is as expected. Thus, the preferred formation of the syn isomer in the second example can be traced to a reversal in π-facial stereoselectivity anticipated from secondary binding to the ester carbonyl as shown. In both transition state representations **1** and **2**, coordination of the indium atom to the aldehyde carbonyl provides a suitable means for activation of this reagent and for S_N' delivery of the allyl fragment. An important consequence of an added flanking carbonyl substituent available for chelation is to reverse the π-surface that is amenable to nucleophilic attack.

The heavy predominance of syn adduct formation by the bromo crotonate when 2-PyCHO and glyoxylic acid are involved requires that a different reaction trajectory be kinetically favored. The only reasonable alternative is that intermolecular chelation as shown in **3** and **5** operates to a high level in these coupling processes. A simple change in the locus of the bromine atom is seen to be adequate to transform 2-PyCHO almost entirely into the anti product. These findings again dispel the notion that steric effects alone control the stereochemical course of allylations performed in aqueous environments. Indeed, it does appear that chelation of indium to the pyridine nitrogen is sufficient to override intramolecular ligation to the ester carbonyl. The selectivity associated with the final entry in Table IV reveals that coordination to the free carboxyl of glyoxylic acid (see **4**) is less important than that involving the ester as in **6** and **7**.

By every indicator we have uncovered, chelation to indium(III) is seen to operate in water where possible, with resultant control of reaction diastereoselectivity. Erosion of selectivity is sometimes encountered, but the causative factors are extrinsic to the chelation phenomenon. In this connection, a comparative analysis of additions involving crotylindium and 3-bromoallylindium to 2-hydroxypropanal is particularly informative (Scheme 2) (*16*). In the case where R is methyl, the proportion of the *syn,syn* and *syn,anti* isomers is maximized at 5.6:1, a value closely comparable to that for simple allylation (*syn/anti* = 7.5:1). When R = Br, the maximum value is only 2:1. In fact, the latter process qualifies as a fully nonselective process. The prevailing hypothesis is that this behavior is linked to facile geometric isomerization in the allylindium species as noted for the Grignard (*17*), potassium (*18*), and lithium derivatives (*19*).

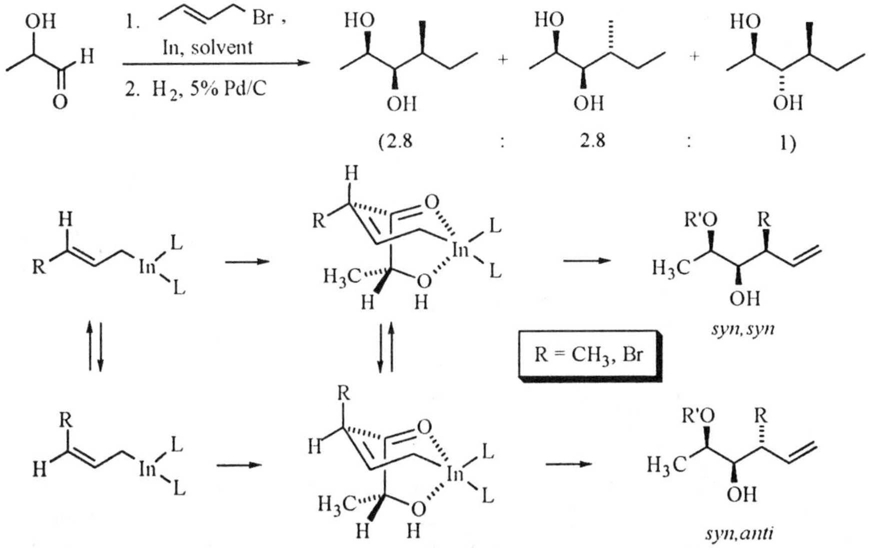

Scheme 2

Coupling Involving Geometrically Biased Allylic Bromides

If the significant dropoff in stereocontrol that materializes in certain cases is due to facile *E/Z* equilibration within the organometallic, a return to synthetically useful levels of asymmetric induction should be seen when stereochemically well defined allylic bromides are involved. Indeed, experimental tests of the ability to set three contiguous stereogenic centers in this manner under aqueous conditions have shown considerable promise (*20*).

For example, the Z form of methyl 2-(bromomethyl)-2-butenoate has been demonstrated to possess greater thermodynamic stability than the E isomer (Scheme 3) (21). Should this ordering carry over to the allylindium derivative and a high Felkin-Anh transition-state preference be exercised, the 3,4-*syn*;4,5-*anti* product should be dominant. The illustrated examples reveal that this is so. Only a modest diminution in coupling diastereoselectivity is observed as the relative size of the aldehyde substituent is increased from methyl to the phenyl and cyclohexyl levels (20).

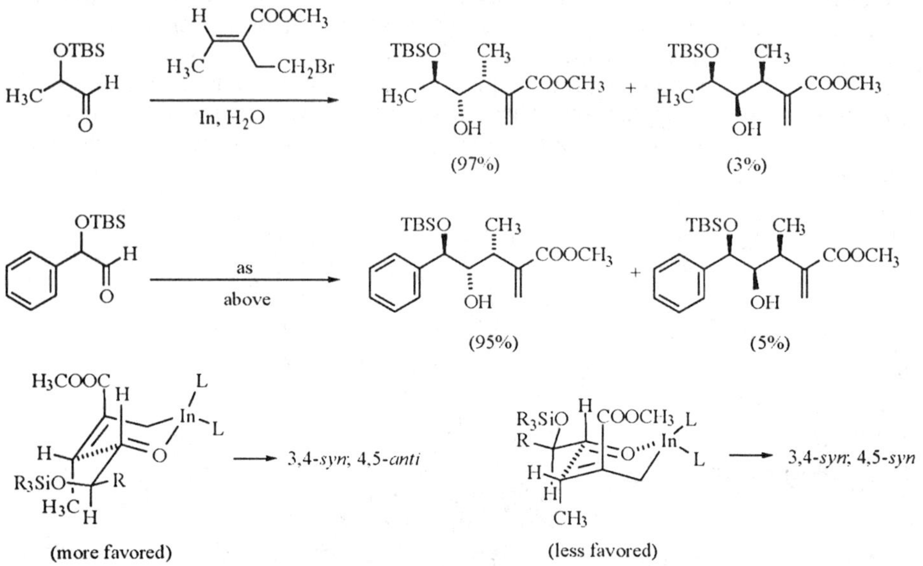

Scheme 3

When the halide is Z-configured as in cinnamyl bromide, the sense of asymmetric induction is reversed and the 3,4-*anti*;4,5-*anti* diastereomer is formed as the major product. Once again, this outcome can be rationalized in terms of the Felkin-Anh paradigm (Scheme 4). In this example, the disfavored transition state is significantly less sterically congested than the predominating alternative, and is adopted at a relatively higher percentage (28%) of the global reaction. Notwithstanding, persistent structural organization within the allylindium reagent is clearly an important determinant of stereocontrolled 1,2-addition to chiral aldehydes.

1,4-Asymmetric Induction

The cooperative interdependencies noted above can be relied upon to be conducive to effective long-range asymmetric stereoinduction in water under the proper circumstances (22,23). When the allyl bromide carries a bulky *tert*-butyldimethylsilyloxy substituent in close proximity to the double bond, the resulting level of *syn*-1,4-stereocontrol is impressively high (Scheme 5). A dropoff in π-facial discrimination occurs with decreasing size and increasing chelating capability of the group on the oxygen. The methoxy and hydroxy derivatives are representative examples where the relative importance of the favored transition state can recognizably be manipulated by proper positioning of the OH or OCH$_3$ group in the allylic halide reactant.

The simple expedient of shifting the oxygenated center in the bromide to a site one atom more distant from the reaction center has proven to be very informative. In these systems, the *O*-silylated derivatives exhibit only modest anti stereoselectivity while the hydroxy bromide gives rise to elevated levels of syn product (*24*). The governing interactions are considered to be fundamental to aqueous indium chemistry. In the first series where a nonchelating oxygen is featured, the stereocontrol element consists of nonbonded steric interactions. The entirely different diastereocontrol that is operational when the flanking oxygen is unprotected is chelative in origin.

Diastereoselective 1,2-Additions to α-Oxygented Ketones

The stereochemical course of the capture of several 2-methoxycyclohexanones and tetrahydrofuranspiro-(2-cyclohexanones) by several allylmetal reagents has been elucidated (*25*). In the simplest methoxy case illustrated here, equatorial attack is generally favored but only to minimal levels (Table V). Recourse to indium resulted in the formation of significantly increased proportions of this diastereomer. In the spirocyclic substrate, steric effects favor predominant attack trans to ether oxygen except when indium is involved. The increased ability of In(III) to anchor onto the neighboring heteroatom is unmistakingly revealed.

Table V. Facial Selectivity in Nucleophilic Additions.

reagent	solvent (T, °C)	chelate/non-chelate ratio	yield, %	reagent	solvent (T, °C)	chelate/non-chelate ratio	yield, %
$CH_2=CHCH_2MgCl$	THF, 0	2.3 : 1	94	$CH_2=CHCH_2MgCl$	THF, 0	1.2 : 1	88
$CH_2=CHCH_2MgCl$, $CeCl_3$	THF, 0	4.5 : 1	82	$CH_2=CHCH_2MgCl$, $CeCl_3$	THF, 0	1 : 2.4	96
$CH_2=CHCH_2Br$, $CrCl_2$	THF, 0	1.2 : 1	78	$CH_2=CHCH_2Br$, $CrCl_2$	THF, 0	1 : 2.6	90
$CH_2=CHCH_2Br$, In	THF, 25	9.0 : 1	84	$CH_2=CHCH_2Br$, In	THF, 25	5.6 : 1	82
$CH_2=CHCH_2Br$, In	THF-H_2O (1:1), 25	14.1 : 1	93	$CH_2=CHCH_2Br$, In	THF-H_2O(1:1), 25	3.9 : 1	95
$CH_2=CHCH_2Br$, In	H_2O, 25	12.5 : 1	95	$CH_2=CHCH_2Br$, In	H_2O, 25	2.7 : 1	81
$CH_2=CHCH_2Br$, In	THF, 25	12.2 : 1	83	$CH_2=CHCH_2Br$, In	THF, 25	3.4 : 1	72

The extent to which cooperation between the α-oxygen atom and control of π-facial nucleophilic attack reaches a maximum (>97:3) is when the system has been rigidified conformationally and the 2-methoxy and 4-*tert*-butyl substituents are both equatorially oriented. Since allylation reactions performed with indium under aqueous conditions are generally far more stereoselective toward α-alkoxy cyclohexanones than are other organometallics under anhydrous conditions, this chemistry merits consideration as a synthetically useful operation.

2-Hydroxycyclohexanones undergo allylindation rapidly and reflect exceptionally high stereocontrol irrespective of whether the unprotected OH is equatorially or axially disposed (*26*). In contrast, 3-hydroxycyclohexanones exhibit none of these characteristics. A systematic investigation involving the allylindium reagent and 6-

substituted-2-hydroxy-1-tetralones demonstrated levels of stereochemical bias that were closely in line with the normal predilection of 2-cyclohexenones for axial attack (27). Although electronic effects were made evident in the context of competition studies, the minimization of torsional effects appears to be the dominant force.

Following the discovery that the indium-promoted addition of functionalized allyl bromides to N-benzyl-2,3-azetidinedione under aqueous conditions can be highly diastereoselective (28), attempts were made to define long-range stereocontrol elements (29). Two azetidinediones carrying (S)-α-methylbenzyl and (R)-α-(1-naphthyl)ethyl residues at the nitrogen center were found to exhibit diastereofacial selectivity directly linked to their R or S configuration (Scheme 6). This

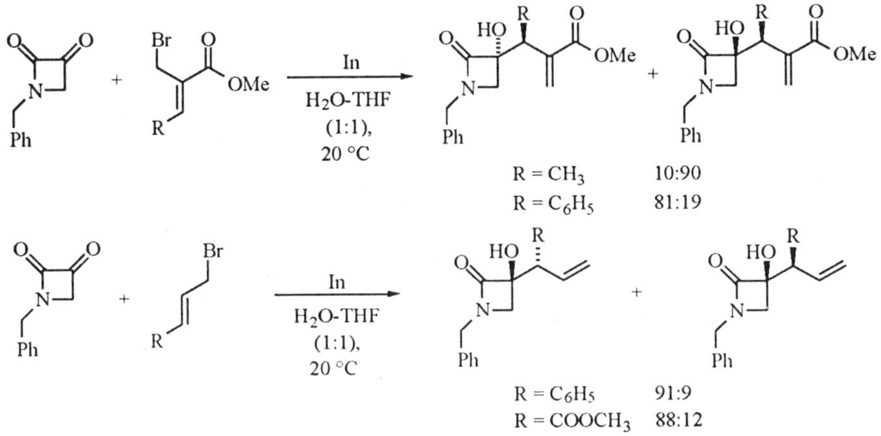

Scheme 6

crossover provides experimental evidence in strong support of cyclic transition states having the capacity for multi-point steric compression as shown in Scheme 7.

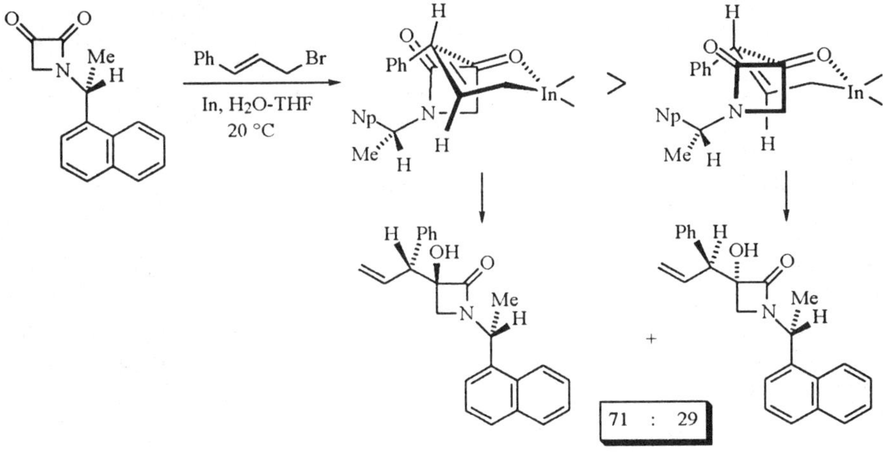

Scheme 7

An Indium-Promoted Alternative to the Knoevenagel Condensation

The Knoevenagel reaction is a carbon-carbon bond-forming reaction having many applications in organic synthesis. A major restriction involves an inability to arrest the coupling to aliphatic aldehydes at the monoaddition stage since these intermediates are highly reactive Michael acceptors in their own right. A simple protocol that skirts this complication has recently been defined (Scheme 8) (29). The two-step process involves initial coupling to methyl (E)-4-bromo-3-methoxycrotonate with indium in the presence of water. The resulting β-hydroxy ester is then subjected to mild acidic hydrolysis. While formaldehyde fares well in the initial allylation step, the subsequent dehydrative conversion to methyl methyleneacetoacetate failed due to the high reactivity of this product.

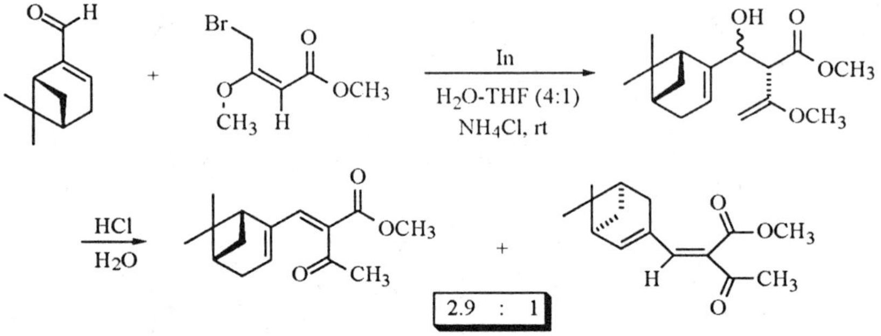

Scheme 8

Formation of α-Methylene-γ-lactones Fused to Medium and Large Rings

The widespread occurrence of α-methylene-γ-lactones in nature goes unquestioned. Many of these are found fused to carbocyclic rings ranging from six- to fourteen-membered. Although methods are available to access this class of compounds, a simpler and more convenient tactic was considered desirable. The new route begins with two ω-unsaturated aldehydes of the same or different chain length (Scheme 9) (30).

The first aldehyde is coupled to methyl acrylate under Baylis-Hillman conditions and the resulting alcohol is activated as the rearranged allylic bromide. The indium-promoted step that follows is fully convergent, permits selection of the desired chain length, and sets the stage for intermolecular allylation and acid-promoted cyclization. The cis and trans lactone isomers are chromatographically separated before advancing into the ring-closing olefin metathesis step. It will be recognized that the order in which the aldehydes are deployed dictates the relative positioning of the intracyclic double bond in the product.

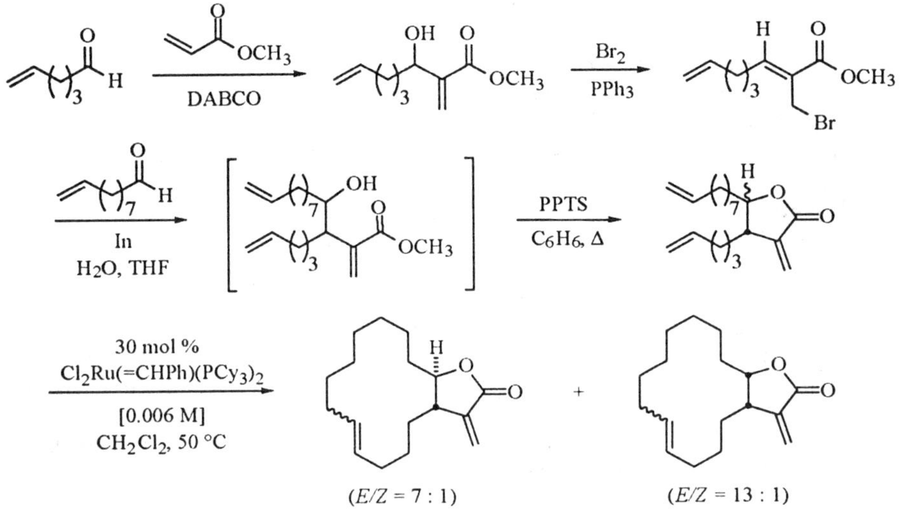

Scheme 9

Summary

The indium-promoted nucleophilic carbonyl addition reaction in water has emerged as a premier method for effecting C-C bond formation under aqueous conditions. The widespread adoption of this technology would eliminate a significant fraction of emissions arising from organic solvent evaporation and go far to offset this aspect of environmental pollution. High levels of stereoselectivity are attainable, protecting groups may be unnecessary, and operational safety arising from greatly reduced flammability concerns is greatly increased. In every instance, the spent indium metal can be recovered totally by simple electrolytic methods. Our experience with indium(0) recovered in this way is to observe greater reactivity than the commercial powdered form.

The benefits associated with this technology will, of course, be demonstrated once it is adopted at production-level scales. The results obtained by our research team and the several others active in this field have laid a rich foundation for bringing this important objective to ultimate realization.

Acknowledgments

The research described in this chapter was funded sequentially by the Emissions Reduction Research Center at the New Jersey Institute of Technology, The US Environmental Protection Agency, and the Paquette Research Fund.

Literature Cited

1. Li, C.-J. *Chem. Rev.* **1993**, *93*, 2023-2035.
2. Lubineau, A.; Augé, J.; Queneau, Y. *Synlett* **1995**, 1087-1096.
3. Li, C.-J. *Tetrahedron* **1996**, *52*, 5643-5668.
4. Li, C.-J.; Chan, T.-H. *Organic Reactions in Aqueous Media*; John Wiley: New York, 1997.
5. Paquette, L. A. In *Green Chemistry: Frontiers in Benign Chemical Synthesis and Processing*; Oxford University Press: Oxford, England, 1998; pp 250-264.
6. Cintas, P. *Synlett* **1995**, 1087-1096.
7. Prenner, R. H.; Binder, W. H.; Schmid, W. *Liebigs Ann. Chem.* **1994**, 73-78.
8. Paquette, L. A.; Mitzel, T. M. *J. Am. Chem. Soc.* **1996**, *118*, 1931-1937.
9. Paquette, L. A.; Mitzel, T. M.; Isaac, M. B.; Crasto, C. F.; Schomer, W. W. *J. Org. Chem.* **1997**, *62*, 4293-4301.
10. Chen, X.; Hortelano, E. R.; Eliel, E. L.; Frye, S. V. *J. Am. Chem. Soc.* **1990**, *112*, 6130-6131; **1992**, *114*, 1778-1784.
11. Cherest, M.; Felkin, H.; Prudent, N. *Tetrahedron Lett.* **1968**, 2199-2204.
12. Ahn, N. T.; Eisenstein, O. *Nouv. J. Chim.* **1977**, *1*, 61-70.
13. Chan, T. H.; Li, C.-J. *J. Chem. Soc., Chem. Commun.* **1992**, 747-748.
14. Carda, M.; Castillo, E.; Rodriquez, S.; Murga, J.; Marco, J. A. *Tetrahedron: Asymmetry* **1998**, *9*, 1117-1120.
15. Paquette, L. A.; Rothhaar, R. R. *J. Org. Chem.* **1999**, *64*, 217-224.
16. Paquette, L. A.; Mitzel, T. M. *J. Org. Chem.* **1996**, *61*, 8799-8804.
17. Whitesides, G. M.; Nordlander, J. E.; Roberts, J. D. *J. Am. Chem. Soc.* **1984**, *84*, 2010-2011.
18. Schlosser, M.; Hartmann, J. *J. Am. Chem. Soc.* **1976**, *98*, 4674-4676.
19. Bates, R. B.; Beavers, W. A. *J. Am. Chem. Soc.* **1974**, *96*, 5001-5002.
20. Isaac, M. B.; Paquette, L. A. *J. Org. Chem.* **1997**, *62*, 5333-5338.
21. Buchholz, R.; Hoffmann, H. M. R. *Helv. Chim. Acta* **1991**, *74*, 1213-1220.
22. Maguire, R. J.; Mulzer, J.; Bats, J. W. *J. Org. Chem.* **1996**, *61*, 6936-6940.
23. Paquette, L. A.; Bennett, G. D.; Chhatriwalla, A.; Isaac, M. B. *J. Org. Chem.* **1997**, *62*, 3370-3374.
24. Paquette, L. A.; Bennett, G. D.; Isaac, M. B.; Chhatriwalla, A. *J. Org. Chem.* **1998**, *63*, 1836-1845.
25. Paquette, L. A.; Lobben, P. C. *J. Am. Chem. Soc.* **1996**, *118*, 1917-1930.
26. Paquette, L. A.; Lobben, P. C. *J. Org. Chem.* **1998**, *63*, 5604-5616.
27. Lobben, P. C.; Paquette, L. A. *J. Org. Chem.* **1998**, *63*, 6990-6998.
28. Paquette, L. A.; Isaac, M. B. *Heterocycles* **1998**, *47*, 107-110.
29. Paquette, L. A.; Kern, B. E.; Méndez-Andino, J. *Tetrahedron Lett.*, **1999**, *40*, 4129.
30. Paquette, L. A.; Méndez-Andino, J. *Tetrahedron Lett.*, **1999**, *40*, 4301.

Biocatalysis and Biosynthesis

Chapter 10

Biocatalytic Production of 5-Cyanovaleramide from Adiponitrile

Robert DiCosimo[1], Eugenia C. Hann[1], Amy Eisenberg[1], Susan K. Fager[1], Neal E. Perkins[1], F. Glenn Gallagher[1], Susan M. Cooper[1], John E. Gavagan[1], Barry Stieglitz[2], and Susan M. Hennessey[2]

[1]DuPont Central Research and Development, [2]DuPont Agricultural Products, Experimental Station, P.O. Box 80328, Wilmington, DE 19880-0328

A biocatalytic process for the hydration of adiponitrile to 5-cyanovaleramide has been demonstrated which can be run to higher conversion, produces more product per weight of catalyst, and generates significantly less waste products than alternate chemical processes. Immobilization of *Pseudomonas chlororaphis* B23 cells in calcium alginate beads produced a catalyst having high nitrile hydratase activity, and excellent stability when recycled in consecutive batch reactions. A total of 13.6 metric tons of 5-cyanovaleramide was produced in 93% yield and 96% selectivity, with a catalyst productivity of 3,150 kg of 5-cyanovaleramide per kg *P. chlororaphis* B23 cells (dry cell weight).

The first step in the manufacture of a new crop protection chemical required the conversion of adiponitrile (ADN) to 5-cyanovaleramide (5CVAM). The hydration of nitriles to amides can be accomplished using a variety of chemical catalysts (*1-5*), and manganese dioxide (*5*) produced 5CVAM with the highest yield and regioselectivity. When a stoichiometric amount of water was reacted with neat ADN at 130 °C using manganese dioxide as catalyst, 5CVAM could be produced at 80 % selectivity if the reaction was run to only ca. 25 % conversion. At higher conversions, the selectivity to 5CVAM decreased significantly as the selectivity to byproduct adipamide (ADAM) increased. Product isolation required dilution of the hot product mixture with toluene and filtration using a filter aid, where the insoluble ADAM remained in the spent manganese dioxide filter cake. Cooling the resulting hot toluene filtrate precipitated 5CVAM, and the large amount of unconverted ADN could subsequently be recovered for recycle by concentration of the remaining toluene solution. Approximately 1.25 kg of catalyst waste (manganese dioxide and filter aid) was generated for each 1 kg of 5CVAM produced, so only small quantities of 5CVAM were prepared for pilot studies using this procedure. The cost and environmental impact of disposal or recycle of the spent, deactivated manganese dioxide catalyst led to the development of an alternate, green process for 5CVAM production.

The nitrile hydratase (EC 4.2.1.84) produced by a variety of bacteria and fungi can readily catalyze the hydration of aliphatic nitriles to the corresponding amides (*6-11*). For the conversion of ADN to 5CVAM, a regioselective nitrile hydratase was required to limit the further conversion of 5CVAM to ADAM. The absence of amidase activity in the microbial cell catalyst was also important to prevent the subsequent hydrolysis of the amide to a carboxylic acid. Bacterial strains of *Bacillus* (*12*), *Bacteridium* (*12*), *Brevibacterium* (*12*), *Micrococcus* (*12*), *Pseudomonas* (*13*), and *Acinetobacter* (*14*) each produce a nitrile hydratase capable of regioselective hydration of aliphatic nitriles, and a variety of these strains were evaluated for the conversion of ADN to 5CVAM. The biocatalytic process which has been developed and commercialized for the production of 5CVAM (*15*) can be run to higher conversion, produces more product per weight of catalyst, eliminates the use of toluene for purification of CVAM, and generates less than 1 % of the catalyst waste produced in the chemical process.

Results and Discussion

Microbial catalysts having nitrile hydratase activity were obtained from publicly-held culture collections, or were isolated from soil samples using one of several aliphatic nitriles or amides as carbon and/or nitrogen source. Of the whole-cell catalysts that were initially screened, two which exhibited high regioselectivity for the hydrolysis of ADN to 5CVAM (Scheme 1) were *Pseudomonas putida* 3LG-1-5-1A (ATCC 55736; 97% regioselectivity) (*13*), and *Rhodococcus sp.* A4 (formerly *Brevibacterium sp.* R312 strain A4, Technische Universiteit Delft, LMD#79.2; 97% regioselectivity) (*12,16*). The stability of the nitrile hydratase activity of each of these whole-cell catalysts was temperature dependent, with the greatest stability under reaction conditions occurring at 5 °C; increasing reaction temperatures above 5 °C resulted in a more rapid loss of nitrile hydratase activity. Whole cells of 3LG (ca. 4,000 ADN IU/g wet cell weight (wcw)) or A4 (ca. 24,000 ADN IU/g wcw) were initially immobilized in polyacrylamide gel (PAG) (*17*) and the resulting catalyst used in consecutive batch reactions with catalyst recycle. Under identical reaction conditions, the 3LG/PAG catalyst lost a higher percentage of nitrile hydratase activity in consecutive batch reactions than the A4/PAG catalyst, therefore A4 was chosen for further development.

Scheme 1

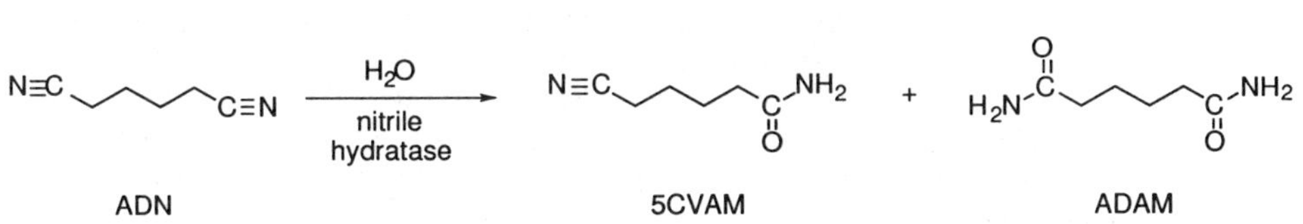

Light Activation of A4 Whole Cells

Prior to use as catalyst for the conversion of ADN to 5CVAM, or for the preparation of an immobilized whole-cell catalyst, A4 cells required irradiation by UV light to activate the nitrile hydratase enzyme which converts ADN to 5-CVAM (*18-20*). The optimum wavelengths for activation are 280 nm and 370 nm, with a slower rate of activation occurring at 370 nm (*18*). When stored in the absence of light, the light-activated enzyme subsequently lost activity, but storage of the cells or enzyme in the presence of 20 mM sodium butyrate stabilized the active form of the nitrile hydratase (*18*). A determination of the parameters which affected the light activation of A4 cell suspensions was performed, and the activation of intracellular nitrile hydratase activity was found to be dependent on the amount of time the cells were suspended in aqueous butyrate buffer before light activation, the ratio of cells to butyrate (wt/wt) in the suspension, the weight percent of cells in the suspension, the surface area/volume ratio of the cell suspension during irradiation, and the flux of UV light at 370 nm/cells/time.

A4 cells having active nitrile hydratase were prepared by first growing the cells by fermentation, then pumping the resulting cell suspension through a 3 kilowatt UV lamp (370 nm) prior to separation of the cells by centrifugation, and storage of the resulting cell paste at - 80 °C. Light-activated A4 cells were stored for over one year at - 80 °C with no significant loss of nitrile hydratase activity. Frozen A4 cell paste was subsequently thawed and suspended in 20 mM sodium butyrate buffer at pH 7.0 and 5 °C, then mixed with a sodium alginate solution, and the resulting cell suspension dripped into 0.20 M calcium chloride/23 mM sodium butyrate at pH 7.0 and 5 °C to produce calcium-crosslinked alginate beads containing entrapped A4 cells (*21*).

Comparison of A4 and *Pseudomonas chlororaphis* B23

Shortly before the planned commercialization of the process using immobilized A4 cells, an additional microbial catalyst was identified which produced 5CVAM with high regioselectivity. *Pseudomonas chlororaphis* B23 was first isolated and characterized by H. Yamada and coworkers (*22-24*), and subsequently employed by the Nitto Chemical Industry Co. (now merged with Mitsubishi Rayon Co.) for the commercial production of acrylamide from acrylonitrile (*25-28*). The measured specific activity of B23 cells was ca. 44,000 ADN IU/g wet cell weight. The hydrolysis of 0.40 M ADN to 5CVAM by either A4/alginate beads or B23/alginate beads was compared by running reactions in a fixed-bed, stainless steel column at 5 °C (Figure 1). The column feed also contained 23 mM sodium butyrate, both to prevent the loss of A4 nitrile hydratase activity in the absence of light, and to stabilize the activity of both A4 and B23 nitrile hydratase (*18, 22*). The selectivity to 5CVAM in a fixed-bed packed column was found to be lower than when running a stirred-batch reaction due to an increase in the hydration of 5CVAM to ADAM in the fixed bed configuration, but the fixed-bed column more efficiently demonstrated catalyst stability and productivity over long reaction times.

A slow decrease in the conversion of ADN to 5CVAM by A4/alginate beads was observed to occur from the very start of the fixed-bed column reaction, where by 700 hours of continuous operation at a constant feed rate, ca. 20 % of ADN remained

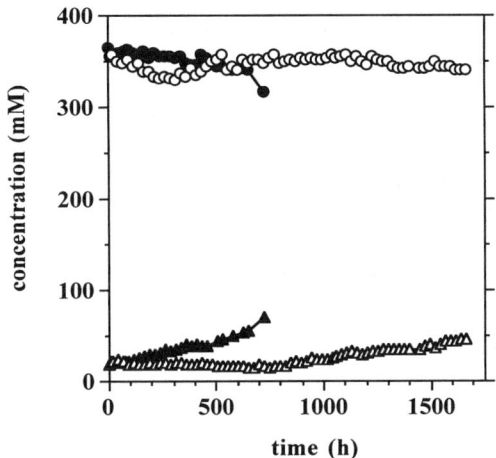

Figure 1. Stability of A4 or B23/alginate beads (20 % wcw, 2.75 % sodium alginate) in a fixed-bed column reactor at 5 °C, feeding 0.4 M ADN in 23 mM sodium butyrate/5 mM calcium chloride (pH 7.0): ●, *5CVAM(A4);* ○, *5CVAM(B23);* ▲, *ADN(A4);* △, *ADN(B23).*

unconverted. Over the same period of time, the B23/alginate catalyst lost no significant activity. At 5 °C, the average yields of 5CVAM and ADAM produced over the first 800 h of continuous operation with the B23/alginate catalyst were 85 % and 10 %, respectively, with a 95 % conversion of ADN (5CVAM:ADAM = 8.5:1); after 800 h, the catalyst began to slowly lose activity. The B23 catalyst did not require light activation prior to immobilization, and was superior to A4 in specific activity, enzyme stability, and 5CVAM productivity (kg 5CVAM/kg biocatalyst/h) in both fixed-bed column and stirred batch reactions, and was therefore substituted for A4 prior to commercialization. Because of the lower selectivity to 5CVAM in the fixed-bed column reaction, and the possibility of precipitation and plugging of the column by the poorly-soluble ADAM byproduct, the stirred batch reaction was selected for scale-up.

Optimization of reaction conditions using immobilized B23 cells

B23 cells must be rapidly separated from the product mixture after complete conversion of ADN to prevent the continued conversion of 5CVAM to ADAM. Filtration or centrifugation for catalyst separation and recovery could not be performed on a commercial scale in a sufficiently short time. Several methods of cell immobilization were examined to produce a catalyst which could be readily separated from the product mixture, and which would also improve the stability of the whole cells against lysis in a two-phase mixture of ADN and aqueous buffer. An immobilization method was required that was simple to perform on a large scale, and produced an immobilized catalyst that had a high specific activity (IU/g catalyst), and was stable to the reaction conditions employed. Immobilization by entrapment in agarose (29) or carrageenan (30) could not be utilized, since the nitrile hydratase activity of B23 cells was rapidly deactivated at the temperatures required for immobilization in these matrices (35-50 °C). Immobilization in either PAG or calcium alginate produced a catalyst with high specific activity, but under identical reaction conditions the B23/alginate catalyst was more stable, and produced more product per weight of catalyst than B23/PAG catalyst.

The dependence of specific activity (ADN IU/gram catalyst) of the B23/alginate beads on pH at 5 °C was determined. The specific activity decreased markedly below pH 7.0, and was fairly constant from pH 7.5 to 9.0. As the optimum pH stability for B23 nitrile hydratase was between pH 6.5 and 7.5, reactions were typically run in a pH range of from 6.8 to 7.2. The dependence of B23/alginate bead specific activity on temperature was also measured; an increase in reaction temperature from 5 °C to 10 °C resulted in an increase in catalyst specific activity (and for a given catalyst loading, reaction rate) of ca. 60 %. Specific activity also increased with decreasing diameter of the B23/alginate beads (measured over a range of from 3.5 mm to 2.0 mm diameter), indicating that the reaction rate was limited at least in part by the rate of diffusion of ADN into the alginate bead.

B23/alginate beads (ca. 2.5 mm dia.) were used as catalyst for the production of 1.5 M 5CVAM in up to 60 consecutive batch reactions with catalyst recycle at 5 °C and in 23 mM butyrate/5 mM calcium chloride (pH 7.0). After the first reaction, the final concentration of 5CVAM in subsequent batch reactions exceeded 1.7 M, as consecutive reactions were run by decanting the product mixture from the previous reaction, and adding fresh buffer and ADN to the catalyst and remaining product mixture from the previous batch reaction (ca. 16 wt % of the initial reaction mixture).

The ratio of 5CVAM to ADAM with alginate beads made using frozen/thawed B23 cells decreased from 36:1 to ca. 24:1 after the first several catalyst recycles, while alginate beads prepared using B23 cells that had not undergone a freeze/thaw cycle produced a final ratio of 5CVAM to ADAM of ca. 15:1 under these same conditions (Figure 2). This difference in selectivity was most likely due to an increase in permeability of the cells to ADN and 5CVAM, caused by the freezing and thawing of the cells prior to immobilization (*31, 32*). The improvement in 5CVAM selectivity was an additional benefit of storing B23 cell suspensions frozen at -80 °C prior to immobilization in alginate beads.

In stirred-batch reactions which produced 1.5 M 5CVAM using B23/alginate beads at 5 °C, the yields of 5CVAM and ADAM were typically 95 % and 3.5 % respectively, with 1.2 % ADN remaining at the reaction endpoint. If the reaction was allowed to continue past the point of complete conversion of ADN (Figure 3), 5CVAM was further converted to ADAM, which would precipitate from the reaction mixture after reaching a concentration of ca. 110-120 mM; this ADAM concentration was more than twice the measured solubility limit for ADAM under these reaction conditions, and may indicate that the product mixture was supersaturated with ADAM. When reactions were run to produce > 1.5 M 5CVAM, no precipitation of ADAM was observed in these reaction mixtures if the catalyst was immediately separated from the product mixture at the end of the reaction. Some B23/alginate beads recovered after use in 60 consecutive batch reactions to produce > 1.5 M 5CVAM had ADAM precipitated within the beads, but a significant decrease in catalyst activity for these catalyst beads was not observed.

Commercial-scale preparation of 5CVAM.

Scale-up of production of B23/alginate catalyst beads was accomplished as follows. A 100-gallon reactor was charged 140 L of 23 mM sodium butyrate buffer (pH 6.8 - 7.2), then 6.42 kg of sodium alginate (Pronova Protanal® LF 10/60, 2.75 % w/w final concentration) was added and the resulting mixture heated to 80 °C for 1 h with stirring. The resulting solution was cooled to 10 °C, then 87.2 kg of frozen *P. chlororaphis* B23 cell slurry (54 % wcw/slurry weight, 12% dry cell weight/slurry weight) was added with stirring while maintaining the temperature between 3 - 7 °C. After the final addition of frozen cell slurry, the suspension was stirred until all the cells were suspended, then the cell suspension was degassed under partial vacuum for an additional 2 h. The resulting cell slurry was maintained at 3 - 7 °C while pumping through a 230 micron and 140 micron filter, then through a 7-inch diameter stainless steel nozzle equipped with 288 20-gauge hollow stainless steel tubes (3.5 cm in length) attached to the top of a 300-gallon reactor containing 757 L of 0.20 M calcium chloride/23 mM sodium butyrate hardening buffer at 5 °C. After the dropwise addition of the cell slurry was complete, the resulting B23/alginate beads were stirred for an additional 2h, then the stirring was stopped and the hardening buffer decanted. An equivalent volume of fresh hardening buffer at 5 °C was added to the catalyst beads and the mixture stirred an additional 16 h at 5 °C. The hardening buffer was again decanted, and the catalyst beads washed twice with ca. 400 L of 23 mM sodium butyrate/5 mM calcium chloride buffer (pH 7.0) at 5 °C to yield 201 kg of B23/alginate beads. The beads were stored in the wash buffer at 5 °C until needed.

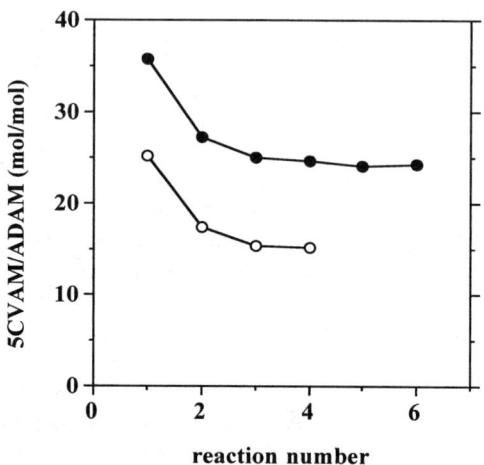

Figure 2. Ratio of 5CVAM/ADAM produced in consecutive batch reactions with recycle of B23/alginate beads prepared using frozen (●) or unfrozen (○) B23 cells. Reactions were run to produce 1.5 M 5CVAM in two-phase reaction mixtures containing ADN and 7.5 % (w/w) B23/alginate beads in 23 mM sodium butyrate/5 mM calcium chloride (pH 7.0) at 5 °C.

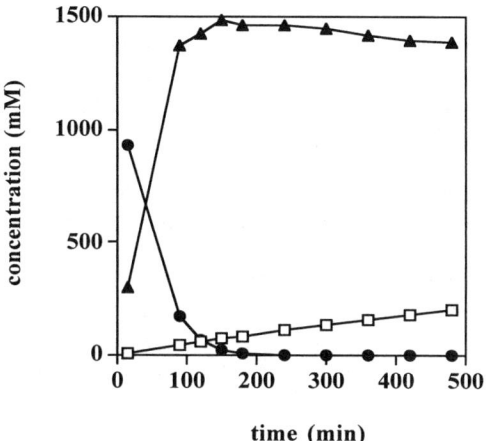

Figure 3. Time course for the production of 5CVAM in a two-phase reaction mixture initially containing 16.2 wt % ADN and 10 wt % B23/alginate beads in 23 mM sodium butyrate/5 mM calcium chloride (pH 7.0) at 5 °C: ●, *ADN;* ▲, *5CVAM;* ❑, *ADAM.*

For production of 5CVAM, a 500-gallon reactor was charged with 1007 L of 23 mM sodium butyrate/5 mM calcium chloride buffer (pH 7.0) at 5 °C and 55.4 kg of B23/alginate beads, followed by 218 kg (2,015 mol) of adiponitrile. The resulting mixture was stirred at 5 °C under nitrogen, and samples were withdrawn at regular intervals and analyzed by HPLC to determine the progress of the reaction. When the conversion of adiponitrile was > 97 % (ca. 4 h initial reaction time), the stirring was stopped, the catalyst beads immediately settled to the bottom of the reactor, and ca. 90 % of the product mixture was decanted away from the catalyst. The reactor was immediately charged with 1007 L of reaction buffer and 218 kg (2,015 mol) of adiponitrile, and the reaction repeated. After running thirteen consecutive batch reactions to establish a reaction baseline, a final addition of 20.4 kg of B23/alginate beads was added to the fourteenth consecutive batch reaction (5.8 wt % final catalyst weight /weight reaction mixture). Each decanted product mixture was briefly heated to 40 °C to deactivate any catalyst activity that was present. The reaction time for reaction number 56 was allowed to extend for four hours past the endpoint of the reaction, then two more reactions were run to determine if the increased production and precipitation of ADAM in reaction number 56 produced a loss of catalyst activity in subsequent reactions; no increase in reaction time for the final two batch reactions was observed.

A total of fifty-eight consecutive 400-gallon batch reactions were run, using 12.7 metric tons of ADN; the reaction times and concentrations of 5CVAM, ADN and ADAM for each reaction are illustrated in Figure 4. The yields of 5CVAM and ADAM from the combined product mixtures were 93 % and 4 % respectively, with 3 % recovery of unconverted ADN. The reaction time increased from 3.5 h to 6 h over the course of the fifty-eight reactions. At 97 % overall conversion of ADN, the combined yield of recovered 5CVAM was 13.6 metric tons. The catalyst productivity was 3,150 kg of 5CVAM produced per kg of B23 cells (dry cell weight). The total weight of product mixture produced was 70.7 metric tons, which contained 19.2 wt % 5CVAM, 0.99 wt % ADAM, and 0.52 wt % ADN. The 5CVAM was recovered from the product mixture by removal of water by distillation under reduced pressure, then dissolution of the resulting oil at > 65 °C in methanol, which precipitated the byproduct ADAM as well as calcium and butyrate salts. After removal of the byproducts by filtration, the methanolic 5CVAM solution was used directly in the next process step for herbicide synthesis.

Conclusions

As an alternative to chemical catalytic methods for the production of 5CVAM from ADN, a biocatalytic process has been developed and commercialized. Immobilization of *Pseudomonas chlororaphis* B23 cells in calcium alginate beads produced a catalyst having high nitrile hydratase activity, and excellent stability when recycled in consecutive batch reactions. The biocatalytic reaction was run to higher conversion, produced more product per weight of catalyst, and generated significantly smaller amounts of byproducts and waste products than alternate chemical processes which were examined. In particular, the ratio of catalyst waste to 5CVAM produced in the biocatalytic process was very low; only 0.006 kg of catalyst waste was produced per kg 5CVAM, and this catalyst waste was 93% water by weight. The

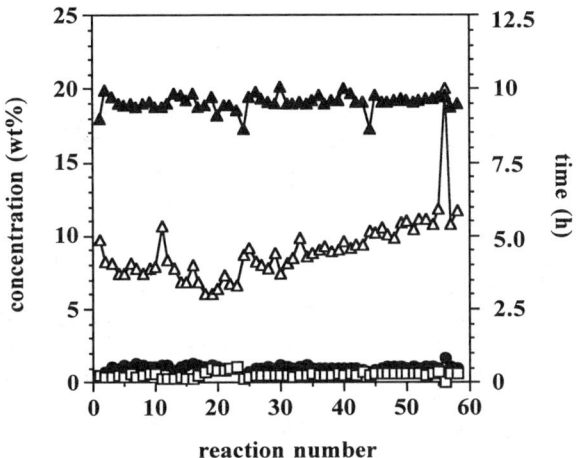

Figure 4. Reaction time (Δ) and wt % 5CVAM (▲), ADAM (●) and ADN (□) for fifty-eight consecutive batch reactions with catalyst recycle to produce a total of 13.6 metric tons of 5CVAM. Reactions were run in two-phase reactions containing ADN and 5.8 wt % (final weight catalyst/weight reaction mixture) B23/alginate beads in 23 mM sodium butyrate/5 mM calcium chloride (pH 7.0) at 5 °C.

reaction produced ca. 190 g 5CVAM per liter of reaction volume, and purification and recovery of the product for use in the next process step was readily performed.

Acknowledgments

The authors gratefully acknowledge the contributions of L. Winnie Wagner, Robert D. Fallon, Mark J. Nelson, David L. Anton, Ron Grosz, Umesh Hattikadur, John Freudenberger, Rafael Shapiro, Onorato Campopiano, and the Agricultural Products Semiworks (DuPont), Coreen R. B. Reed (Pronova Biopoylmer, Inc.), and Tetsuro Horinouchi and Katsumi Nakamura (Mitsubishi Rayon Co.). Sections of the Results and Discussion which describe optimization of reaction conditions using immobilized B23 cells, and commercial-scale preparation of 5CVAM, are reprinted from reference 15 (Copyright 1999) with permission from Elsevier Science Ltd.

Literature Cited

1. Izumi, Y. *Catal. Today* **1997**, *33*, 371 - 409.
2. Larock, R. C., *Comprehensive Organic Transformations: A Guide to Functional Group Preparations*; VCH: New York, 1989; p 993.
3. Schaefer, F. C., In *The Chemistry of the Cyano Group*; Rappoport, Z., Ed.; Interscience: New York, NY 1970; Chapter 6, pp 256 - 262.
4. Sugiyama, K.; Miura, H.; Nakano, Y.; Suzuki, H.; Matsuda, T. *Bull. Chem. Soc. Jpn.* **1987**, *60*, 453 - 456.
5. Onuoha, N. I.; Wainwright, M. S. *Chem. Eng. Commun.* **1984**, *29*, 1 - 12.
6. Holland, H. L. *Curr. Opin. Chem. Biol.* **1998**, *2*, 77 - 84.
7. Sugai, T.; Yamazaki, T.; Yokoyama, M.; Ohta, H., *Biosci. Biotech. Biochem.* **1997**, *61*, 1419 - 1427.
8. Meth-Cohn, O.; Wang, M.-X. *J. Chem. Soc., Perkin Trans. 1* **1997**, 3197-3204.
9. Crosby, J.; Moilliet, J.; Parratt, J. S.; Turner, N. J. *J. Chem. Soc., Perkin Trans. 1* **1994**, Issue 13, 1679-1687
10. De Raadt, A.; Klempier, N.; Faber, K.t; Griengl, H. *J. Chem. Soc., Perkin Trans. 1* **1992**, 137 - 140.
11. Yokoyama, M.; Sugai, T.; Ohta, H. *Tetrahedron: Asymmetry* **1993**, *4*, 1081 - 1084
12. Andresen, O.; Godtfredsen, S. E. European Patent 178,106B1, 1993.
13. DiCosimo, R.; Stieglitz, B.; Fallon, R. D. U.S. Patent 5,728,556, 1998.
14. Saito, O.; Kawakami, K. Jpn. Kokai Tokkyo Koho JP 02154692 A2 900614.
15. Hann, E. C.; Eisenberg, A.; Fager, S. K.; Perkins, N. E.; Gallagher, F. G.; Cooper, S. M.; Gavagan, J. E.; Stieglitz, B.; Hennessey, S. M.; DiCosimo, R. *Bioorg. Med. Chem.* **1999**, *7*, 2239-2245.
16. Bernet, N., Arnaud, A., Galzy, P. *Biocatalysis* **1990**, *3*, 259-267.
17. Skryabin, G. K.; Koshcheenko, K. A. *Methods Enzymol.* **1987**, *135*, 198 -216.
18. Honda, J., Kandori, H., Okada, T., Nagamune, T., Shichida, Y., Sasabe, H., Endo, I. *Biochemistry* **1994**, *33*, 3577-3583.
19. Tsujimura, M., Odaka, M., Nagashima, S., Yohda, M., Endo, I. *J. Biochem.* **1996**, *119*, 407-413.

20. Scarrow, R. C.; Strickler, B. S.; Ellison, J. J.; Shoner, S. C.; Kovacs, J. A.; Cummings, J. G.; Nelson, M. J. *J. Am. Chem. Soc.* **1998**, *120*, 9237-9245.
21. Bucke, C. *Methods Enzymol.* **1987**, *135*, 175-189.
22. Nagasawa, T.; Nanba, H.; Ryuno, K.; Takeuchi, K.; Yamada, H. *Eur. J. Biochem.* **1987**, *162*, 691-698.
23. Yamada, H.; Ryuno, K.; Nagasawa, T.; Enomoto, K.; Watanabe, I. *Agric. Biol. Chem.* **1986**, *50*, 2859-2865.
24. Asano, Y.; Yasuda, T.; Tani, Y.; Yamada, H. *Agric. Biol. Chem.* **1982**, *46*, 1183-1189.
25. Ashina, Y.; Suto, M. *Bioprocess Technol.* **1993**, *16*, 91-107.
26. Kobayashi, M.; Nagasawa, T.; Yamada, H. *Trends Biotechnol.* **1992**, *10*, 402 – 408.
27. Nagasawa, T.; Ryuno, K.; Yamada, H. *Experientia* **1989**, *45*, 1066 - 1070.
28. Ryuno, K.; Nagasawa, T.; Yamada, H. *Agric. Biol. Chem.* **1988**, *52*, 1813 - 1816.
29. Nilsson, K.; Brodelius, P.; Mosbach, K. *Methods Enzymol.* **1987**, *135*, 222 - 230.
30. Chibata, I.; Tosa, T.; Sato. T.; Takata, I. *Methods Enzymol.* **1987**, *135*, 189 - 198.
31. Felix, H. *Anal. Biochem.* **1982**, *120*, 211 - 234.
32. Oliveira, D. E.; Santos Neto, L. C.; Panek, A. D. *Anal. Biochem.* **1981**, *113*, 188 -192.

Chapter 11

Bioconversion of Toluene to *p*-Hydroxybenzoate Using a Recombinant *Pseudomonas putida*

Edward S. Miller, Jr., and Steven W. Peretti[1]

Department of Chemical Engineering, North Carolina State University, Raleigh, NC 27695-7905

Toluene bioconversion to *p*-hydroxybenzoate (HBA) by resting cell suspensions of recombinant *P. putida* EM2878 showed greater than 99% selectivity for *para*-hydroxybenzoate production. The first step in the pathway, toluene-4-monooxygenase (T4MO) conversion of toluene to *p*-cresol, significantly constrained the carbon flux through the pathway giving a maximum rate of HBA production of 1.61 ± 0.15 nmol min.$^{-1}$ mg total protein^{-1} with a yield of 11 to 15 mg L^{-1} in shake flask studies. Substrate range studies of *P. putida* EM2878 revealed that T4MO conversion of substituted toluenes was more sensitive to steric hindrance than the rest of the pathway enzymes. *P. putida* EM2878 will also catalyze the conversion of *o*-xylene, *m*-xylene, 3-chlorotoluene, 3,4-dimethylphenol, 2-chloro-4-methylphenol, and 2-amino-4-methylphenol to products which maintain the 1-4 carboxyl to hydroxyl group symmetry important in the production of liquid crystal polymers (LCP's).

HBA is representative of a class of regio-specifically oxidized aromatic molecules that possess a 1-4 symmetry of carboxyl to hydroxyl groups. It is the chief monomer used in LCP's due to its structural symmetry, availability, and favorable pricing relative to the more expensive multiple ring monomers derived from naphthalene. HBA also finds extensive use as an intermediate in the manufacturing of dyes, pesticides, and pharmaceuticals (*1*). In addition, the n-alkyl esters of HBA (parabens) are widely used as preservatives in food (e.g. in fruit juices), cosmetics, pharmaceuticals, and technical products (e.g. antifreeze agents) due to their effective bacteriostatic and fungiastatic properties (*2*).

Current HBA synthesis constitutes a 7000 ton yr.$^{-1}$ industry. The state of the art for HBA production is the two step Kolbe-Schmitt carboxylation of phenol, a process discovered in the early 1860's (*3*). The Kolbe-Schmitt process is both energy intensive, requiring strong forcing conditions, Figure 1, and lacks absolute regiospecificity. Only 48% of the available potassium phenolate is converted to HBA, while 12% is converted to *o*-hydroxybenzoic acid and 4-hydroxyisophthalic acid

[1]Corresponding author.

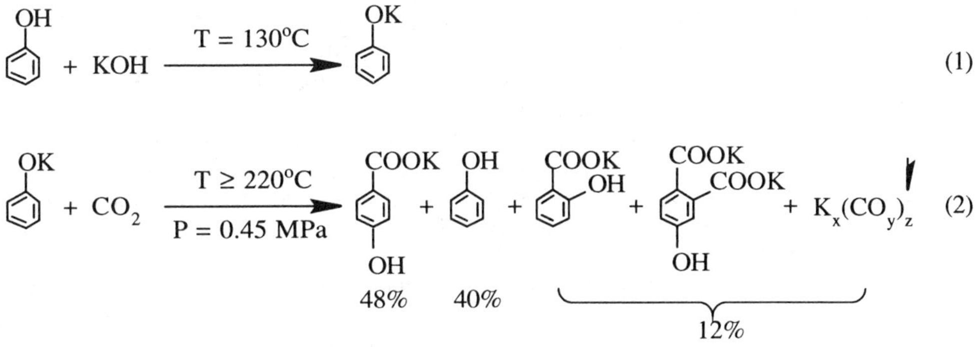

$$\text{PhOH} + \text{KOH} \xrightarrow{T = 130°C} \text{PhOK} \quad (1)$$

$$\text{PhOK} + CO_2 \xrightarrow[P = 0.45 \text{ MPa}]{T \geq 220°C} \underset{48\%}{\text{4-HO-C}_6\text{H}_4\text{-COOK}} + \underset{40\%}{\text{PhOK}} + \underset{}{\text{salicylate}} + \underset{}{\text{dicarboxylate}} + K_x(CO_y)_z \quad (2)$$

Figure 1. Reactions of the Kolbe-Schmitt process.

byproducts (*3*). The remaining potassium phenolate (40%) is oxidized back to phenol in the second reaction step, Figure 1. Phenol is a water soluble, EPA priority pollutant under the Clean Air and Water Act, which must be recovered from gas and waste water streams (*4*). Other process emissions include the production of metal salts and heavy residues which are becoming increasingly expensive to land-fill and incinerate, and CO_2 gas emissions, the principle cause of global warming (*5*).

The need for the development of a cleaner technology for HBA production is fueled by the increasing demand for HBA brought on by the growth and expansion of the LCP and skin care markets (*6-9*). Toluene bioconversion to HBA represents a move toward the development of a pollution source reduction technology for the production of HBA, and structurally-related specialty chemicals. The high degree of specificity of enzymatically catalyzed reactions can reduce the generation of unwanted byproducts, lessening the amount of material that needs to be disposed of or treated. In the case of toluene bioconversion to HBA, this could account for the reduction of 1,750 ton yr.$^{-1}$ of unwanted side products that are generated by the Kolbe-Schmitt process. Increased levels of safety for both the environment and human health can be achieved due to the mild reaction conditions that are generally afforded by bioprocess routes. It is estimated that through the use of such conditions, 1.8×10^{10} Btu yr.$^{-1}$ in energy usage could be recovered by switching from the Kolbe-Schmitt process. Furthermore, the mild reaction conditions employed by biocatalytic HBA production could lessen the immediate environmental and human dangers associated with the high pressures and temperatures used in the existing HBA process, and minimize handling of potentially explosive materials (potassium phenolate) and reactive catalysts (alumina powder). It is this latter point that OSHA investigators determined to be the cause of the explosion that occurred on April 21, 1995, at Napp Technologies, Inc.'s HBA and parabens operation in Lodi, NJ, which killed 4 workers, completely destroyed the plant and surrounding offices, and resulted in the evacuation of more than 400 people from the surrounding community. The resulting fire released sulfur dioxide and nitrogen compounds into the air, and 'significant volumes' of phenol into the Saddle River (*10*).

HBA production can now be achieved by a single processing step using the proposed bioconversion process. By dispensing with the need for multiple step syntheses, often required to produce hydroxy-substituted aromatic products using traditional chemistry, the extreme reaction conditions and generation of potentially toxic waste products that are associated with these types of syntheses can be avoided. Concomitant with the reduced source generation of byproducts, comes the more efficient management of existing petrochemical resources. Potential emissions resulting from the application of this technology could be minimized, and would include waste water and spent biomass. Since toluene exhibits poor water solubility (515 mg L^{-1} at 25°C (*11*)) and low boiling point (110.6°C (*12*)), cost effective methods for its recovery and recycling could be easily implemented, eliminating possible VOC emissions. In addition, the biomass could be dried, ground up, and given to farmers to be used as fertilizer, providing a positive environmental impact.

The work presented here describes initial fundamental research efforts aimed at the generation of a co-metabolic process intended to replace the Kolbe-Schmitt carboxylation of phenol. At present a prototype organism has been constructed utilizing enzymes from multiple aromatic catabolic pathways encoding for toluene and *p*-cresol metabolism (*13,14*). Recombinant *P. putida* EM2878 catalyzes the conversion of toluene to HBA in a single processing step. Central to the conversion of toluene to HBA in *P. putida* EM2878 is the T4MO system from *P. mendocina* KR1, a 4-component monooxygenase that catalyzes the conversion of toluene to *p*-cresol by regioselective incorporation of 1 atom of O_2 into the aromatic ring *(13,15)*. Transformation of *p*-cresol to *p*-hydroxybenzaldehyde (HBZ) is catalyzed by *p*-cresol methylhydroxylase A (PCMH), a flavocytochrome protein from *P. putida* NCIB9869 that incorporates the oxygen from water into the methyl group of *p*-cresol to give *p*-hydroxybenzyl alcohol, followed by dehydrogenation to give HBZ *(16)*. The final step, dehydrogenation of HBZ to HBA, is catalyzed by *p*-hydroxybenzaldehyde dehydrogenase (PHBZ), an NAD^+ dependent aromatic aldehyde dehydrogenase from *P. putida* NCIB9866 *(17)*. The proposed reaction sequence is given in Figure 2.

In this paper we describe the initial characterization of a novel biocatalytic process for the production of HBA from toluene using resting cell suspensions of recombinant *P. putida* EM2878. Bottlenecks to carbon flux in the pathway are identified, and dependence of HBA production on toluene and induction of T4MO are quantified. In addition, the substrate range of the organism is presented, further demonstrating the utility of this technology for the clean, highly specific synthesis of molecules with regiospecific features important to the production of LCP's.

Materials and Methods

Generation of a non-HBA Degrading *P. putida* Mutant.

Initial studies of *p*-hydroxybenzoate hydroxylase, PobA, Figure 3, activity in *P. putida* KT2440, the parent strain of *P. putida* EM2878, revealed that HBA turnover was rapid in the presence of glutamate as a primary carbon source, Table I. Disruption of PobA activity was obtained by introducing point mutations into the *pobA* DNA sequence using N-methyl-N'-nitro-nitrosoguanidine chemical mutagenesis *(18)*. Following mutagenesis and 3 rounds of D-cycloserine/piperacillin enrichment, approximately 3000 colonies were screened for loss of PobA activity. 195 colonies were initially isolated that showed no growth on solid HBA minimal media and positive growth on benzoate minimal media. Of these, 4 colonies were shown to be deficient in PobA activity following assay of crude cell extracts as described by Entsch et al. *(19)*; Table 1. Both HBA and HBZ were used as inducers to distinguish between true loss of PobA activity and loss of the HBA transport system *(20)*. One Pob⁻ mutant from the original 4 was selected at random to be the host of the system and was designated *P. putida* EM2839.

Construction and Organization of Pathway Genes.

The five essential genes encoding T4MO, *tmoABCDE*, and 2 genes encoding the cytochrome c and flavoprotein subunits of PCMH, *pchCF*, were obtained from published sources *(21,22)*. The single 1.5 kb gene encoding PHBZ, *phbz*, was cloned by PCR from a preparation of *P. putida* NCIB9866 genomic DNA *(23)*. The Tn5 constructs used in this work are based on the pCNB5 and pCNB4 mini-Tn5 transposon systems developed by deLorenzo et al. *(24)*. pUC18Not was used to add flanking *Not*I ends to the structural gene cassettes prior to insertion into the unique

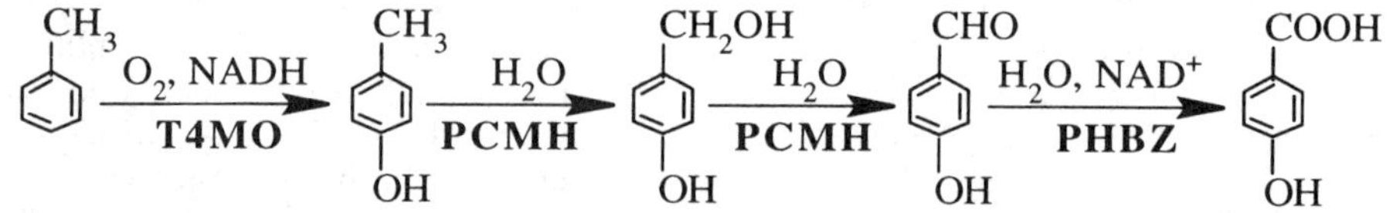

Figure 2. Successive reactions of the proposed biocatalytic pathway. (1), Toluene; (2), p-cresol; (3), p-hydroxybenzyl alcohol; (4), HBZ; (5), HBA. (Reproduced with permission from reference 26. Copyright 1999 The Royal Society of Chemistry.)

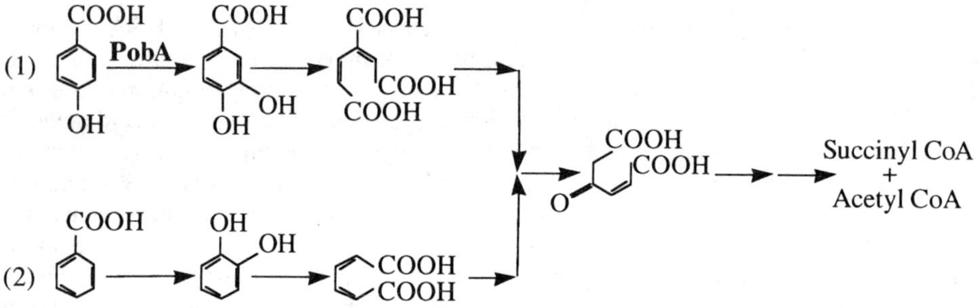

Figure 3. Converging pathways leading into the β-ketoadipate pathway. (1), Protocatechuate Branch; (2), Catechol Branch. The role of PobA is shown for clarity.

NotI site in the mini-Tn5 systems (25). Mobilization of mini-Tn5 constructs into P. putida EM2839 derivatives and selection of putative transconjugants were carried out as described (20,26).

Table I. PobA Enzyme Activity

Strain	Inducer	Specific Activity[a]
P. putida KT2440	10 mM HBA	0.294 ± 0.054
	1.5 mM HBZ	0.046 ± 0.011
P. putida EM2839	10 mM HBA	ND[b]
	1.5 mM HBZ	ND

[a] Specific activity units are µmol min.$^{-1}$ mg total protein^{-1}.
[b] ND, none detected.
SOURCE: Reproduced with permission from reference 26. Copyright 1999 The Royal Society of Chemistry.

The pathway for toluene conversion was set up in two regulatory units in order to minimize the inefficient transcription of distal genes, and to allow for the optimization of expression of key steps in the pathway. Figure 4 depicts the plasmid constructions and the reactions catalyzed by the products of cloned gene expression. pESM21 contains the five structural genes necessary for T4MO conversion of toluene to p-cresol under salicylate inducible Psal promoter control. pESM23 contains the structural genes necessary for the two step PCMH conversion of p-cresol to HBZ, and PHBZ conversion of HBZ to HBA. In pESM23, structural genes are expressed from a single IPTG-inducible Ptrc promoter with pchCF being transcribed before phbz.

Batch Transformation Experiments Using Resting Cell Suspensions of P. putida EM2878.

Resting cell suspensions were prepared from 200 mL of late exponential phase induced cultures of P. putida EM2878 grown in L-Salts minimal media (27) with 0.6% glutamate as the carbon and energy source (26). Upon resuspension, 500 µg mL^{-1} chloramphenicol was used to inhibit further protein synthesis in all rate determination experiments. Batch transformation experiments were conducted in 250 mL screw top flasks sealed with Teflon backed septa through which 1 mL samples were taken periodically using a syringe. Following centrifugation, bioconversion products and intermediates in supernatant fractions were separated by high performance liquid chromatography (26). Products were identified using a Waters 990 Photodiode Array Detector (Waters Chromatography Division, Milford, MA) by retention time and spectral comparison with known standards. HBA, p-cresol, p-hydroxybenzyl alcohol, and HBZ were quantified using previously determined correlations. Cell pellets obtained from the 1 mL samples taken during the course of the batch bioconversion experiments were analyzed for total protein using a bicinchoninic acid kit (Pierce, Rockford, IL) following ultrasonic disruption as described previously (26). Activity values reported in text are the average ± the standard deviation of at least two independent determinations.

(a)

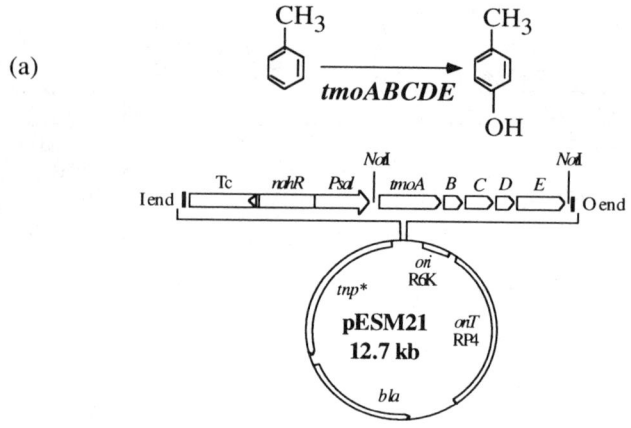

(b)

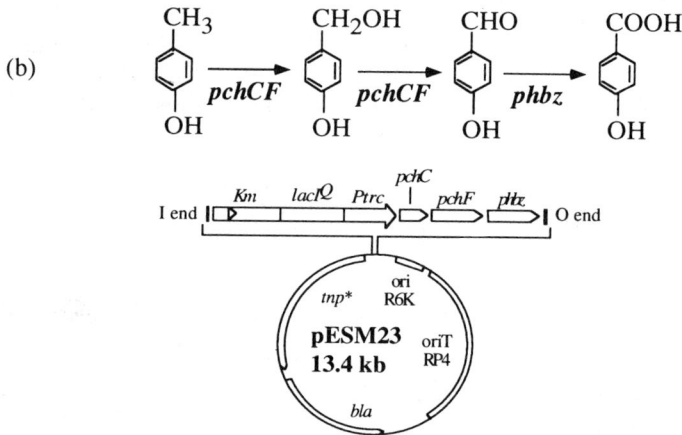

Figure 4. Construction of plasmids harboring mobile expression elements containing structural genes for the conversion of toluene to p-cresol, and p-cresol to HBA. (a) pESM21 and (b) pESM23. The structure of the mini-Tn5 transposons is emphasized in the figure for clarity. The DNA sequences coding antibiotic resistance markers aphA and tet, regulatory proteins lacI^Q and nahR, and promoters Ptrc and Psal are indicated. The mobile units are present in the delivery plasmid pUT as XbaI-EcoRI restriction fragments (44).

Results and Discussion

Toluene Conversion to HBA by Resting Cells of *P. putida* EM2878.

Resting cell suspensions prepared from exponentially growing cultures provided the highest rate of toluene conversion, 1.61 ± 0.15 nmol min.$^{-1}$ mg total protein^{-1}. Cultures prepared from cells in stationary phase were not as active, giving a maximal rate of toluene conversion of 0.65 nmol min.$^{-1}$ mg total protein^{-1}. The difference in activity may be attributed to increased proteolytic activity that is known to occur as cells enter stationary phase, decreased *de novo* synthesis of enzyme, and depletion of cofactor.

Analysis of pathway activity in *P. putida* EM2878 revealed two bottlenecks to carbon flux. The first, low T4MO specific activity, results in the slow turnover of toluene to *p*-cresol. To illustrate this, identical cultures of resting cells of *P. putida* EM2878 were spiked with varying levels of *p*-cresol during toluene bioconversion. The HBA profiles generated are shown in Figure 5a. Upon addition of *p*-cresol, an increase in the rate of HBA production results due to excess enzyme activity in the lower part of the pathway. The maximum HBA production rate from toluene was measured to be 1.61 ± 0.15 nmol min.$^{-1}$ mg total protein^{-1}, and was independent of whether expression of *tmoABCDE* was from the *Psal* or *Ptrc* promoters, data not shown.

Conversion of *p*-cresol to HBA resulted in higher rates of HBA production than from toluene. However, at high rates of *p*-cresol turnover, transient accumulation of *p*-hydroxybenzyl alcohol appeared during batch transformation, Figure 5b. This is caused by the increased selectivity with which PCMH catalyzes the conversion of *p*-cresol to *p*-hydroxybenzyl alcohol, $V_{max}/K_m = 5.8$ μM^{-1} s^{-1}, *vs.* *p*-hydroxybenzyl alcohol to HBZ, $V_{max}/K_m = 1.7$ μM^{-1} s^{-1}. (Calculation assumes MW$_{PCMH}$ = 114 kDa; data for V_{max} and K_m taken from Keat and Hopper (28), and McIntire et al. (29).) Thus the slower rate of PCMH conversion of *p*-hydroxybenzyl alcohol to HBZ causes a second rate limiting step in the pathway.

HBA production from *p*-cresol was nonlinear with a maximum rate of 17 ± 0.75 nmol min.$^{-1}$ mg total protein^{-1} while *p*-cresol disappearance was measured to be 26 ± 5.1 nmol min.$^{-1}$ mg total protein^{-1}. From the above data, it is apparent that HBA production from *p*-cresol proceeds by at least an order of magnitude faster than from toluene. Furthermore, since T4MO conversion of toluene cannot supply *p*-cresol at a rate that exceeds the rate of PCMH conversion of *p*-hydroxybenzyl alcohol to HBZ, the transient accumulation of *p*-hydroxybenzyl alcohol is never observed during toluene conversion to HBA. As such, under the conditions employed in this study, HBA production from toluene is dependent on the level of T4MO induction and toluene concentration.

To quantify this dependence, the intrinsic kinetic parameters characterizing this strain were determined using resting cell suspensions in shake flask studies. Hanes regression of the initial rate data yielded Monod saturation dependence of HBA production on toluene, giving $V_{max} = 1.61 ± 0.15$ nmol min.$^{-1}$ mg total protein^{-1} and $K_m = 58 ± 5$ μM. Monod dependence of HBA production on salicylate induction of T4MO was also observed giving $V_{max} = 1.69 ± 0.09$ nmol min.$^{-1}$ mg total protein^{-1} and $K_m = 208 ± 29$ μM. Furthermore, quantification of SDS-PAGE gels using the

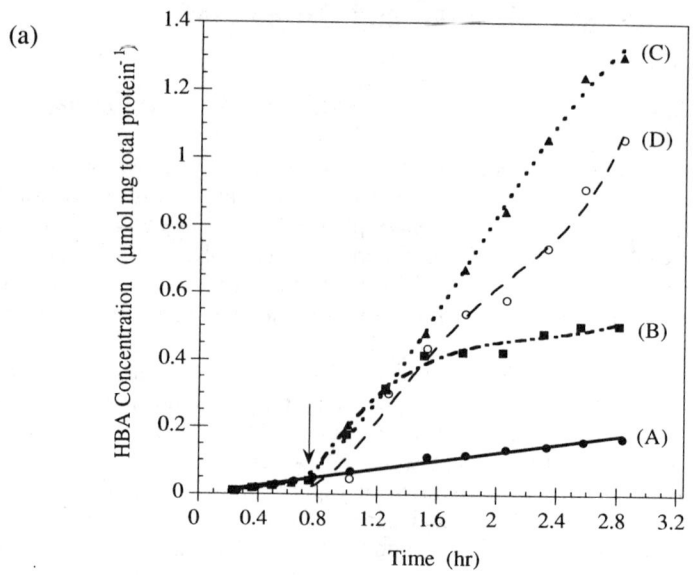

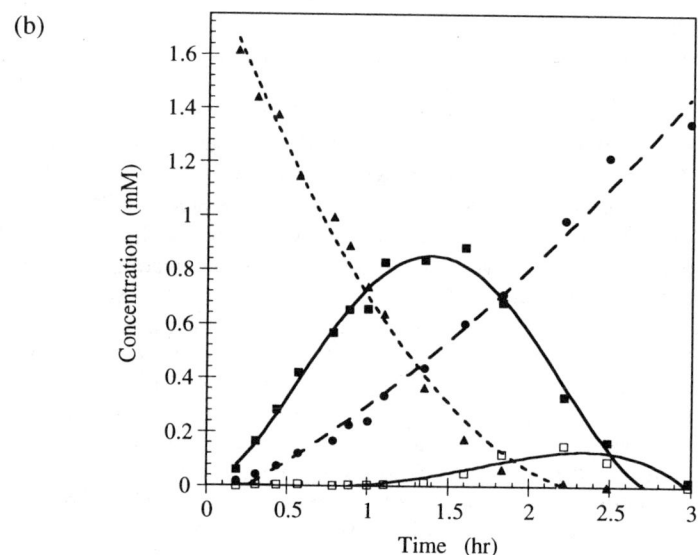

Figure 5. Determination of carbon flux bottlenecks in the toluene bioconversion pathway using resting cells of P. putida EM2878. (a) Identification of T4MO rate limiting step. Arrow denotes time at which p-cresol was added to culture. Curves indicate amount of p-cresol added; (A), 0 ppm; (B), 10 ppm; (C), 50 ppm; (D), 150 ppm. (b) Transient accumulation of p-hydroxybenzyl alcohol in conversion of p-cresol to HBA. (▲) p-cresol; (●) HBA; (□) HBZ; (■) p-hydroxybenzyl alcohol. Lines drawn to indicate trends. (Reproduced with permission from reference 26. Copyright 1999 The Royal Society of Chemistry.)

NIH Image software (*30*) revealed that expression of TmoA and TmoE monomers become saturated at salicylate concentrations roughly equal to 0.5 mM, Figure 6. The maximum expression levels of TmoA and TmoE in these studies is estimated to be equal to 1.3% to 1.9% of total protein.

The maximum yield of HBA realized in the batch transformation studies was 11 to 15 mg L^{-1} and usually achieved in about 4 hours. Conversion of toluene to HBA was stoichiometric up to initial toluene concentrations of 0.1 mM indicating no degradative losses of product due to cellular metabolism. However, above this concentration, stoichiometric conversion of toluene to HBA was not observed, giving only 6% conversion at concentrations that were saturating for enzyme activity. Enzyme instability and inactivation (*31*) and lack of *de novo* protein synthesis due to chloramphenicol inhibition may explain the phenomena. The selectivity of toluene bioconversion to HBA by batch cultures of *P. putida* EM2878 was 99+% for *para*-hydroxybenzoate product. The presence of *m*-cresol, a possible byproduct of the reaction, was undetectable by HPLC analysis of culture supernatants. This peak would appear at 16.78 min. in the spectra in Figure 7. The high degree of selectivity is attributed to the absolute structural requirement by PCMH for a 1-4 juxtaposition of methyl(ene) group to hydroxyl group for activity (*32*), and 96% selectivity of toluene to *p*-cresol conversion by T4MO (*33*).

Substrate Range of *P. putida* EM2878.

In addition to HBA, two other types of mononuclear aromatic molecules are of interest in commercial LCP production: (1) those with longer carbon side chains that when polymerized, can add flexibility to the polyester chain, and (2) molecules which can increase the spacing between polymer chains, such as those having one or two side groups in addition to the 1-4 juxtaposition of hydroxyl and carboxyl groups necessary for polymerization. To further illustrate the utility of the proposed technology, we investigated the biocatalytic production of additional compounds with unique structural features that may have potential application in the production of LCP's. Table II presents a summary of the results of substrate range experiments using overnight incubations of *P. putida* EM2878.

Mono-Substituted Benzenes with Increasing Side Chain Length.
HPLC data indicates that ethylbenzene is a substrate for bioconversion, as would be expected based on previous data (*29,34*). Conversion of 100 ppm ethylbenzene is estimated at less than 7%, following an 18 hour incubation, with 4-ethylphenol detected as a major product. Another major peak was detected in HPLC chromatograms with a retention time that would be consistent with an aromatic alcohol (*35*). This compound is expected to be 2-(4'-phenyl)ethanol based on the results of McIntire et al. (*29*); however, its identity could not be conclusively determined using our analytical procedures.

Overnight incubations of *P. putida* EM2878 with *n*-propylbenzene and 4-propylphenol, the expected intermediate formed by T4MO conversion of *n*-propylbenzene and a substrate for PCMH, produced no detectable products using our HPLC assay. This suggests that both T4MO and PCMH enzymes do not recognize substrates with side chains of greater than 2 carbon atoms in length. However, these findings contradict previously reported data of McIntire et al. (*29*) who have shown

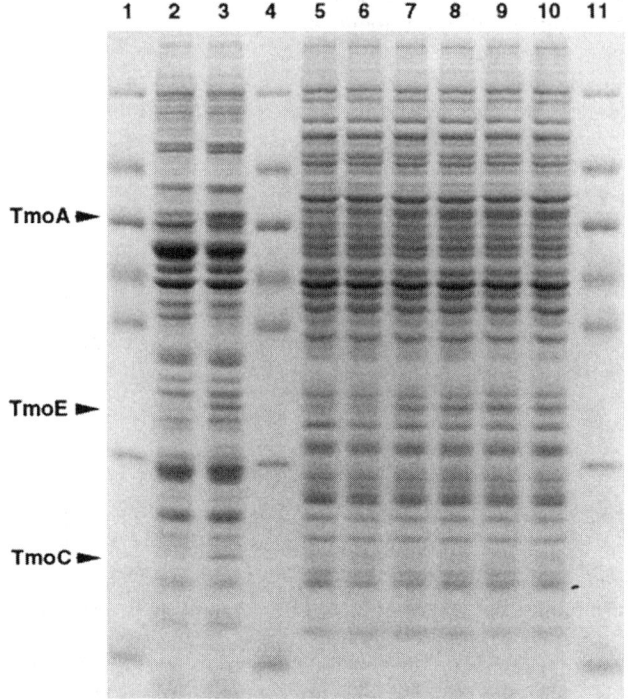

Figure 6. SDS PAGE analysis of selected T4MO polypeptides as a function of salicylic acid induction level. Lanes (1,4, and 11) Molecular weight markers. Lane (2) E. coli/pUC18 induced with 2 mM IPTG; negative control. Lane (3) E. coli/pUC18 with EcoRI/SspI fragment containing tmoABCDE genes induced with 2 mM IPTG; positive control. Lane (5) Uninduced P. putida EM2878. Lane (6) P. putida EM2878 induced with 2 mM IPTG without salicylate; negative control. Lane (7) P. putida EM2878 induced with 0.1 mM salicylate and 2 mM IPTG. Lane (8) P. putida EM2878 induced with 0.5 mM salicylate and 2 mM IPTG. Lane (9) P. putida EM2878 induced with 3 mM salicylate and 2 mM IPTG. Lane (10) P. putida EM2878 induced with 10 mM salicylate and 2 mM IPTG. SDS PAGE was performed as described by Laemmli (45).

137

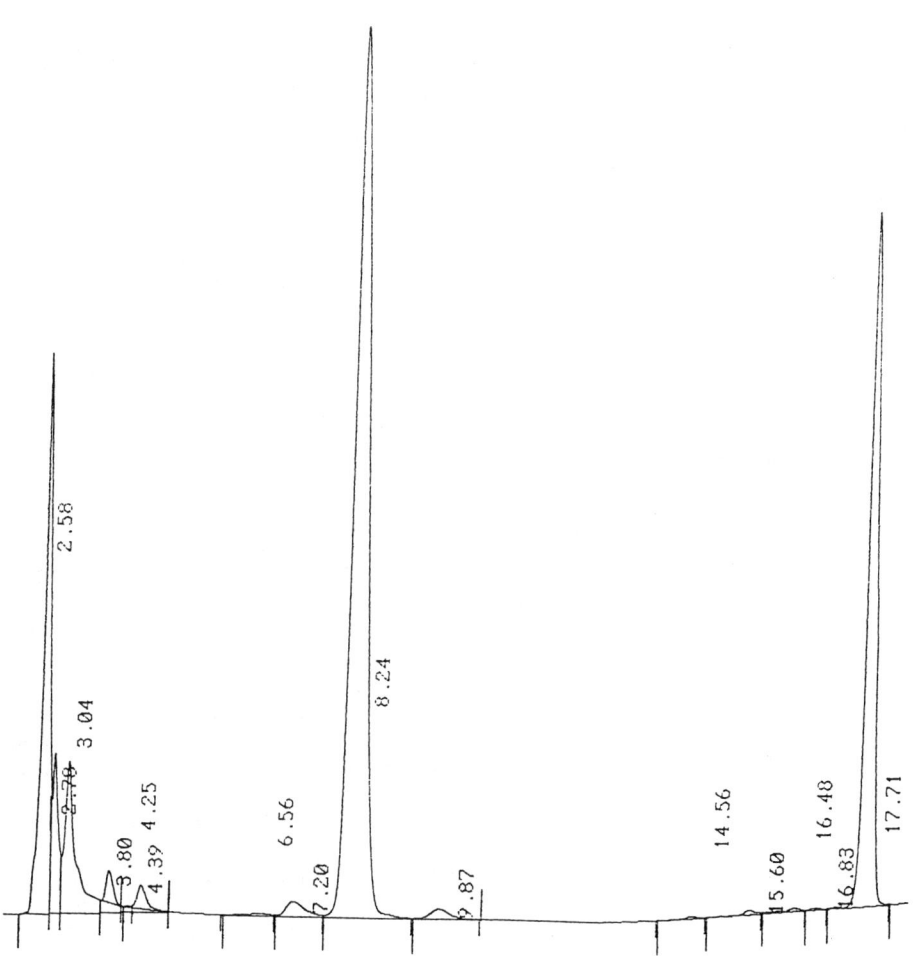

Figure 7. HPLC spectra of toluene bioconversion to HBA. HBA retention time, 8.24 min.; salicylate retention time, 17.71 min.; and m-cresol retention time, 16.78 min. All other peaks detected in the chromatogram are associated with cell growth and glutamate metabolism. (Reproduced with permission from reference 26. Copyright 1999 The Royal Society of Chemistry.)

PCMH to be active with 4-propylphenol giving $V_{max} = 0.839$ μmol min.$^{-1}$ mg protein^{-1} and $K_m = 0.281$ mM. Furthermore, the substrate range of T4MO is known to include nitrobenzene, and phenylacetic acid (34,36); compounds with side groups that are larger ($MW_{Nitro} = 46$, and $MW_{Aceto} = 73.1$) than the propyl side group ($MW_{Propyl} = 43$). In view of previous data and steric arguments, it stands to reason that the present construct should be active with these compounds. We attribute our observations to the greater sensitivity of the assay techniques used in the previous studies, and the poor water solubility of n-propylbenzene and 4-propylphenol.

Table II. Results of Substrate Range Experiments[a]

Substrate	Reactivity[b]	Major Products Detected
ethylbenzene	+	4-ethylphenol
		2-(4'-phenyl)ethanol[c]
n-propylbenzene	-	
4-propylphenol	-	
o-xylene	+/-	4-hydroxy-2-methylbenzoate
3,4-dimethylphenol	+++	4-hydroxy-2-methylbenzyl alcohol[c]
		4-hydroxy-2-methylbenzaldehyde[c]
		4-hydroxy-2-methylbenzoate
2-chlorotoluene	-	
m-xylene	++	4-hydroxy-3-methylbenzoate[d]
3-chlorotoluene	+/-	3-chloro-4-hydroxybenzoate
2,4-dimethylphenol	+++	4-hydroxy-3-methylbenzoate[d]
2-chloro-4-methylphenol	+++	3-chloro-4-hydroxybenzoate
3-nitrotoluene	-	
4-methyl-2-nitrophenol	-	
2-amino-4-methylphenol	+++	3-amino-4-hydroxybenzoate

[a] Products obtained were based on retention time and spectral comparison with known standards unless otherwise indicated.
[b] Reactivity based on estimated percent of conversion of 100 ppm starting substrate concentration in overnight incubations with *P. putida* EM2878. -, no conversion; +/-, conversion estimated at less than 1%; +, conversion estimated at less than 10%; ++, conversion estimated between 20% and 30%; +++, conversion estimated at greater than 80%. Overnight incubations with *P. putida* EM2839 served as negative controls.
[c] Suspected product based on previous work (29), but not conclusively verified in this study.
[d] Standard for comparison obtained from conversion of 4-hydroxy-3-methylbenzaldehyde by *P. putida* EM2839.

Mono-Substituted Toluenes with Substitution at the C-2 Position.
Of the C-2-substituted toluenes tested, o-xylene and 2-chlorotoluene, only o-xylene was found to undergo bioconversion, giving 4-hydroxy-2-methylbenzoate as a major product. Less than 1% of the initial 100 ppm o-xylene was found to undergo bioconversion, whereas 6% conversion of toluene, and 86% conversion of 3,4-dimethylphenol, the predicted product of o-xylene conversion by T4MO, were

observed. The low level of *o*-xylene conversion may be due to steric hindrance caused by the presence of the second methyl group at the C-2 position in the active site of the T4MO enzyme. Such effects would be magnified in the case of 2-chlorotoluene due to the larger size of the chloro side group, molecular weight$_{Cl}$, (MW)$_{Cl}$, = 35 *vs*. MW$_{Methyl}$ = 15, explaining the observed inactivity of the strain with 2-chlorotoluene. The steric hindrance argument is further strengthened by the fact that chlorobenzene is a substrate for T4MO conversion (*34*), thus ruling out inactivity of T4MO on 2-chlorotoluene due to possible electronegative effects of the chloro side group.

Mono-Substituted Toluenes with Substitution at the C-3 Position.
3-Substituted toluenes serving as substrates for bioconversion included both *m*-xylene and 3-chlorotoluene, giving 4-hydroxy-3-methylbenzoate and 3-chloro-4-hydroxybenzoate as major products, respectively. Roughly 26% of the starting 100 ppm *m*-xylene was converted to 4-hydroxy-3-methylbenzoate, whereas less than 1% of the available 3-chlorotoluene was converted to 3-chloro-4-hydroxybenzoate. The effects of steric limitation caused by the larger chloro group size may explain the lower conversion of 3-chlorotoluene. In contrast, conversion of 2-chloro-4-methylphenol and 2,4-dimethylphenol to 3-chloro-4-hydroxybenzoate and 4-hydroxy-3-methylbenzoate, respectively, was nearly 100%, indicating that downstream enzymes from T4MO were not subject to the same steric constraints.

Overnight incubations of *P. putida* EM2878 with 3-nitrotoluene and 4-methyl-2-nitrophenol, the expected intermediate formed by T4MO conversion of 3-nitrotoluene, produced no detectable products as determined by our HPLC assay. Steric hindrance caused by the large size of the nitro group (MW$_{Nitro}$ = 46) of 3-nitrotoluene is suspected to be the cause of inhibition of T4MO activity. This is consistent with the decrease in activity observed when the size of the side group at the C-3 position was increased from methyl, MW$_{Methyl}$ = 15, to chloro, MW$_{Cl}$ = 35. Furthermore, Haigler et al. have shown the T4MO system to be effective in oxidizing nitrobenzene to a mixture of 3-nitrophenol and 4-nitrophenol (*36*). In this instance, the strong electron withdrawing nature of the nitro group was found to stabilize more than one transition intermediate during O$_2$ addition, thus reducing the *para* specificity of the reaction.

In agreement with the results of our study, previous study has also shown PCMH to be inactive with 2-nitro-4-methylphenol (*29*). However, 4-methyl-3-nitrophenol has been shown to be a substrate for PCMH conversion, suggesting that strong electronic interactions of the side group with the aromatic ring does not affect the ability of the enzyme to form the quinone methide intermediate. This argument is strengthened by the fact that we have also observed nearly 100% conversion of 2-amino-4-methylphenol to 3-amino-4-hydroxybenzoate, where the amino group constitutes a strong electron donating side group. We speculate that the observed PCMH activity with 4-methyl-3-nitrophenol and not 4-methyl-2-nitrophenol is due to the close proximity of the nitro group to the hydroxyl group in the latter compound which would allow for hydrogen bonding to take place between an nitro group oxygen atom and the phenolic hydrogen atom. This bond would disrupt the flow of electrons in formation of the quinone methide intermediate (*16*) by stabilizing the phenolic hydrogen atom against removal during catalysis.

Conclusions

HBA is representative of a class of regio-specifically oxidized molecules for which existing chemistry is ill-suited for efficient production. The goal of this research is the development of an alternative technology that can address many of the environmental concerns and synthetic limitations of the Kolbe-Schmitt process used for the production of HBA. Bioconversion of toluene to HBA capitalizes on the two inherent traits of microbial based chemicals production; namely, mild reaction conditions and improved catalyst specificity. By exploiting these two benefits, process environmental impact can be greatly reduced by minimizing the generation of waste products and lowering overall energy consumption.

The results presented in this work provide 'proof of concept' of the utility of toluene bioconversion in the production of HBA, and merely serve as the starting point in the evolution of an efficient biocatalytic process. We have observed HBA yields in unoptimized fed-batch cultures of 135 mg L^{-1} (unpublished results), and though this represents a 10-fold improvement over yields obtained in this study, this is still not sufficient. Assuming an HBA selling price between one and ten dollars per kilogram (2), HBA concentrations in the fermentation medium need to reach ~50 g L^{-1} for an economically attractive process to be considered. However, the potential savings achieved by changing feedstocks from phenol to toluene, estimated at \$315,000 yr.$^{-1}$ for a 500 ton yr.$^{-1}$ producing plant, also needs to be considered. At any rate, further work is needed to eliminate bottlenecks that restrict carbon flux through the pathway, to improve enzyme stability and expression, and to optimize fermentation conditions and reactor design.

While not optimal, the recombinant system used in these studies captures many of the important details concerning the development of the toluene bioconversion process and provides a reference point from which further strain and process improvements can be initiated. T4MO conversion of toluene to *p*-cresol was shown to be the rate limiting step in the toluene bioconversion pathway. A 15-fold improvement in T4MO activity could easily be realized through co-expression of the terminal oxido-reductase gene, *tmoF*, not present in our current constructs, with the *tmoABCDE* gene cluster, and through the use of plasmid-based expression systems (*37*). HBA yields approaching 2 g L^{-1} could immediately be realized from these modifications, making toluene bioconversion to HBA competitive with other proposed biocatalytic routes to this compound (*38,39*).

Further improvements in T4MO specific activity could be gained by increasing the stability of the iron centers in TmoC and TmoH subunits using site-directed mutagenesis (*40*), eliminating upstream DNA sequences from the ATG start codon in *tmoA* (*21*), modifying amino acid codon usage to effect more efficient translation in a given host, and mutagenesis of promoter and ribosomal binding sites to effect more efficient transcription and translation of the T4MO mRNA. Furthermore, advances in protein engineering using error-prone PCR and/or gene shuffling techniques (*41,42*) can be used for the *in vitro* evolution of T4MO with improved specific activity. Combined use of these techniques has led to a 100-fold improvement in evolved pNB esterase specific activity in 30% dimethylformamide over that of the parent enzyme (*43*). Application of these techniques coupled with fermentation optimization studies could be used to reach economically viable HBA yields.

Acknowledgments

We are grateful to Dr. William S. McIntire (Veterans Affairs Medical Center and University of California at San Francisco, San Francisco, CA) for providing p16*Sma*I, Dr. Kenneth Timmis (GBF-National Research Centre for Biotechnology, Braunschweig, FRG) for providing pUC18Not, mini-Tn5-Tc and pCNB5 transposon systems, and *E. coli* S17-1λpir, Dr. Thomas K. Wood (University of Connecticut, Storrs, CT) for providing pCNB4, and Dr. Burt D. Ensley (Phytotech) for providing pKMY402.

This work was supported by Hoechst Celanese, as part of the Hoechst Celanese and Kenan Fellows Program at North Carolina State University; the National Science Foundation, grant #BCS-9157913-01, and the Department of Chemical Engineering at North Carolina State University.

Literature Cited

1. Ritzer, E.; Sunderman, R. In *Ullmann's Encyclopedia of Industrial Chemistry*; Elvers, B., Hawkins, S., Russey W., and Schulz, G., Eds.; New York, 1993; Vol. A13; pp 519-526.
2. Erickson, S. H. In *Kirk-Othmer Encyclopedia of Chemical Technology*, 4th ed.; Grayson, M., Ed.; John Wiley and Sons: New York, 1982; Vol. 20; pp 500-524.
3. Lindsey, A. S.; Jeskey, H. *Chem. Rev.* **1957**, *57*, 583-620.
4. Keith, L. H.; Telliard, W. A. *Environ. Sci. Technol.* **1979**, *13*, 416-423.
5. Raloff, J. *Science News* **1997**, *152*, 277.
6. Lerner, M. *Chemical Marketing Reporter* **1995**, *247*, 14-16.
7. McChesney, C.; Dole, J. *Machine Design* **1997**, *69*, 108-113.
8. Ouellette, J. *Chemical Market Reporter* **1997**, *251*, 24-26.
9. Vernyi, B. *Plastics News* **1996**, *8*, 10-16.
10. Fattah, H. *Chemical Week* May 3, 1995, pp 7-8.
11. McAuliffe, C. *J. Phys. Chem.* **1966**, *70*, 1267-1275.
12. *CRC Handbook of Chemistry and Physics*, 67th ed.; CRC Press, Inc.: Boca Raton, FL, 1986, pp 2406.
13. Whited, G. M.; Gibson, D. T. *J. Bacteriol.* **1991**, *173*, 3017-3020.
14. Hopper, D. J.; Taylor, D. G. *J. Bacteriol.* **1975**, *122*, 1-6.
15. Pikus, J. D.; Studts, J. M.; Achim, C.; Kaufman, K. E.; Munck, E.; Steffan, R. J.; McClay, K.; Fox, B. G. *Biochem.* **1996**, *35*, 9106-9119.
16. Hopper, D. J. *Biochem. J.* **1978**, *175*, 345-347.
17. Hewetson, L.; Dunn, H. M.; Dunn, N. W. *Genet. Res., Camb.* **1978**, *32*, 249-255.
18. Carlton, B. C.; Brown, B. J. In *Manual of Methods for General Bacteriology*; Gerhardt, P., Ed.; American Society for Microbiology: Washington, DC, 1981; pp 222-242.
19. Entsch, B. In *Hydrocarbons and Methylotrophy*; Lidstrom, M. E., Ed.; Academic Press, Inc.: San Diego, CA, 1990; Vol. 188; pp 138-147.
20. Harwood, C. S.; Nichols, N. N.; Kim, M.-K.; Ditty, J. L.; Parales, R. E. *J. Bacteriol.* **1994**, *176*, 6479-6488.

21. Yen, K.-M.; Karl, M. R.; Blatt, L. M.; Simon, M. J.; Winter, R. B.; Fausset, P. R.; Lu, H. S.; Harcourt, A. A.; Chen, K. K. *J. Bacteriol.* **1991**, *173*, 5315-5327.
22. Kim, J.; Fuller, J. H.; Cecchini, G.; McIntire, W. S. *J. Bacteriol.* **1994**, *176*, 6349-6361.
23. Miller, E. S. Ph.D. thesis, North Carolina State University, Raleigh, NC, 1999.
24. deLorenzo, V.; Herrero, M.; Jakubzik, U.; Timmis, K. N. *J. Bacteriol.* **1990**, *172*, 6568-6572.
25. Herrero, M.; deLorenzo, V.; Timmis, K. N. *J. Bacteriol.* **1990**, *172*, 6557-6567.
26. Miller, E. S.; Peretti, S. W. *Green Chemistry* **1999**, *3*, 143-152.
27. Thomas, S. M.; Peretti, S. W. *Biotechnol. Bioeng.* **1998**, *58*, 1-13.
28. Keat, M. J.; Hopper, D. J. *Biochem. J.* **1978**, *175*, 649-658.
29. McIntire, W.; Hopper, D. J.; Singer, T. P. *Biochem. J.* **1985**, *228*, 325 - 335.
30. *NIH Image,* URL http://rsb.info.nih.gov/nih-image/.
31. Arnold, F., Personal communication.
32. Hopper, D. J.; Taylor, D. G. *Biochem J.* **1977**, *167*, 155-162.
33. Whited, G. M.; Gibson, D. T. *J. Bacteriol.* **1991**, *173*, 3010.
34. Yen, K.-M.; Blatt, L. M.; Karl, M. R. U.S. Patent 5,171,684, 1992.
35. Miller, E. S.; Peretti, S. W. Department of Chemical Engineering, North Carolina State University.
36. Haigler, B. E.; Spain, J. C. *Appl. Environ. Microbiol.* **1991**, *57*, 3156-3162.
37. Yen, K.-M.; Karl, M. R. *J. Bacteriol.* **1992**, *174*, 7253-7261.
38. Frost, J. W., Personal communication.
39. Draths, K., Personal communication.
40. Murdock, D.; Ensley, B. D.; Serdar, C.; Thalen, M. *BIO/Technol.* **1993**, *11*, 381-386.
41. Kuchner, O.; Arnold, F. H. *TIBTECH* **1997**, *15*, 523-530.
42. Zhao, H.; Arnold, F. H. *Nucleic Acids Res.* **1997**, *25*, 1307-1308.
43. Moore, J. C.; Jin, H.-M.; Kuchner, O.; Arnold, F. H. *J. Mol. Biol.* **1997**, *272*, 336-347.
44. deLorenzo, V.; Eltis, L.; Kessler, B.; Timmis, K. N. *Gene* **1993**, *123*, 17-24.
45. Laemmli, U. K. *Nature* **1970**, *227*, 680-687.

Chapter 12

Genetically Engineered *Saccharomyces* Yeasts for Conversion of Cellulosic Biomass to Environmentally Friendly Transportation Fuel Ethanol

Nancy W. Y. Ho, Zhendao Chen, Adam P. Brainard, and Miroslav Sedlak

Laboratory of Renewable Resources Engineering (LORRE), Purdue University, 1295 Potter Center, West Lafayette, IN 47907–1295

Ethanol is an effective, environmentally friendly, nonfossil, transportation biofuel that produces far less pollution than gasoline and contributes essentially no net carbon dioxide to the atmosphere. Furthermore, unlike crude oil for the production of gasoline, ethanol can be produced from plentiful, domestic, renewable, cellulosic biomass feedstocks. However, a major obstacle in this process is that cellulosic biomass contains two major sugars, glucose and xylose. *Saccharomyces* yeasts, traditionally used for large scale industrial production of ethanol from glucose, is unable to ferment xylose to ethanol. This makes the use of the safest, most effective microorganism for conversion of cellulosic biomass to ethanol economically unfeasible. In the fall of 1993, we achieved a historic breakthrough in the successful development of genetically engineered recombinant *Saccharomyces* yeast that can effectively ferment both glucose and xylose to ethanol. This paper provides an up-to-date overview of the design, development, and continuous innovative perfection of our recombinant *Saccharomyces* yeast that is widely regarded as the microorganism which will make the conversion of cellulosic biomass to ethanol commercially possible.

Introduction

Numerous studies have proven that ethanol as a transportation fuel produces less air pollutants than gasoline. This environmentally friendly liquid fuel can be used directly as a neat fuel (100 %) or as a blend with gasoline at various concentrations. The raw material used for the production of ethanol fuel is renewable and abundantly available domestically. Thus, the use of ethanol to supplement or replace gasoline not only reduces air pollution and ensures a cleaner environment, but also reduces the dependency of our nation on imported foreign oil, protects our nation's energy security, and reduces our nation's trade deficit due to imported oil for the production of gasoline.

Ethanol has been produced by fermenting glucose-based food crops, such as cane sugar, corn starch, and other starch-rich grains, using yeasts, particularly *Saccharomyces* yeasts, which remain the only microorganisms used for large scale industrial ethanol production since the pre-industrial age. However, these agricultural crops are expensive and in limited supply.

Cellulosic biomass, which includes agriculture residues, waste streams from agricultural processing, sugarcane bagasse, maunicipal solid wastes, yard and wood wastes, wastes from paper mills, etc. is an attractive feedstock for ethanol-fuel production by fermentation because cellulosic biomass is not only renewable and available domestically but also available at low cost and in great abundance. However, there are problems which must be solved for such ideal feedstocks to be economically converted to ethanol. One serious problem is that the major sugars derived from cellulosic biomass include not only glucose but also xylose with a ratio of glucose to xylose approximately 2 or 3 to 1. It is generally agreed that unless both glucose and xylose from the cellulosic biomass could be fermented, the economics of converting cellulosic biomass to ethanol are not favorable. Yeasts, particularly *Saccharomyces*, which have traditionally been used and are still the only microorganisms used by industry for fermenting glucose-based feedstocks to ethanol, are unable to ferment xylose or utilize xylose for growth. It has been found that these yeasts are missing the enzyme(s) that are responsible for the conversion of xylose to xylulose (Figure 1). Furthermore, there are no other naturally occurring microorganisms known that are capable of effectively converting both glucose and xylose to ethanol. Thus, the major obstacle that prevents the economical conversion of cellulosic biomass to ethanol has been the lack of microorganisms, particularly yeasts, that can effectively convert both glucose and xylose to ethanol.

Two decades ago, a great deal of efforts worldwide were devoted to finding new yeasts, particularly *Saccharomyces* yeasts, capable of effectively fermenting both glucose and xylose. However, none were found. Nevertheless, a few other yeasts such as *Pichia stipitis*, *candida shehatae*, and *Pachysolen tannophilus* were discovered which could ferment xylose and utilize it for growth. Unfortunately, these yeasts are not effective fermentative-microorganisms and are not suitable to industry for large scale ethanol production because they are not user-friendly.

More than a decade ago, after exhausting all other means, the scientific community worldwide turned to genetic engineering to develop effective *Saccharomyces* yeasts for fermenting xylose to ethanol. Nearly ten research groups (at least five of them in the United States, including our group) initiated a worldwide attempt to genetically engineer *Saccharomyces* yeasts for xylose fermentation. Without exception, each group initially cloned a xylose isomerase gene from different bacteria into the *Saccharomyces* yeasts to make the latter ferment xylose (Figure 1). However, all efforts to express a bacterial xylose isomerase gene in the *Saccharomyces* yeasts failed to produce a genetically engineered *Saccharomyces* yeast that could

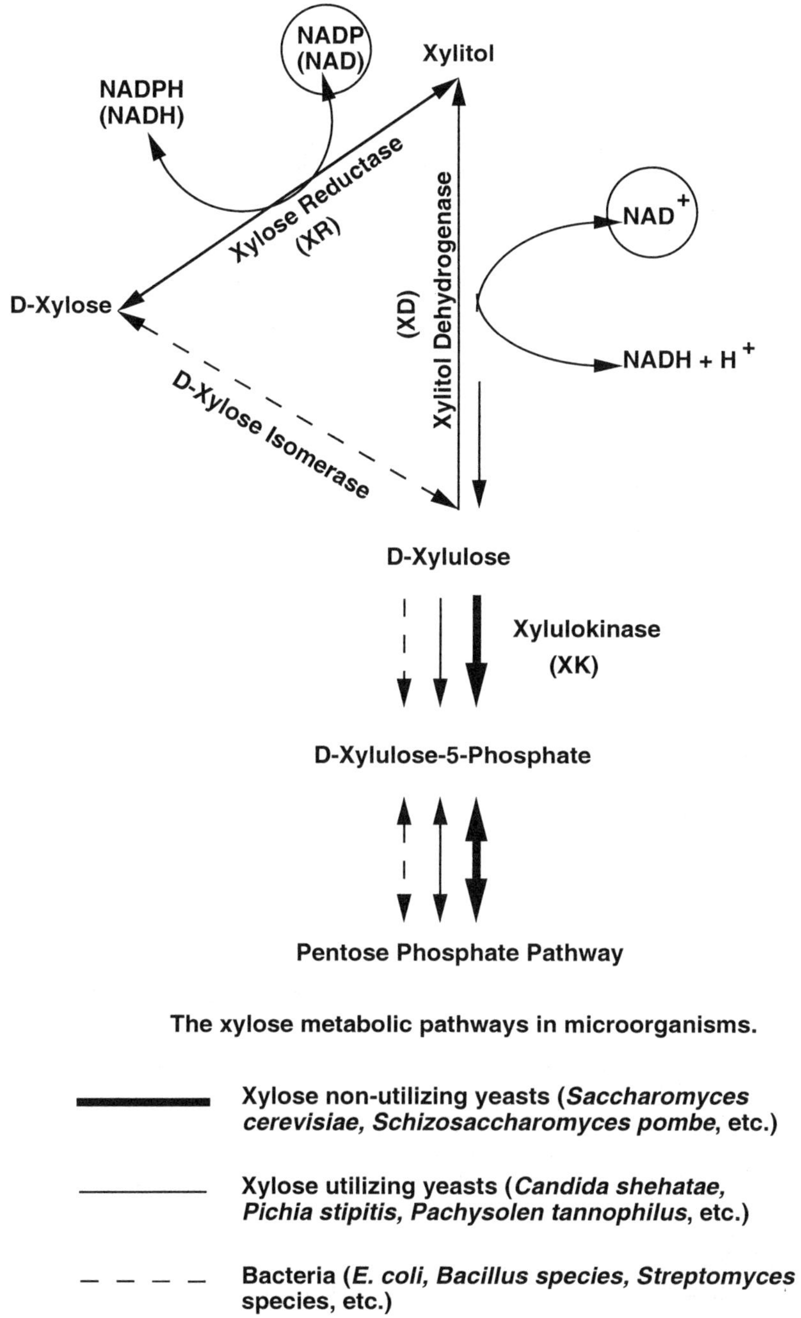

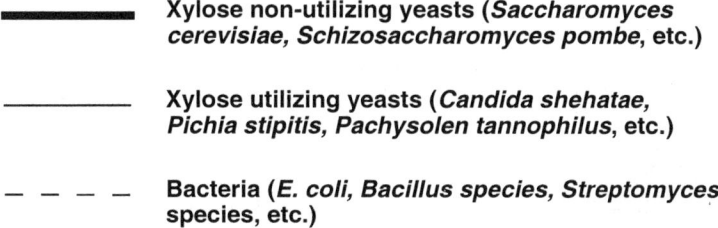

Figure 1. Xylose metabolic pathways in different types of microorganisms.

ferment xylose. After these unsuccessful attempts to clone a xylose isomerase gene in yeast, all U. S. groups except our group at Purdue University gave up on this difficult undertaking. Most experts by then had proclaimed that perhaps it was not possible to genetically engineer any *Saccharomyces* yeasts to effectively ferment xylose to ethanol. They reasoned that the only remaining way to genetically engineer the *Saccharomyces* yeast to ferment xylose was to transfer the genes for the conversion of xylose to xylulose; namely the xylose reductase gene (*XR*), encoding xylose reductase (XRtase), and the xylitol dehydrogenase gene (*XD*), encoding xylitol dehydrogenase (XDnase), from *Pichia stipitis*, the best natural xylose-fermenting, non-*Saccharomyces* yeast (Figure1). As shown in Figure 1, the Pichia as well as all the natural xylose-fermenting non-*Saccharomyces* yeast do not have an adquate system to provide cofacters to the XRtase and XDnase (*1*). On one hand, XRtase can use both NADPH and NADH as its cofactor but has much stronger affinity towards NADPH than NADH. On the other hand, XDnase requires NAD for its sole cofactor. In yeast there is no proper enzyme system to convert NAD to NADPH or vice versa, particularly under anaerobic conditions. Thus, most scientists predicted that xylose fermentation with such a genetically engineered yeast would stop due to lack of the proper cofactors. Furthermore, it was known that the XDnase catalyzes the interconversion between xylulose and xylitol but favors the formation of xylitol (the reverse reaction), which is compounded by the fact that most *Saccharomyces* yeasts have very low levels of xylulokinase activity (*2*). Thus, most experts concluded that even if one succeeded in cloning the XR and XD genes from *P. stipitis* into a *Saccharomyces* yeast, the engineered yeast would produce xylitol rather than ethanol.

Instead of abandoning this important project we carefully planned our approach to not only overcome all the above problems but also to design our engineered yeasts with additional properties that would make them ideal for fermenting glucose and xylose present in cellulosic biomass to ethanol. This paper presents an up-to-date overview of the design, development, and continuous innovative perfection of our genetically engineered recombinant *Saccharomyces* yeasts that can not only effectively ferment xylose to ethanol but also simultaneously coferment glucose and xylose to ethanol.

Design and Development of an Ideal Yeast for Effective Cofermentation of Glucose and Xylose to Ethanol

The uniqueness of our approach is that the yeasts engineered to not only ferment xylose but also (i) to effectively direct the metabolic flux towards the production of ethanol rather than the production of byproducts such as xylitol, (ii) to overcome the natural barrier "glucose effect," making it possible for the resulting engineered yeast to effectively coferment both glucose and xylose simultaneously so that the mixed sugars will be fermented as fast as possible, (iii) to easily convert most *Saccharomyces* strains, particularly the superior glucose-fermenting industrial strains to coferment xylose in addition to glucose, (iv) to use rich medium for growth and fermentation to make the engineered yeast grow and ferment sugars faster, and (v) to solve the potential waste problems by recycling used yeast cells for the production of crude yeast extracts for culturing new yeast cells. Furthermore, the final genetically engineered yeasts are stable and can be used in either batch or continuous process for ethanol production without the use of special chemicals as selection pressure to maintain the cloned genes.

In order to accomplish all of the above, we cloned not only a xylose reductase gene (XR) and a xylitol dehydrogenase gene (XD) but also a third gene, the xylulokinase gene (XK), even though all the *Saccharomyces* yeasts already contain a functional XK. We also replaced the signal sequences that control the expression of the three cloned genes with sequences that control the expression of yeast glycolytic genes.

After over ten years of dedicated research, our group at Purdue University achieved a historic breakthrough in the fall of 1993. As the first and still the only research group in the world to overcome the aforementioned obstacles, we succeed in the development of genetically engineered *Saccharomyces* yeasts that effectively ferment both glucose and xylose present in cellulosic biomass to ethanol. This was accomplished by first developing several stable plasmids such as pLNH32 and pLNH33 (Figure 2) that contain the three cloned xylose-metabolizing genes XR, XD, and XK, all fused to glycolytic promoters and subsequently designated as AR (or $A*R$), KD, and KK (for details see ref. *3*). These recombinant plasmids were constructed by closely following the design outlined above. They performed as expected. For example, these plasmids are able to transform various xylose non-utilization yeasts, particularly the *Saccharomyces* yeasts, and convert them to genetically engineered yeasts that are able to effectively coferment both glucose and xylose to ethanol. Yeast strain 1400 is one of the *Saccharomyces* yeasts that have been transformed with pLNH32 and pLNH33. The resulting genetically engineered yeasts, designated 1400(pLNH32) and 1400(pLNH33) (also referred to as LNH32 and LNH33), can effectively coferment glucose and xylose to ethanol. For example, 1400 (pLNH32) can ferment 8% glucose and 4% xylose mostly to ethanol in 48 hr as shown in Figure 3. The parent yeast, strain 1400, can ferment only glucose but not xylose to ethanol (Figure 4) *(3)*.

Our genetically engineered *Saccharomyces* yeasts have outstanding properties that make them ideal for the industrial production of ethanol. For instance, they can effectively ferment glucose and xylose both separately and simultaneously to ethanol as shown in Figure 3. This is an extremely important property for economical production of ethanol from biomass. Furthermore, our engineered yeasts are very stable and can use rich medium for fermentation and growth. They do not require the presence of expensive or undesirable chemicals, such as antibiotics, to maintain the cloned genes for fermentation of xylose. In addition, our genetically engineered yeasts produce very little byproduct such as xylitol (Figure 3) and can grow very well in medium using either glucose or xylose as the sole carbon source, as shown in Table 1.

The Effect of Cloning and Overexpression of the Xylulokinase Gene

One of the major differences between our genetically engineered *Saccharomyces* yeasts and those reported by others (Kotter et al., Walfridsson et al., and Tantirungkij et al.) *(4-6)* is that in addition to cloning the XR and XD genes, we also cloned and overexpressed the XK genes, even though nearly all *Saccharomyces* yeasts have an active xylulokinase gene and produce an active xylulokinase enzyme in the presence of

148

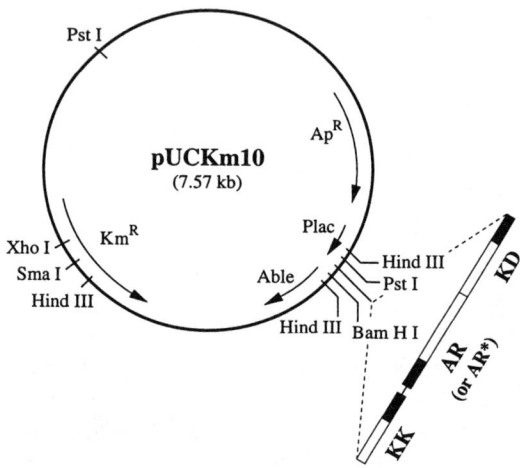

Figure 2. Restriction map of the PLNH plasmids. (Reproduced from reference 3. Copyright 1998 American Society for Microbiology).

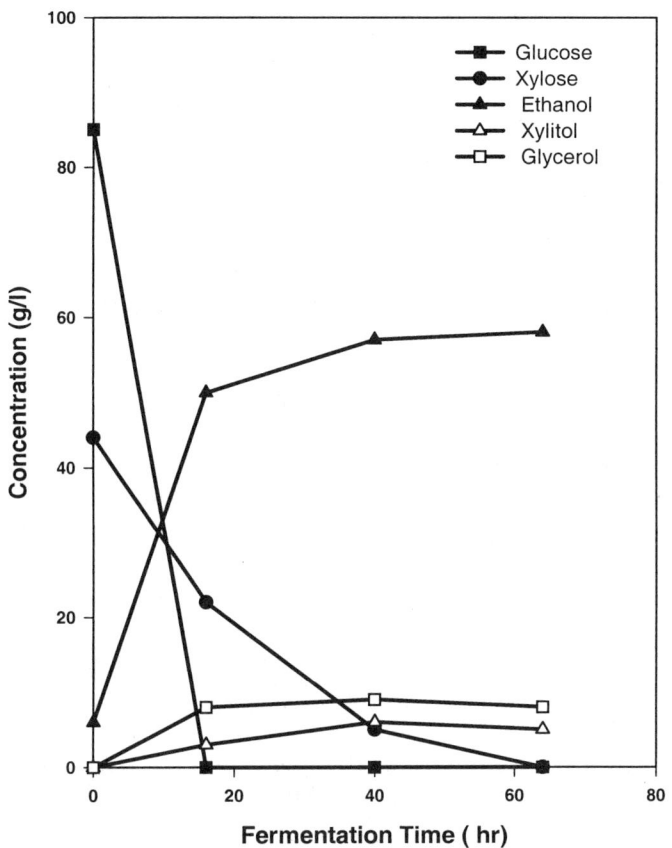

Figure 3. Cofermentation of glucose and xylose by genetically engineered yeast 1400(pLNH 32). (Reproduced from reference 3. Copyright 1998 American Society for Microbiology.)

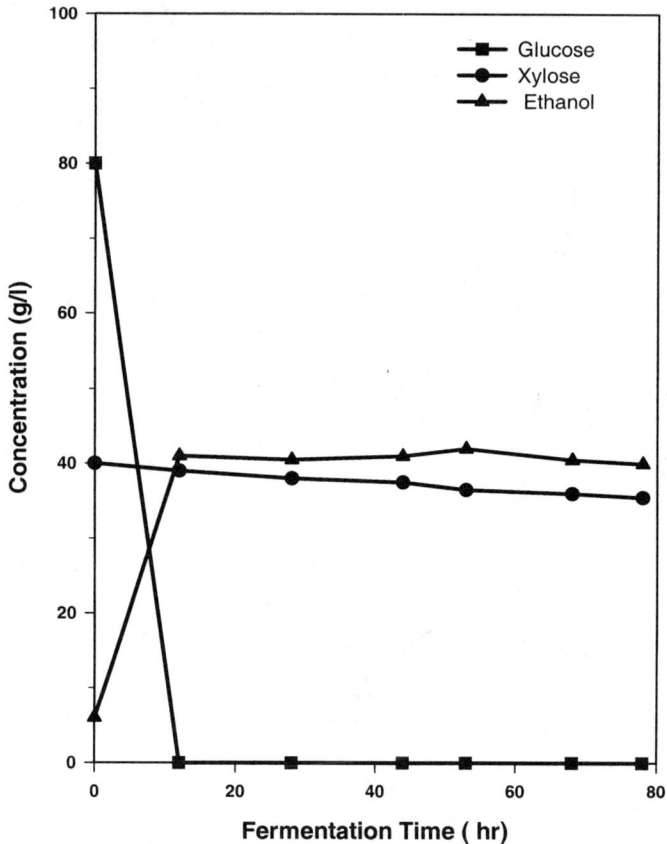

Figure 4. Cofermentation of glucose and xylose by the parent yeast strain 1400. (Reproduced from reference 3. Copyright 1998 American Society for Microbiology.)

Table 1. Comparison of the growth of Recombinant *Saccharomyces* and the Parent Yeast in Glucose and Xylose Media.

hr		0	2	4	6	8	11	15
Strains	Medium				KU^1			
LNH32 &	YEPD[2]	21	22	44	120	289	455	472
LNH33	YEPX[3]	18	21	37	68	122	500	550
LNH-ST(1)	YEPD	17	19	35	87	205	400	425
	YEPX	17	18	30	59	116	482	520
Parent Yeast	YEPD	17	17	39	120	290	433	460
	YEPX	54	66	72	75	75	79	79

1. KU = Klett Units, the optical density units measured by Klett-Summerson photoelectric colorimeter.
2. 1% yeast extract, 2% peptone, 2% glucose.
3. Same as 2 except 2% xylose instead of glucose.

xylose. As a result, our genetically engineered yeast can effectively utilize xylose for growth and ferment xylose to ethanol. On the contrary, those developed by Kotter et al. *(4)* and Walfridsson et al. *(5)* are neither able to utilize xylose for growth nor able to ferment xylose to ethanol. Although Tantirungkij et al. *(6)* reported that their engineered yeast can grow with xylose as the sole carbon source, their yeast still cannot ferment xylose to ethanol. It appears that overexpression of the xylulokinase gene makes it possible to overcome the problems of cofactor imbalance and the favoring of xylitol production as described above.

To demonstrate the effect of cloning *XK* on metabolizing xylose by the *Saccharomyces* yeast, we constructed a new plasmid pUCKm10-XR-XD which is identical to pLNH33 except that it lacked the cloned XK gene. *Saccharomyces* strain 1400 transformants containing pUCKm10-XR-XD utilized xylose for growth at much slower rates, less than one-fifteenth the rate of the same yeast strain containing pLNH33. Furthermore, the genetically engineered 1400 yeast containing pUCKm10-XR-XD, designated 1400(pXR-XD), could only convert xylose to xylitol, but not ethanol as illustrated in Figures 5A and 5B. Preliminary results of this work have been published elsewhere *(7)*.

The Effect of Replacing the Original Promoters of the cloned *XR, XD,* and *XK* Genes with *Saccharomyces* Glycolytic Promoters

Nearly all naturally occurring microorganisms, including the Saccharomyces yeasts, only metabolize glucose when a mixture of sugars, such as glucose and xylose, are present in their media. Furthermore, microorganisms require the induction of new enzymes to use another sugar, for example xylose, after glucose is depleted from their media. Thus, there usually is a lag period after glucose has been depleted before the second sugar can be actively used by a microorganism, a phenomenon generally known as the "glucose effect". In our view, the time required for microorganisms, including yeasts, to convert a feedstock containing mixed sugars to ethanol would be greatly shortened if the microorganisms were made free from the glucose effect and also the requirement that xylose be present for the induction of the xylose-specific enzymes. We determined that this could be accomplished by replacing the regulatory sequences controlling gene expression (present within the 5' noncoding sequences) of those genes specifically required for metabolism of xylose with regulatory sequences not subject to the glucose effect. Besides the *XR* and *XD* genes, the XK gene is also subject to the glucose effect and requires the presence of xylose for induction of the synthesis of its protein. Thus, this is another reason why it is necessary to clone the xylulokinase gene, even if the parent yeasts contain high levels of xylulokinase activity.

In order to prove that cloning KD, AR or A*R, and KK (XD, XR, and XK structural genes fused to glycolytic promoters) is extremely important for cofermentation of feedstocks (or media) containing mixtures of glucose and xylose, we compared the fermentation of a mixture of glucose and xylose by 1400(pLNH 32) and *Pichia stipitis*. As described above, the Pichia yeast is a naturally occurring xylose-fermenting yeast that can ferment xylose with similar efficiency at the laboratory scale under well controlled conditions as the genetically engineered 1400(pLNH32). However, the expression of *Pichia's XD, XR,* and *XK* genes is subject to glucose

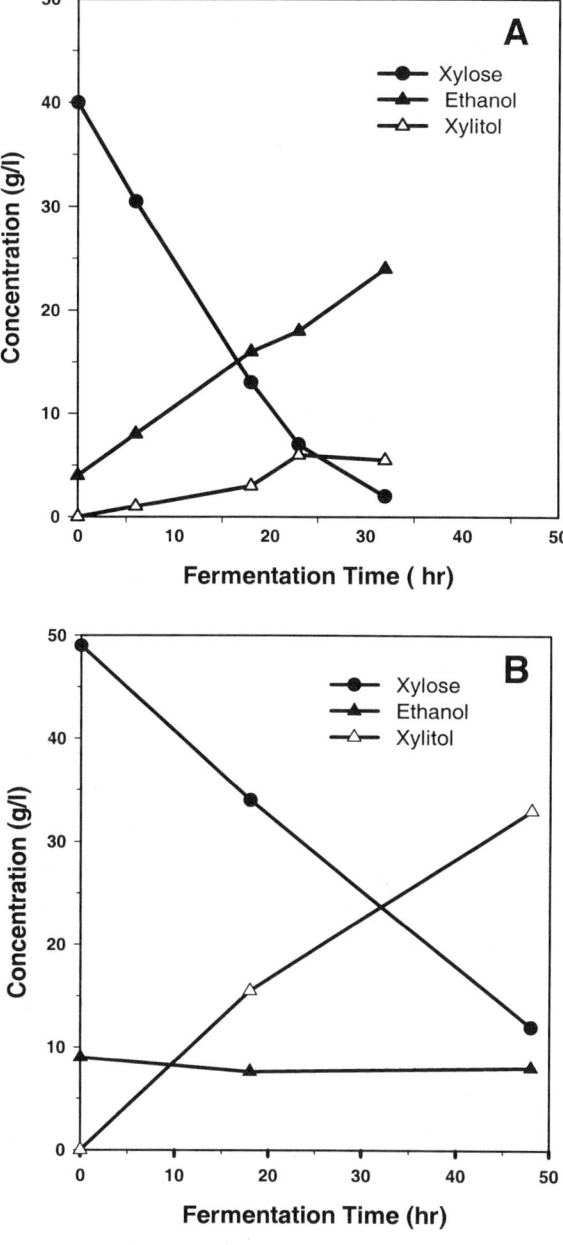

Figure 5. Comparison of fermentation of xylose under identical conditions as described in the Methods of reference 3 by (A) genetically engineered Saccharomyces yeast strain 1400(pLNH32) which contains the cloned and genetically modified XR, XD, and XK genes and by (B) 1400(pXR-XD) which contains only the same cloned XR and XD genes, but not the cloned XK gene.

inhibition and requires the presence of xylose for induction. As shown in Figures 6A and 6B, when these yeasts were cultured in YEPX and used to ferment a mixture of glucose and xylose, 1400(pLNH32) effectively cofermented both glucose and xylose to ethanol, whereas *P. stipitis* fermented glucose but not xylose to ethanol. Furthermore, our engineered yeast cultured in 50 ml YEPD medium with 2 ml of YEPX pregrown cells as the seed culture is also able to ferment a mixture of glucose and xylose to ethanol with similar efficiency as shown in Figure 6A. Thus, our results demonstrate that in order for microorganism (yeast) to be effective in cofermenting xylose and glucose to ethanol, it must be able to synthesize xylose-metabolizing enzymes in the presence of glucose. Preliminary results of this work have been published (7).

Effective New Gene-Integration Technique for the Development of Superstable Genetically Engineered *Saccharomyces* Yeasts

The 1400(pLNH32) and related plasmid-mediated genetically engineered xylose-fermenting *Saccharomyces* yeasts are very stable and their xylose-fermenting ability can be maintained by an ideal selection mechanism (3). However, a more perfect yeast for fuel ethanol production should be completely stable without the need for selection pressure to maintain its xylose-fermenting ability at any stage of growth or fermentation. In 1995, we successfully developed the first superstable strain of genetically engineered glucose-xylose-cofermenting *Saccharomyces* yeast, 1400(LNH-ST), which requires no selection pressure to maintain its ability to ferment xylose to ethanol (data not shown) or to utilize the sugar for growth as shown in Table 1. Furthermore, this superstable xylose-fermenting genetically engineered yeast coferments glucose and xylose to ethanol with equal or greater effectiveness than the plasmid-mediated recombinant yeasts such as 1400(pLNH32). This superstable genetically engineered *Saccharomyces* yeast contains multiple copies of the same three modified XYL genes, AR, KD, and KK, integrated into the host chromosome. It is generated by means of a much improved technique for integrating multiple copies of genes into the yeast chromosome, a method that is actually much easier to accomplish than all methods previously described in the literature. At this time, we have completed the development of three stable glucose-xylose-cofermenting *Saccharomyces* yeast strains, 1400(LNH-ST), 259A(LNH-ST), and 424A(LNH-ST). They are derived from three different parent strains of *Saccharomyces* yeasts, and they are all efficient in co-fermenting glucose and xylose to ethanol. The use of this unique gene-integration technique allows one to easily control the copy-number of the integrated genes which thereby facilitates development of the ideal *Saccharomyces* yeasts for ethanol fuel production from cellulosic biomass as envisioned years ago (see Figure 7). According to our original design, *Saccharomyces* yeasts that are superior for ethanol production from traditional glucose-based feedstocks, or that have strong tolerance to ethanol, high temperature, or inhibitors present in hydrolysates of the cellulosic biomass, should all be successfully genetically engineered by our technology to effectively ferment xylose to ethanol. This unique gene-integration technique makes it possible to convert most, if not all, *Saccharomyces* yeasts to effectively coferment glucose and xylose to ethanol. It should also make it possible to further improve our engineered yeasts to be able to coferment other sugars present

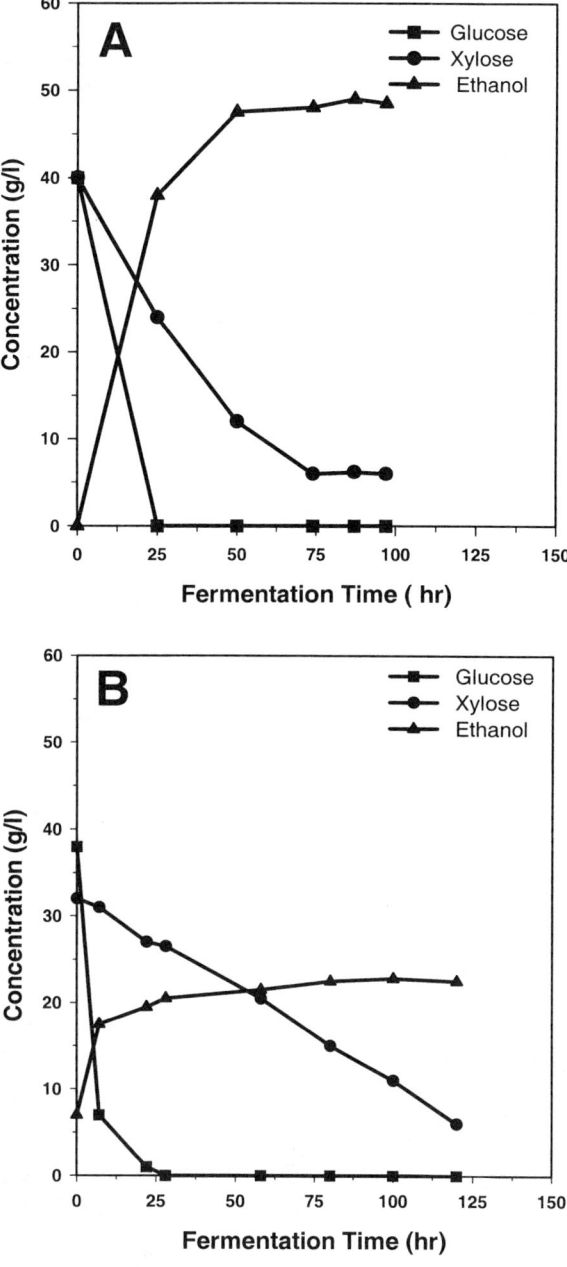

Figure 6. Comparison of cofermentation of glucose and xylose under identical conditions as described in the Methods of reference 3 by (A) genetically engineered Saccharomyces yeast strain 1400(pLNH32) and by (B) Pichia stipitis. Symbols: square, glucose; circle, xylose; triangle, ethanol.

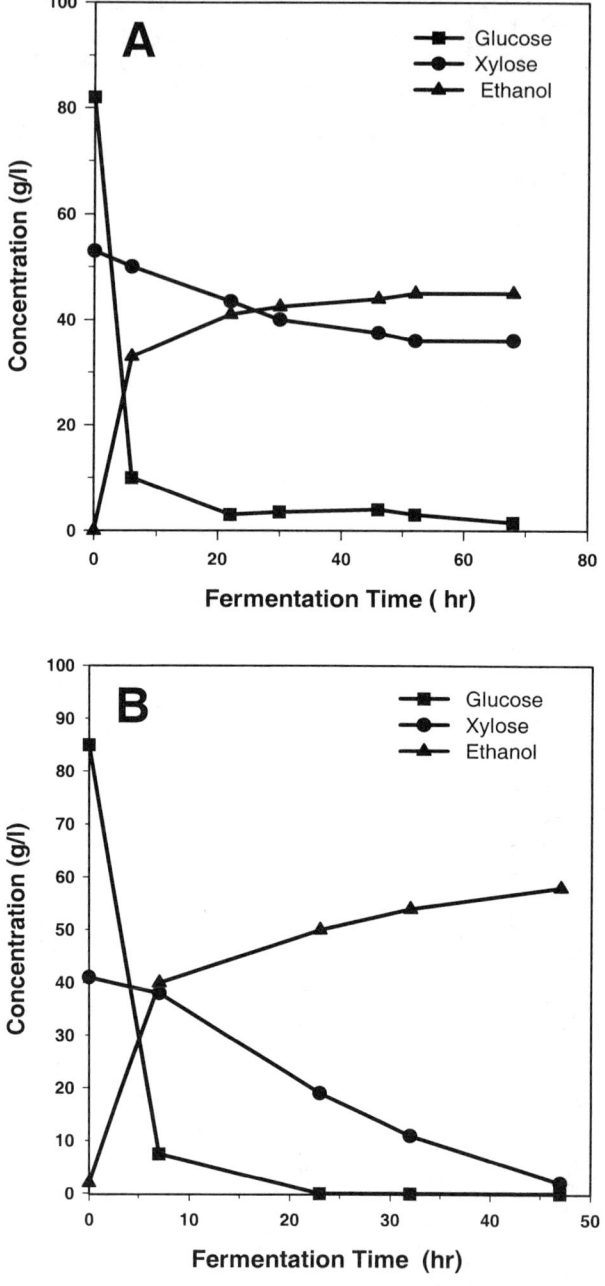

Figure 7. Stepwise integration of multiple copies of the cloned multiple genes into the host chromosome(s). 259A(LNH-ST)(Int 0) (shown in A) contains fewer copies of the integrated xylose-metabolic genes than 259A(LNH-ST)(Int 3) (shown in B)

in the cellulosic biomass hydrolysates, as well as to produce some high-valued byproducts or co-products in addition to ethanol. The development of these stable genetically engineered glucose-xylose-cofermenting *Saccharomyces* yeasts and the method of integrating multiple copies of gene(s) into the yeast chromosome have been briefly described previously *(8)* and the details of the development of these stable yeasts will be reported elsewhere.

Repeated Cofermentation of a mixture of Glucose and Xylose Efficiently to Ethanol

Recently we determined that our engineered glucose-xylose-cofermenting *Saccharomyces* yeasts can repeatedly coferment a mixture of glucose and xylose efficiently to ethanol for numerous cycles as shown in Figure 8. Moreover, our "stable" glucose-xylose cofermenting yeasts can repeatedly coferment sugars present in hydrolysates of cellulosic biomass including glucose, xylose, mannose, and galactose. Our initial results showed that the engineered yeasts can produce 20%-40% more ethanol than the unengineered parent yeast, depending on the initial pH at each cycle of fermentation (data not shown). The latter results may lead to the development of much more cost-effective processes for the production of cheap ethanol from cellulosic biomass.

Genetically Engineered *Saccharomyces* Yeasts Cofermenting Glucose and Xylose Present in Crude Hydrolysates

Our stable strain 1400(LNH-ST) was confirmed to be able to effectively coferment glucose and xylose to ethanol by a continuous process in a 9000-L pilot scale bioreactor for 90 days with both pure sugars and sugars from crude corn biomass hydrolysates as the feedstocks *(9)*. Our genetically engineered *Saccharomyces* yeasts were shown to be able to coferment glucose and xylose present in various other crude cellulosic biomass hydrolysates very effectively *(7, 10)*. Furthermore, our glucose-xylose-cofermenting *Saccharomyces* yeasts can also repeatedly coferment glucose and xylose present in crude hydrolysates of soft wood to ethanol (Ho et al. unpublished results) and undoubtedly these yeasts should also be able to repeatedly coferment glucose and xylose present in crude hydrolysates of other cellulosic biomass.

Conclusions and Future Perspective

In this paper, we reviewed our strategies used to genetically engineer the *Saccharomyces* yeasts, resulting in the development of innovative technology that can successfully transform most, if not all, the *Saccharomyces* yeasts from non-xylose-fermenting to xylose-fermenting as well as to glucose and xylose cofermenting species not subject to control by "glucose" inhibition.

Our unique gene-integration technique that can effectively integrate multiple copies of multiple genes into the yeast chromosome has made it possible for us to develop (the perfect) "stable" genetically engineered yeasts that can effectively

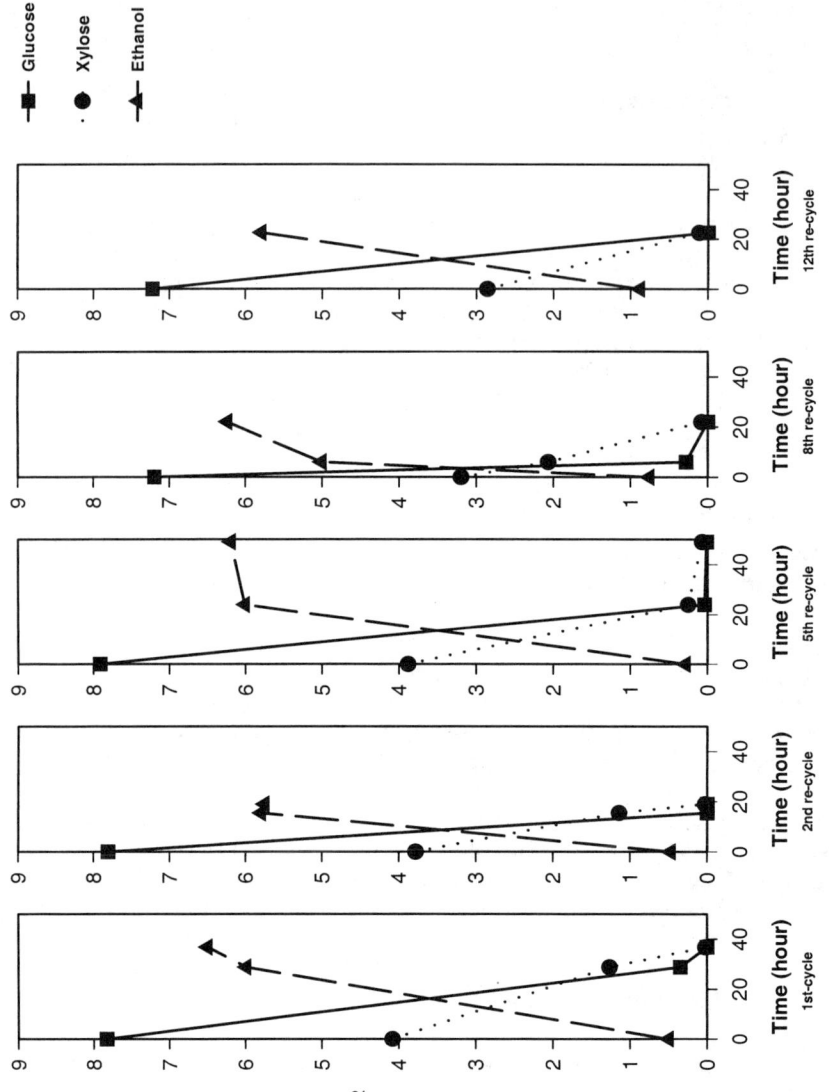

Figure 8. Repeated cofermentation of glucose and xylose to ethanol by the same batch of genetically engineered stable glucose-xylose-cofermenting Saccharomyces yeast 424A(LNH-ST).

coferment glucose and xylose to ethanol for hundreds of generations without requiring the presence of any selection pressure to maintain the three cloned xylose metabolism genes.

With our reliable and effective gene-integration technique coupled with the most advanced recombinant DNA techniques available to yeast manipulation, and the most complete genetic information and data available for the *Saccharomyces* yeasts, our engineered yeasts can further be improved for the coproduction of numerous other high-value products. Because *Saccharomyces* yeasts are the safest, hardiest, and most user-friendly of the industrial microorganisms, there is no doubt that our genetically engineered *Saccharomyces* yeasts should be and will be the microorganisms of choice for the cost-effective production of cheap ethanol from cellulosic biomass, just as traditional *Saccharomyces* yeasts are the microorganisms of choice for the large-scale industrial production of ethanol from glucose-based feedstocks.

Acknowledgments

The authors wish to acknowledge that their current work is supported in part by the U.S. Environmental Protection Agency and the National Renewable Energy Laboratory of the U. S. Department of Energy and that their past work was supported by the U.S. department of Agriculture, the U. S. Department of Energy, and Swan Biomass Company.

Literature Cited

1. Bruinenberg, P. M.; deBot. P. H. M.; van Dijken, J. P.; Scheffers, W. A. *Appl. Microbiol. Biotechnol.* **1984**, *19*, 256-260.
2. Deng, X. X.; Ho, N. W. Y. *Appl. Biochem. Biotechnol.* **1990**, *24/25*, 193-199.
3. Ho, N. W. Y.; Chen, Z.; Brainard, A. P. *Appl Environ. Microbiol,* **1998**, *64*, 1852-1859.
4. Kotter, P.; Amore, R.; Hollenberg, C. P.; Ciriacy, M. *Curr. Genet.*, **1990**, *18*, 493-500.
5. Tantirungkij, M.; Nakashima, N.; Seki, T.; Yoshida, T. *J. Ferment. Bioeng.* **1993**,*75*, 83-88.
6. Walfridsson, M.; Anderlund, M; Bao, X.; Hahn-Hagerdal, B. *Appl. Microbiol. Biotechnol.*, **1997**, *48*, 218-224.
7. Ho, N. W. Y.; Chen, Z.; Brainard, A. P.; Sedlak, M. in Recent Progress in Bioconversion of Lignocellulosics; Tsao, G. T. Ed.; Springer-Verlag: Heidelberg, Germany; in press.
8. Ho. N. W. Y.; Chen, Z. D., Patent pending, PCT patent No. WO97/42307.
9. Toon, S. T,; Philippidis, g. P., Ho, N. W. Y.; Chen, Z. D.; Brainard, A.; Lumpkin, R. E.; Riley, C. J. *Appl. Biochem. Biotechno,.* **1997**, *63-65*, 243-255.
10. Moniruzzaman, M.; Dien, B. S.; Skory, C. D.; Chen, Z. D.; Hespell, R. B.; Ho, N. W. Y.; Dale, B. E.; Bothast, R. J. World *J. Microbiol. Biotechnol.*, **1997**, *13*, 341-346.

Chapter 13

Bioconversion of Sugar Cane Vinasse into Microbial Biomass by Recombinant Strains of *Aspergillus nidulans*

André O. S. Lima and Aline A. Pizzirani-Kleiner

Department of Genetics, University of São Paulo—ESALQ,
Piracicaba-SP, 13418–900, Brazil

Sugar cane broth is widely used in Brazil in fermentation to produce alcohol. This process yields a large amount of a toxic byproduct, vinasse. Due to its high organic matter, law prohibits its disposal into rivers. The microbial conversion of the vinasse into Single Cell Protein (SCP)/microbial biomass, as a protein and/or methane (fuel) resource, is being evaluated world wide. The present study illustrates an approach to microbial biomass production based on strains of the fungus *Aspergillus nidulans* produced through parasexual crossing. Six diploid strains and 23 recombinants were evaluated and compared for their ability to degrade vinasse based on measurement of COD reduction and biomass synthesis. SCP production was based on mycelial dry weight, and indirectly on radial colony growth on Petri dishes. Most of the diploids and certain recombinant strains gave enhanced performance compared to the parental strains.

Sugar Cane Vinasse and Its Application

Vinasse Production

The Brazilian alcohol project began in 1975 as a result of the world oil crisis. At that time, the Brazilian production of ethanol was 320 million liters per year. Production peaked in 1986 rising to nearly 12 billion liters/year, a 37-fold increase. The increase in production was followed by an increase in production of cars that use hydrated alcohol (álcool) as fuel. By 1992, there was a scarity of alcohol, and the country needed to import ethanol to supply the automotive demand. This was followed a decline in interest in, and production of, alcohol burning cars as well as a decline in the production of hydrated ethanol. In 1994, federal regulations required that 22 percents of anhydrous alcohol should be added to the commercial gasoline, consequently anhydrous production rose. In 1996-97, Brazilian alcohol production reached 14 billion liters, with one third being anhydrous (Marcus Farhat, personal communication).

During the alcohol distillation process, vinasse, a byproduct, is formed at approximately 13 parts vinasse to one of alcohol. In the State of São Paulo alone, the largest producing area in the country, this represents 115 billion liters of vinasse per year. Due to its high organic content, vinasse is prohibited from disposal in rivers. Its composition is variable, being influenced by many factors such as the type of sugar, the composition of the broth, and the distillation process *per se*. However, some features of this byproduct are standard. It is an acidic liquid, rich in nitrogen, potassium and organic acids such as succinic and lactic (*1*), and containing amino acids, such as leucine, isoleucine and valine (*2*). It is poor in sugars with just 10 percent of the total amount of organic matter composed of saccharose and fructose (*3*). The pollution index of Chemical Oxygen Demand (COD) is 35 to 60 grams of oxygen per liter, and there is also a large Biochemical Oxygen Demand (BOD$_5$) which varies from 7 to 20 grams of oxygen per liter.

Vinasse Utilization and Bioconversion

There are few practical ways to dispose of vinasse. The most common application is the use of vinasse on agricultural lands as fertilizer (*4*) providing nitrogen and potassium. The effect of vinasse on the activity of soil microorganisms under a range of conditions shows a temporary increase in their activity (*5, 6*). Diverse studies have compared the optimization of vinasse as a fertilizer, and there are several official recommendations for vinasse application.

Processes aimed at using not just the salts but also the organic content of vinasse are being investigated, for instance, the anaerobic conversion of vinasse to methane (*7, 8, 9*). The aerobic degradation of vinasse to yield microbial Single Cell Protein (SCP) has also attracted attention. In this case, the biomass produced is used as a protein source for animal feed. Bacteria (*10, 2*), yeast (*11, 12, 13, 14*) and filamentous fungi (*14, 15, 16, 17, 18, 19, 20, 21*) have been evaluated for this purpose. Others applications, including silaging (*22*), sugar cane bagasse composting and enrichment (*23*), or burning and use of the ashes as a soil fertilizer (*24*), have also been evaluated at experimental and pilot scales.

The aerobic conversion of vinasse to microbial biomass provides significant environmental benefit as the vinasse's toxicity (BOD$_5$ and COD) is reduced and its organic fraction utilized in the production of SCP. In the late 1960's, Brazil had two plants using this process, each producing about 3000 tons of SCP per year. The filamentous fungi gave good yield due in part to the secretion of a wide range of enzymes and easy recovery of SCP by filtration. Certain species of the fungi *Aspergillus* were identified as especially efficient (*15, 16, 17, 18, 20, 21, 25*).

The focus of this chapter is the aerobic conversion of vinasse by *Aspergillus* into SCP. Specifically, parasexual recombination is evaluated as a genetic tool for improving *Aspergillus*' ability to consume vinasse. The development work followed three steps. First, the optimal growth medium was established; second, the *Aspergillus* species were compared; and third, using *A. nidulans* as the model system, diploids and recombinant strains were constructed and evaluated for vinasse conversion.

Material Methods

Species

Aspergillus niger NRRL 337; *A. oryzae* ATCC 7252; *A. sulphureus* IZ (USP) A186; *A. nidulans* strains: Master Strain E (MSE) [*suA1 adeE20 yA2 adeE20; wA3; galA1; pyroA4; facA303; sB3; nicB8; riboB2*] (*26*); Abnc [*proA1 pabaA6* T (I-II); *bncA*]; PPY [*proA1 pabaA6 yA*]; BM [*biA1; methG1*]; INO [*inoB*]; VE [*veA1*] (*27*). All strains were stocks from ESALQ,-USP from mineral oil or SiO_2.

Medium

The strains were maintained on Complete Medium (CM) for *A. nidulans* (*28*) as modified in (*29*). The vinasse medium was established in this work; its pH was adjusted to 6.8 with 4N NaOH. For plate assays, 1.5% agar was added.

Sugar Cane Molasse and Vinasse

The sugar cane molasses was donated by a distillery from the Companhia Industrial e Agricola Ometto, and the vinasse was obtained at the pilot distillery of the Dept. Ciência e Tecnologia Agroindustrial (ESALQ/USP). Both were stored at 1^0C.

NPK solution

The commercial fertilizer NPK (10:10:10) was crushed, solubilized (1%) in distilled water, filtered (Whatman #1), autoclaved (1 atm, 20min.) and stored at 4^0C.

Radial Growth

The strains were point inoculated on Petri plates containing the appropriate medium for each assay and the colony diameter was measured with the help of a light box.

Dry Biomass Production

The dry biomass production was measured after growing the strain in liquid medium under defined growth conditions - see individual experiments. After cultivation, the mycelium was filtered through nylon screen (95 mesh), washed and dried at 70^0C until constant weight.

Pollution Indexes

The COD as well the BOD$_5$ were performed as described in (*30*). The percent decrease was calculated as a function of the control, uninoculated sample.

Parasexual Crosses, Diploid and Recombinant strains

The parasexual crosses (a process that results in the formation of recombinant nuclei) were conducted as described in (*28*), in which haploid spores from two genetically contrasting strains (different auxotrophs) were cultivated in minimal medium. By this method, diploid cells containing genetic material of both parents were isolated. Recombinant haploid strains were then obtained by the exposure of the diploid to benomyl (benomyl is a agricultural antifungal agent that disrupts the formation of spindle microtubules, inhibiting mitosis and inducing non-disjunction). The diploid condition was confirmed by the ability to grow in minimal medium, by the spore size (data not shown) that is bigger, and by the capacity for the haploid to segregate from the diploid during growth on the plate.

Results and Discussion

Medium Establishment

Nutrient supplementation of vinasse has been evaluated by other researchers to assess enhancement of biomass productivity (*11, 12, 16, 20, 25, 31, 32*). In our initial trials we attempted to optimize the culture media by using a randomly picked strain of *A. nidulans* (BM) and evaluating its radial growth using vinasse plus different added compounds.

Molasses

The effects of the addition of sugar cane molasses in increasing concentrations were evaluated in terms of growth and cost of molasses. According to the results (Table 1), optimal relation between radial growth and the cost of the supplement was obtained with the addition of two percent molasses per liter of vinasse (w:v).

NPK

The addition of a commercial fertilizer with salts of nitrogen, phosphorous and potassium (10:10:10) was evaluated. The vinasse appears to have adequate amounts of these components for microbial development, since a decline in their radial growth was observed when these salts were added to the media (Table 2), even when reduced (1/5) doses were evaluated (data not shown). As a control, fertilizer was replaced with the NPK salts from *A. nidulans* complete medium. Thus in further assays, vinasse medium was standardized as vinasse supplemented with two percent molasses (w:v).

Table 1. The Effect on the Radial Growth of *A. nidulans* by the Addition of Sugar Cane Molasses to Vinasse [1]

	Colony Diameter (mm)	
Molasses [2]	24 h	48 h
0	41 a [3]	079 a
1	51 b	112 b
2	54 bc	129 c
3	58 c	134 c
4	59 c	142 d
5	60 c	144 d

[1] Experimental outline: completely random, split-plot in time (24 and 48 hours), 10 repetitions. Cultivation at 37°C. [2] Molasses (g/100mL of vinasse). [3] Letters represent the groups formed by the Tukey test (5% confidence); the same letters = no significant variation.

Table 2. The Effect on the Radial Growth of *A. nidulans* by the Addition of NPK solution [1]

	Colony Diameter (mm)	
NPK (%)	43 hours	67 hours
0	103 a [3]	162 a
2	72 b	137 b
4	65 c	126 b
6	51 d	104 c
8	49 d	107 c
10	47 d	102 cd
12	50 d	093 d
CM	74 b	152 a

[1] Cultivation conditions: 37°C, medium vinasse + 2% molasses (w:v). NPK % (v:v). CM = 6 g $NaNO_3$ + 1.5 g KH_2PO_4 + 0.5 g KCl * L^{-1}. [2] Experimental outline: completely random, split-plot in time (43 and 67 hours), 16 repetitions. [3] Letters represent the groups formed by the Tukey test (5% confidence), same letters represent no significant variation.

Vinasse Biodegradation by *Aspergillus* species

The next stage was to compare the model *A. nidulans* (Abnc) with other *Aspergillus* species that had been assessed to be efficient in vinasse degradation. These species were *A. niger* (9,12,7), *A. oryzae* (4, 9, 12, 24) and *A. sulphureus* (Dr. Choit Kiyan, personal communication).

Temperature

The first trial was to determine a common temperature for growth of all 4 species. For that purpose we measured the radial growth of the species on solid vinasse medium at three different temperatures (27, 32 and 37°C) twice (after 35 and 70 hours of cultivation). All the species grew well at 32°C (Table 3). This temperature was subsequently used.

Table 3. Effect of temperature on the radial growth of *Aspergillus* species [1]

Cultivation (h)	°C	Colony Diameter (cm)			
		A. nidulans	*A. niger*	*A. oryzae*	*A. sulphureus*
35	27	0.63 c [2]	1.56 c	1.70 b	0.49 c
	32	0.95 b	2.40 b	2.46 a	0.77 a
	37	1.10 a	2.53 a	0.65 c	0.64 b
70	27	1.96 c	4.27 c	4.19 b	1.24 b
	32	2.90 b	6.17 a	5.63 a	1.34 a
	37	3.23 a	5.95 b	1.97 c	1.26 ab

[1] Experimental outline: completely random factorial (temperature X species), two measurements (35 and 70 hours of incubation), 6 repetitions. Cultivation on vinasse medium. [2] Letters represent the groups formed by the Tukey test (5% confidence), same letters represent no significant variation.

Biomass Production and Pollution Reduction

After establishing the optimal medium and the growth temperature, the 4 species were evaluated for biomass production and COD and BOD_5 reduction. The species were cultured in liquid vinasse medium at 32°C with agitation. The results varied significantly between species (Table 4). Two groups were separated by the Tukey test for biomass production and BOD_5 reduction. As radial growth measurements had low correlation to the dry biomass data, they could not be used as an indicator for others assays in liquid medium; however, the correlation of 0.7 between COD and BOD_5 reduction, enabled the estimation of BOD_5 reduction by that of COD.

With the highest ratio between dry biomass produced and the COD/BOD_5 consumed (conversion index [%]), *A. nidulans* was identified as the best converter (Table 4). However, the aim of the process was not only to produce biomass, but also to consume or degrade vinasse. Thus, a good converter may not be the best producer or degrader. As an example, with *A. niger* and *A. sulphureus*, the first showed more productivity and reduction of vinasse than the second, which had better conversion indexes (Table 4).

Table 4. Biomass Production and Pollution Reduction of Vinasse by *Aspergillus species* [1]

Species	Dry Biomass (mg)	Productivity (g/L/hour)	COD Reduction (%) [2]	COD Conversion Index (%) [3]	BOD$_5$ Reduction (%) [4]	BOD$_5$ Conversion Index (%) [3]
A. nidulans	507 a [5]	0.267	31.12 c	3.06	55.85 ab	6.58
A. niger	493 a	0.260	48.27 a	1.92	61.88 a	5.78
A. oryzae	369 b	0.194	43.42 b	1.60	57.96 ab	4.61
A. sulphureus	407 b	0.215	34.73 c	2.21	50.51 b	5.85

[1] Incubation Conditions: 7×10^6 spores, 50 mL vinasse medium (Erlenmeyer 250mL), 32°C, 40 hours, agitation of 120 rpm, 3 replicates. [2] Average from 2 digestions from a compost sample constituted by the 3 replicates. [3] Conversion index: biomass produced (g) / COD or BOD$_5$ consumed (g) * 100. [4] Average from 4 digestions (2 by each dilution) from a compost sample constituted by the 3 replicates. [5] Letters represent the groups formed by the Tukey test (5% confidence), same letters represent no significant variation.

Vinasse Conversion by Recombinant Strains of *A. nidulans*

The main aspect of this study was the development and evaluation of some diploid and recombinant strains of *A. nidulans* for vinasse utilization.

Diploids, Recombinants and Genealogy

The diploid strains were constructed by parasexual crosses using four genetically contrasting strains as parents, namely M (MSE), P (PPY), B (BM) and I (INO). The combination of these strains resulted in 6 diploid strains: M//P, M//B, M//I, P//B, P//I and B//I. Four recombinants were isolated by haploidization from each diploid, for example Rec MP1, Rec MP2, etc. The prototrophic haploid strain VE was used as a control.

Growth Kinetics

Next, a determination was made of the growth kinetics and the time required for the culture to attain the stationary phase of development. We evaluated biomass production (Figure 1) and radial growth (data not shown) of five strains (M, P, B, M//P, M//B). The results were analyzed by logistic regression (data not shown). The growth on solid medium was similar to the that observed on liquid vinasse. In both cases, the results suggest that the prototrophic diploid strains started their log phase of growth earlier than the auxotrophic haploids. The control VE was not assayed. No correlation was observed between the strain performance and spore viability (data not shown). The time yielding an average of 90% of the maximum biomass production for all strains was 43 hours. This culture period was used for further cultivation.

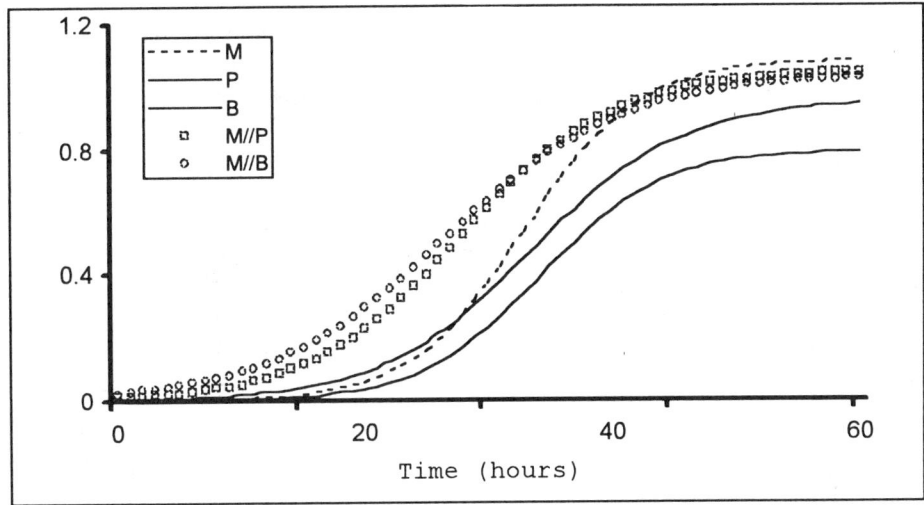

Figure 1. Growth Kinetics of A. nidulans strains cultivated on liquid vinasse medium. Approximately 9×10^6 spores (3 replications) were grow on 100mL of vinasse medium at 37 °C with agitation of 150 rpm. Samples were taken every 12 hours for 60 hours. The results were evaluated by logistic regression analysis.

COD Reduction and Biomass Production

To compare SCP production and COD reduction four kinds of *A. nidulans'* strains: parental, diploid, recombinant and a control, were grown in liquid vinasse medium. Figure 2, a dispersion graph, shows SCP production and COD reduction for the four strains. While recombinants showed a wide range of performance; nearly all diploid strains, and a few recombinants, showed activity over that of the parents. Analyzing the results, using the Student Newman Keuls test ($\alpha=0.05$), in which a different letter is assigned for each statistically different group gave the following results. For COD reduction, diploids (group A) were superior, the parentals and the control (group B) were less active, and the recombinants (group C) showed the least activity. For biomass production, the diploids (group A) were again the most active, followed by parentals (group A, B), control (group B) and recombinants (group C).

Analyzing some specific strains, Parent I had the best performance of all parent strains; its diploids also seem to be superior to all other diploids (not statistically tested). In the different crosses, however, the activity of their recombinants (Rec M//I, Rec B//I, Rec P//I) varied; for example: the cross between M and I yielded the best recombinants for biomass and COD (Student-Newman-Keuls test, $\alpha=0.05$) with an intermediate performance in relation to that of the parents, while the cross between B and I produced recombinants for biomass and COD (Student-Newman-Keuls test, $\alpha=0.05$) that were less efficient than the other recombinants and less efficient than the parents. The greatest increase was detected in the diploid MP, which had a 50% increase in SCP production as compared to parent M, and a 30% increase in COD reduction as compared with parent P.

We conclude that the parent sets used in the crosses have a greater influence on the recombinants than on the diploids as the diploids' overall performance was constant, and they were generally the most efficient. The same was not observed for the recombinants which presented a much greater amplitude of variation. This suggests that enhancement of both measured parameters is more easily obtained by the construction of diploids rather than by recombinants. However, it is important to note that diploids are unstable and tend to segregate to the haploid, undesirable in an industrial process. This condition can be avoided by the use of a recessive lethal mutation, a mutation that when present on a single allele (haploid state) kills the cell and prevents unwanted changes in the fermentation culture.

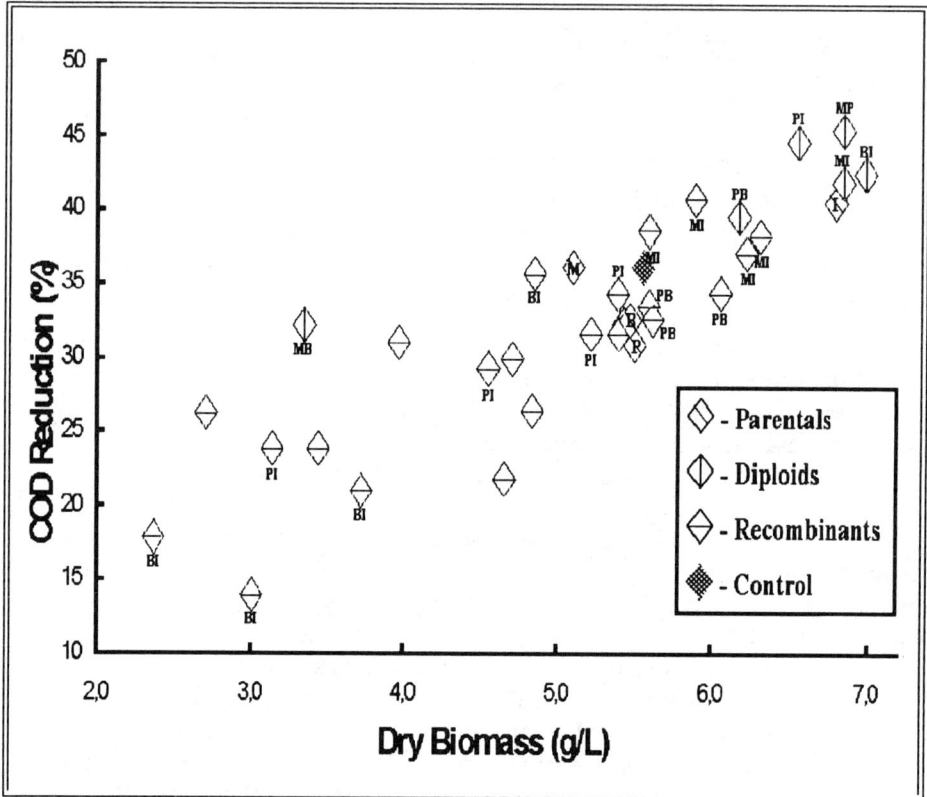

Figure 2. Dry Biomass Production and COD Reduction by A. nidulans strains. Approximately 9×10^6 spores (4 replicates) were cultivated in 100mL of vinasse medium at 37 °C with agitation at 150 rpm for 43 hours.

A correlation of 0.87 between dry biomass production and COD reduction was found from the results presented in Figure 2. To determine if the increased performance of the diploids over that of the recombinants was due to the presence of multiple auxotrophic markers in the recombinants, we made a genetic comparision of the recombinants. No correlation was detected between the quantity of genetic markers and their performance (data not shown).

Final Consideration

Utilization of alcohol industry by-products such as sugarcane bagasse, straw, yeast filter cake and vinasse, has the potential to reduce both the cost of alcohol production, and the cost to the environment due to their disposal. Due to the enormous volume produced per year, vinasse can be used in different systems of conversion. This study indicates that *A. nidulans* can be constructed to more optimally convert vinasse to biomass. Using parasexual recombination, it was possible to increase the indices of pollution reduction, and of SCP production by the construction of diploid and recombinant strains. In so doing, we have demonstrated that *A. nidulans* can be used as a genetic model for the study of vinasse degradation. This approach to the synthesis of microbial biomass from vinasse makes possible the transformation of a known pollutant into a benign resource. By reusing the carbon and
energy inherent in vinasse via an environmentally friendly biological process, *A. nidulans* bioconversion produces a useful, non-toxic agricultural feedstock.

Acknowledgments

Thanks to Dra. Dejanira F. de Angelis (UNESP- Brazil) for the laboratory facilities, to CAPES (Brazil) for the scholarship and to Dr. Douglas E. Eveleigh (Rutgers University – USA), Gavin Swiatek and James K. McCarthy for the English review.

Literature Cited

1. Rotenberg, B.; Antonaccio, L. D.; Lachan, A. *Inform. INT.* **1979**, *12*, 25-28.
2. Martelli, H. L.; Sousa, N. O. *Revta. bras. Tecnol.* **1978**, *9*, 157-164.
3. Dowd, M. K.; Johansen, S. L.; Cantarella, L.; Reilly, P. J. *J. agric. Fd Chem.* **1994**, *42*, 283-288.
4. Pande,H. P.; Sinha, B. K. In *Sugarcane: Agro-industrial Alternatives*; Singh, G. B.; Solomon, S.; Oxford & IBH Publishing Co. Pvt. Ltd: New Delhi, India, 1995; pp 401-413.
5. Tauk, S. M.; Ruegger, M. S. *Revta. Microbiol.* **1987**, *18*, 67-76.
6. Martinez Cruz, A.; Alemán, I.; Bach, T.; Calero, B. J. *Ciencias de La Agricultura.* **1987**, *30*, 118-127.

7. Vazoller, R. F. *International Symposium on Microbial Ecology*; 7; ISME: Santos, Brazil, 1995; pp5.
8. Valdes, E.; Obaya, M. C.; Ramos, J.; Leon, O. L. *Rev. Icidca Sobre Los Derivados de La Cana de Azucar.* **1992**, *26*, 33-36.
9. Bazile, F.; Bories, A. *Bull. Agron.* (Petit Bourg). **1989**, *9*, 20-26.
10. Ferraz, C. A. M.; Aquarone, E.; Krauter, M. *Revta. Microbiol.* **1986**, *17*, 15-21.
11. Costa, A. C. M.; Ceccato-Antonini, S. R.; Tauk-Tornisielo. *Arquiv. Biol. Tec.* **1992**, *35*, 575-584.
12. Delgado, R.; Jiménez, R.; Linares, H.; Gutiérrez, C. *Annls. Technol. agric.* **1975**, *24*, 287-295.
13. Tauk, S. M. *Revta. bras. Microbiol.* **1976**, *7*, 92-97.
14. Araújo, N. Q.; Visconti, A. S.; Castro, H. F.; Silva, H. G. B. *Inform. INT.* **1977**, *10*, 12-19.
15. Ceccato-Antonini, S. R.; Tauk, S. M. *Revta Microbiol.* **1992**, *23*, 43-47.
16. Ruegger, M. J. S.; Tauk-Tornisielo, S. M. *Arquiv. Biol. Tec.* **1996**, *39*, 323-332.
17. Nudel, B. C.; Waehner, R. S.; Fraile, E. R.; Giulietti, A. M. *Biological Wastes.* **1987**, *22*, 67-73.
18. De Lamo, P. R.; Menezes, T. J. B. *Coletânea do Instituto de Tecnologia de Alimentos.* **1978**, *9*, 281-312.
19. Araújo, N. Q.; Visconti, A. S.; Castro, H. F.; Silva, H. G. B. *Bras. Açuc.* **1976**, *88*, 35-45.
20. Kiyan, C. *STAB Açuc, Álcool Subpr.*, **1988**, *7*, 55-58.
21. Venturini Filho, W. G.; Camargo, R. *STAB Açuc, Álcool Subpr*, **1987**, *5*, 38-41.
22. Ortiz Ortiz, G. A.; Perez Lezama, O.; Juarez Lagunes, F. I. *Tecnica Pecuaria en Mexico.* **1994**, *32*, 134-138.
23. Stamford, T. L. M.; Fernandez, Z. F.; Stamford, N. P. *Arquiv. Biol. Tecnol.* **1991**, *34*, 503-507.
24. Deshpande, V. V. *Modernisation of India sugar industry*; Arnold Publishers: New Delhi, India, 1990; pp 146-152.
25. Araújo, N. Q.; Visconti, A. S.; Esteves, A. M. L.; Baggio, C. A.; Castro, H. F.; Silva, H. G. B.; Ferraz, M. H. A.; Salles Filho, M.; Reis, R. F.; Schneiderman, V. M. S.; Almeida, W. R. *Inform. INT.* **1978**, *11*, 3-8.
26. Mc Cully, K. S.; Forbes, E. *Genet. Res.* **1965**, *6*, 352-359.
27. Käfer, E.; Chen, T. L. *Can. J. Genet. Cytol.* **1964**, *6*, 249-254.
28. Pontecorvo, G.; Roper, J. A.; Hemmons, L. M.; McDonald, K. D.; Bufton, A. W. *Advanc. Genet.* **1953**, *5*, 141-238.
29. Azevedo, J. L.; Costa, S. O. P. *Exercícios Práticos de Genética*; Companhia editora Nacional: São Paulo, Brazil, 1973, 288p.
30. Silva, S. A.; Osvaldo, M. *Análise Fisico Química Para Controle das Estações de Tratamento de Esgoto*; CETESB: São Paulo, Brazil, 1977, 225p.
31. Selim, M. H.; Elshafei, A. M.; El-Diwany, A. I. *Bioresource Tech.*. **1991**, *36*, 157-160.
32. Tauk, S. M. *Ciência e Cultura.* **1978**, *31*, 522-530.

Environmentally Benign Catalysis

Chapter 14

Fluorous Biphasic Catalysis: A Green Chemistry Concept for Alkane and Alkene Oxidation Reactions

Jean-Marc Vincent[1,2], Alain Rabion[1,3], and Richard H. Fish[1,4]

[1]Lawrence Berkeley National Laboratory, University of California, Berkeley, CA 94720
[2]Laboratoire de Chimie Organique et Organométallique, UMR CNRS 5802, Université Bordeaux 1, 351 Cours de la Libération, 33405 Talence Cedex, France
[3]Groupement de recherche de Lacq, BP 34, 64170 Artix, France

> Novel and innovative approaches for the selective catalytic homogeneous oxidation of alkanes and alkenes to produce industrially important alcohols, aldehydes, ketones, and epoxides are still urgently needed. Homogeneous catalytic oxidation is a process that is more selective to products and, as well, the reactions are carried out at lower temperatures in comparison to heterogeneous catalytic oxidation reactions. *However, the separation of the homogeneous catalyst from the oxidation products is energy intensive; for example, distillation, as well as possible thermal decomposition of the catalyst during distillation.* We will discuss the separation of the homogeneous catalyst; for example, Ru^{2+}, Co^{2+}, or Mn^{2+} complexes, and the products of oxidation, via a novel two phase homogeneous process called *fluorous biphasic catalysis* (FBC). By using a lower phase fluorocarbon solvent that solubilizes the homogeneous catalyst, modified via the addition of long chain perfluoroalkyl derivatives to their basic structure, designated as fluoroponytails, and a second upper solvent phase that solubilizes the substrate alkane/alkene, the oxidant, and the products of alkane and alkene oxidation, we will demonstrate the viability of this FBC concept. Thus, the oxidation products and the fluoroponytailed, homogeneous catalyst reside in separate solvent phases, and this represents an innovative method to perform homogeneous catalytic oxidation chemistry, and an elegant way to a potentially new environmentally efficient, green chemical, industrial process for the synthesis of important global organic chemicals.

[4]Corresponding author.

The question that homogeneous catalysis chemists have continually asked themselves is as follows: how do we separate the homogeneous catalyst from the substrate and the product of the reaction? The answer was to develop biphasic solvent approaches to better conduct this separation process. More importantly, the process of developing a new *Green Chemistry* paradigm for the separation of a homogeneous catalyst from the substrate and the product must start with the solvent system. Thus, it is important to note that fluorocarbons (C_nF_{2n+1}) are characterized by very strong intramolecular bonds, but very low intermolecular interactions. Therefore, due to the high fluorine electronegativity, and its low polarisability, fluorocarbons exhibit interesting physical properties that make them excellent candidates for an environmentally friendly, industrial process. For example, they are among the most inert organic compounds known, the best solvents for gases (e.g. O_2, CH_4, and CO_2), and present the unique property of being both hydrophobic and lipophobic.

These unusual properties were the basis of the *fluorous biphasic catalysis* process (FBC) first published in 1994 by Horváth and Rábai and demonstrated using hydroformylation chemistry as a pertinent example (*1, 2*); in a 1991 Ph.D. thesis, that was unfortunately not readily available to the homogeneous catalysis community nor published in the open literature, M. Vogt, under the guidance of his Ph.D. advisor, W. Keim, of the Rheinisch-Westfälischen Technischen Hochschule in Aachen, Germany, presented the first conceptual aspects of the FBC approach with an emphasis on oligomerization of alkenes, oxidation of alkenes, hydroformylation of olefins, and telomerization of dienes (*3, 4*).

The FBC concept entailed the use of a fluorocarbon as one phase containing the fluorocarbon soluble catalyst, whereas the second, alkane/alkene phase contains the substrate and the subsequent products, allowing the facile recovery of the catalyst by simple decantation. Moreover, due to their chemical inertness, perfluorinated compounds are environmentally friendly and represents a potential *Green Chemistry* concept for future utilization by industry. This elegant approach has been utilized for many organic reactions such as organotin hydride reductions (*5*), Stille coupling reactions (*6*), and the addition of allyltin derivatives to aldehydes (*7*); recent reviews of this burgeoning FBC technology establish the wide scope of this concept (*8, 9*). In this chapter, we will focus our attention on the FBC concept applied to alkane and alkene oxidation chemistry, which represents an important advancement for the potential catalytic synthesis of organic chemicals of global interest.

Results and Discussion

Alkane and Alkene Oxidation with Perfluorocarbon Soluble Metalloporphyrins as Catalysts

Since the analogues of the heme prosthetic group containing enzymes (e.g. cytochrome P450, peroxidases, chloroperoxidases and catalase) have been implicated in biological oxidation reactions (*10, 11*), then their use as biomimetic, synthetic metalloporphyrins in the area of alkane/alkene oxidation chemistry has shown that they are some of the most efficient catalysts for this purpose. For example, Fe^{3+} and Mn^{3+} porphyrins have been the most widely used catalysts in biomimetic oxidations in conjunction with various oxidants such as PhIO, NaOCl, H_2O_2, O_2, and $KHSO_5$.

The first attempt to solubilize metalloporphyrins in perfluorocarbons was achieved by Pozzi and coworkers (*12*). They modified the tetraarylporphyrin (TPP) nucleus by introducing perfluoroalkylponytails (-C_7F_{15}) linked to the meso-aryl groups through amido bonds (Figure 1). Unfortunately, these ligands were found to be rather insoluble in perfluorocarbons, but soluble, or slightly soluble, in common organic solvents. The four perfluoroalkylponytails (fluorine content 45%) were not sufficient to render these ligands perfluorcarbon soluble, particularly in the presence of very polar dichlorophenyl and amido groups. The Mn^{3+} complexes of ligands **1** and **2** were prepared and found to catalyze the epoxidation of cyclooctene and 1-dodecene using NaOCl as the oxygen donor with 82% (cyclooctene) and 33% (1-dodecene) yields for the best catalyst (**2**). These results are similar to those obtained with the non-perfluoro analog, Mn(TPP).

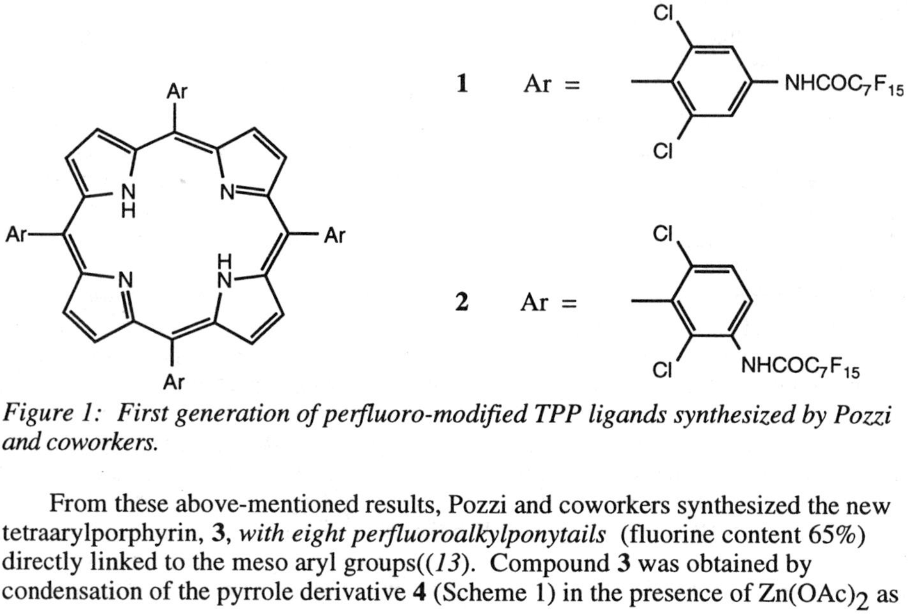

Figure 1: First generation of perfluoro-modified TPP ligands synthesized by Pozzi and coworkers.

From these above-mentioned results, Pozzi and coworkers synthesized the new tetraarylporphyrin, **3**, *with eight perfluoroalkylponytails* (fluorine content 65%) directly linked to the meso aryl groups((*13*). Compound **3** was obtained by condensation of the pyrrole derivative **4** (Scheme 1) in the presence of $Zn(OAc)_2$ as the templating agent. However, the synthesis of **3** requires 7 steps with an overall yield of 2.3%! Both **3** and its Co^{2+} complex (synthesized from $Co(OAc)_2$) were found to be soluble in perfluorocarbons, whereas they were insoluble in common organic solvents. It was found that the cobalt complex of ligand **3** was an efficient catalyst for the fluorous biphasic epoxidation of alkenes using molecular oxygen and the sacrificial 2-methylpropanal as the reducing agent, presumably to form the corresponding percarboxylic acid *insitu* that logically performs the subsequent epoxidation chemistry. Reactions were carried out in the dark in a biphasic mode (CH_3CN/perfluorohexane) at room temperature, and under an atmospheric pressure of O_2. Quantitative conversion of *cis*-cyclooctene into the epoxide (conversion >95%) was obtained with a much higher substrate/catalyst ratio (1000/1) than previously reported for the non-perfluoro analogue of the Co^{2+} complex of **3** (20/1). At the end of the reaction, the catalyst was completely partitioned into the fluorous phase and can be reused for a second run, without decreasing the substrate conversion or the chemoselectivity. However, the major drawback of this elegant and academic approach is the tedious synthesis of the fluorocarbon soluble porphyrin.

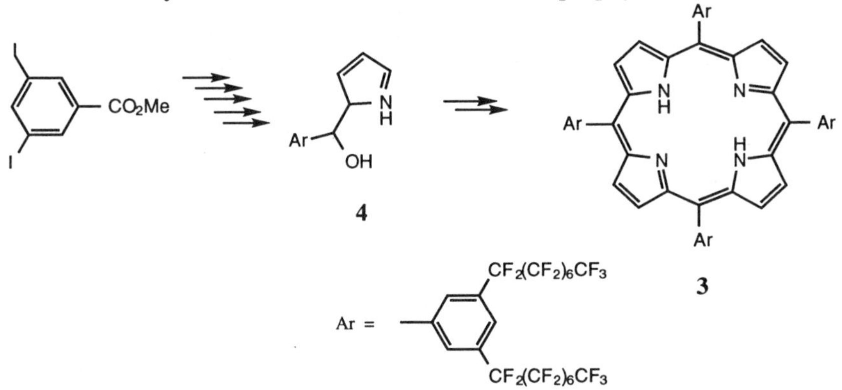

Scheme 1: Key intermediates in the synthesis of the perfluorocarbon soluble porphyrin, 3.

Alkane and Alkene Oxidation with Perfluorocarbon Soluble Non-porphyrin Catalysts

For the reasons mentioned above concerning the tedious synthesis of fluoroponytalied porphyrin ligands, our LBNL group, and others, have been interested in the preparation of perfluorocarbon soluble non-porphyrin ligands. Therefore, azamacrocycles were thought to be excellent candidates as non-porphyrin ligands for two main reasons: (**1**) metal complexes of azamacrocycles were known to catalyze the oxidation of hydrocarbons (*14-16*); and (**2**) perfluoroponytails could be easily introduced by alkylation of the secondary amine groups. In our FBC studies at LBNL, we prepared two types of highly fluorinated 1,4,7-triazacyclononane (TACN) derivatives (Figure 2) (*17*).

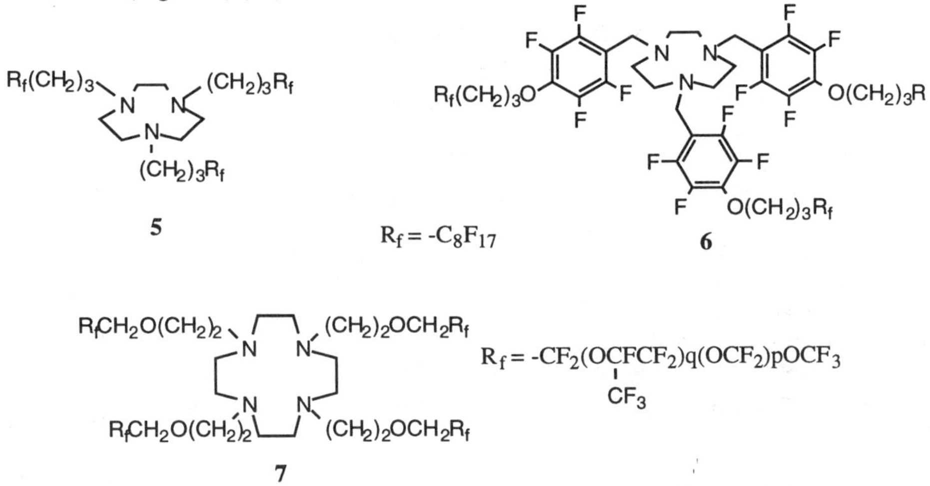

Figure 2: Structures of the synthesized perfluoroponytailed azamacrocycles.

Ligands **5** and **6** were synthesized either by alkylation of TACN with $R_f(CH_2)_3I$ or by nucleophilic substitution on the N-pentafluorophenyl derivative with $R_f(CH_2)_3OH$ (Scheme 2). The three methylene spacer in $R_f(CH_2)_3I$ was introduced to avoid the elimination of HI during the alkylation step, and to further insulate the metal coordinating heteroatom from the strong electron-withdrawing effect of the perfluoroponytail. Ligand **5** was found to be soluble in perfluorocarbons, whereas **6** was not, showing that polar groups, such as pentafluorophenyl, have to be avoided. Pozzi and coworkers used a similar procedure to obtain the perfluorocarbon soluble cyclam derivative, **7** (Figure 2) (*18*).

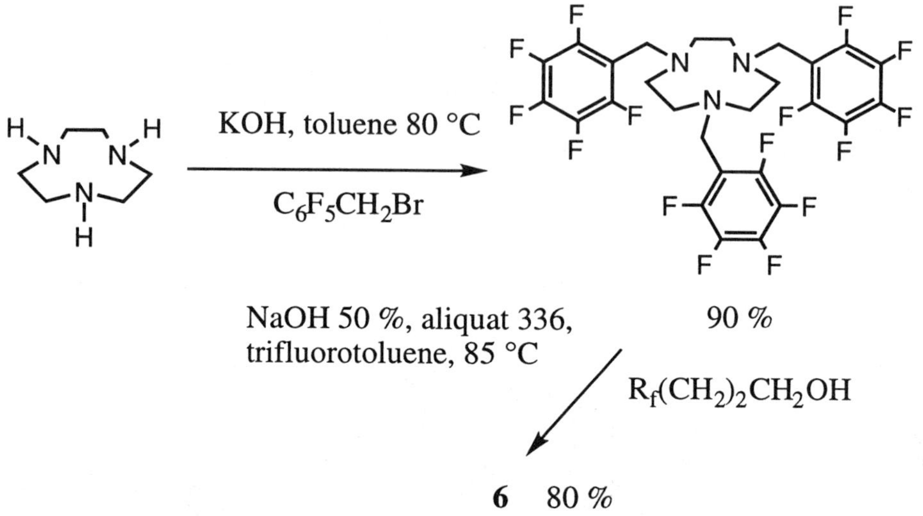

Scheme 2: Reaction pathway for the synthesis of 6.

The use of metal salts such as, $Mn(ClO_4)_2$, $Co(ClO_4)_2$ or $Ni(ClO_4)_2$ did not allow the preparation of complexes soluble in perfluorocarbons, since charged species are probably too polar for the fluorocarbon solvents. To overcome this deficiency, perfluorocarboxylate counterions were used to prepare the following salts: $[Mn(C_8F_{17}CH_2CH_2CO_2)_2]$, $[Co(C_8F_{17}CH_2CH_2CO_2)_2]$, and $[Co(C_7F_{15}CO_2)_2]$. These salts are slightly soluble in fluorocarbons, but they are completely solubilized upon addition of one molar equivalent of ligand **5** or **7**. The CuCl complex has also been solubilized in the presence of **7**, with the chloride anion possibly still ligated to the copper metal center.

The oxidation reactions of alkanes (cyclohexane, cyclooctane) or an alkene (cyclohexene) were carried out in the presence of both t-butyl hydroperoxide (TBHP) and O_2 in biphasic solutions, with the alkane/alkene phase being the substrate itself (Scheme 3). The Cu complex of ligand **7** was found to be the best catalyst for alkane oxidation (cyclooctane) affording cyclooctanol and cyclooctanone in 80% yield (24 h) with respect to TBHP. However the alkane conversion yield remains low (~ 1%); with the Mn^{2+} and Co^{2+} complexes of **5**, or the Co^{2+} complex of **7**, the yields vary from 12% to 30 %. In the cyclohexene oxidation, only products of allylic oxidation are formed in yields ranging from 500% to 750% (based on TBHP utilized). In the absence of O_2 or TBHP, negligible amounts of 2-cyclohexen-1-one and 2-cyclohexen-1-ol were detected, indicative of an autoxidation reaction involving the presumed intermediacy of cyclohexenyl hydroperoxide. At the end of the reaction, the fluorocarbon layer is recovered by simple phase separation and can be reused for several runs with similar catalytic activity (*17*).

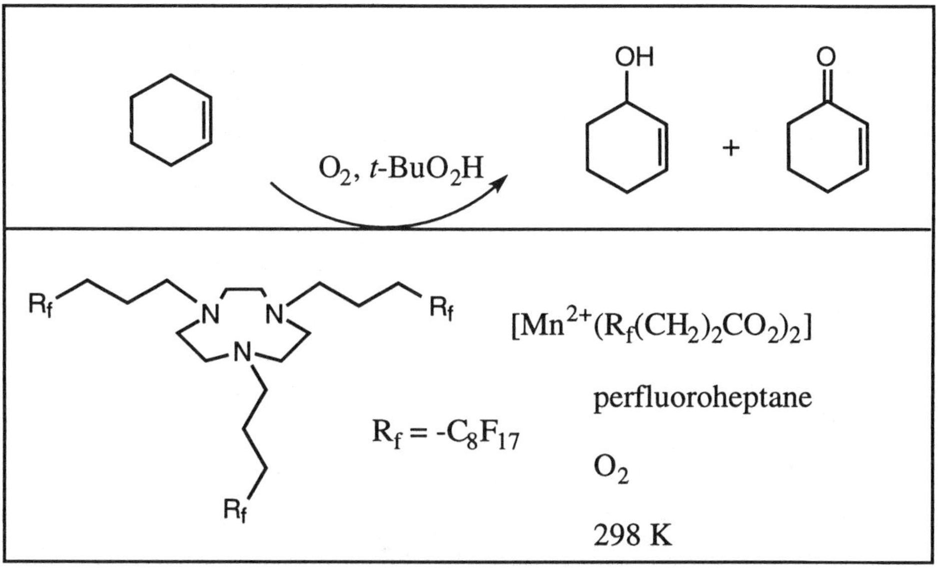

Scheme 3: Fluorous biphasic oxidation catalysis with cyclohexene as the substrate.

It was also shown that the Mn^{2+} complex of ligand **5**, formed during the oxidation reaction, was present mainly in its $Mn^{3+/4+}$ oxidation state. The occurrence of a antiferromagnetically coupled mixed valent Mn^{3+}/Mn^{4+} species was unambiguously confirmed by ESR studies at 9K, with a strong 16-lines signal being observed in the fluorous phase after one hour of reaction. At the end of the reaction, this signal had almost disappeared, suggesting that the Mn^{3+}/Mn^{4+} complex might be involved in the catalytic decomposition of the presumed cyclohexenyl peroxide intermediate.

Epoxidation Reactions

Another FBC approach was developed by Knochel and co-workers to oxidize olefins, sulfides, and aldehydes to epoxides, sulfoxides, sulfones, and carboxylic acids (*19*). They used the Ru and Ni complexes of a perfluorinated 1,3-diketone, obtained in one step in 80 % yield by the condensation of a perfluoromethyl ester and a perfluoromethyl ketone, as catalysts for these oxidation reactions. More importantly, these Ru and Ni complexes were found to be highly soluble in perfluorocarbons (Figure 3).

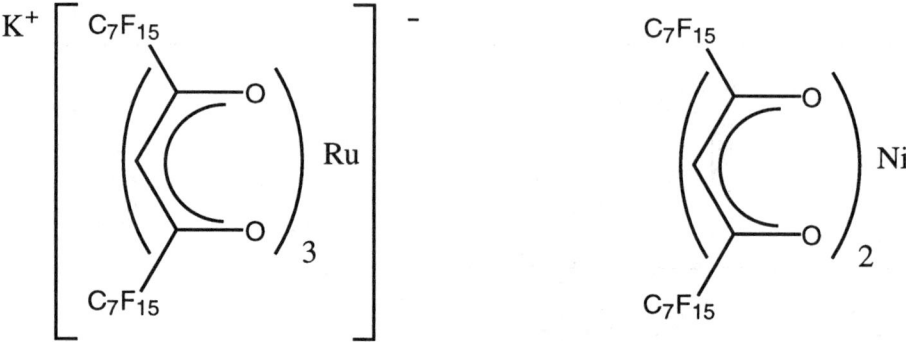

Figure 3: Structures of the perfluorosoluble catalyst developed by Knochel and coworkers.

Aldehydes and sulfides were oxidized with the Ni catalyst (3 mol%) in good yields (60% to 90%) in a single phase solvent system at 64 °C (PhCH$_3$/ perfluorodecalin or C$_8$F$_{17}$Br) under 1 atm of O$_2$ (see Figure 4 for carboxylic acid, sulfoxide/sulfone, and epoxide products). Sulfone to sulfoxide selectivity was obtained by using a larger amount of the sacrificial isobutyraldehyde (5 equiv instead of 1.6). The catalyst was readily recycled by simple decantation at ambient temperature (biphasic), with the reaction yield still 70% after 6 runs. Olefin epoxidation reactions were carried out with the Ru complex under the same conditions and, remarkably, only disubstituted alkenes are selectively converted in good yields (71%-85 %). The very stable ruthenium catalyst was recovered in over 95% by weight from the fluorous phase, after several reaction cycles, making this system, in the authors point of view, suitable for a large scale industrial process.

A most challenging FBC application was developed by Pozzi and co-workers with the synthesis of chiral SALEN-Mn complexes that were found to be soluble in perfluorocarbons (*20*) by introducing two perfluoroalkylponytails on each phenol group (Figure 5). The ligands were obtained in six steps with good yields and their Mn complexes were found to be soluble in perfluorocarbons.

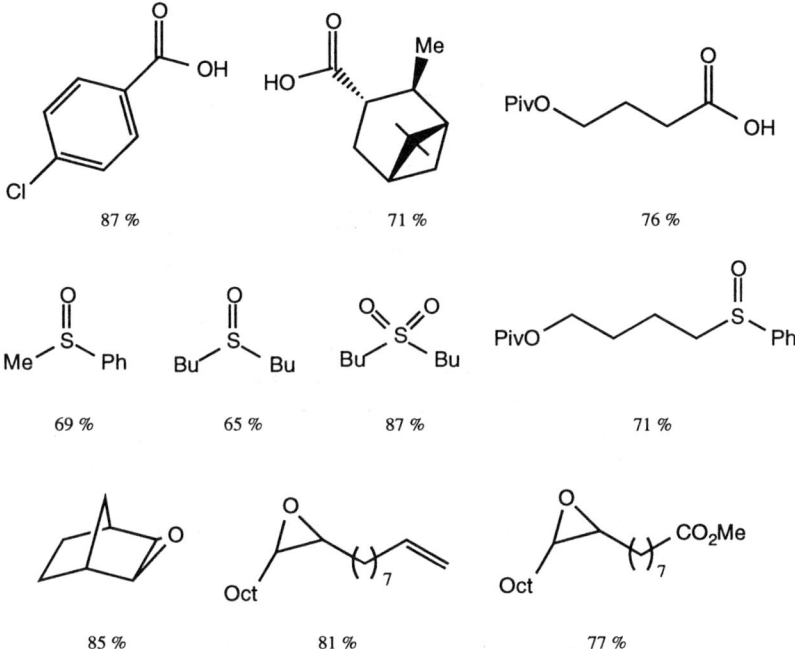

Figure 4: Examples of product yields and selectivity obtained with the Ru and Ni catalysts in Figure 3, under FBS conditions.

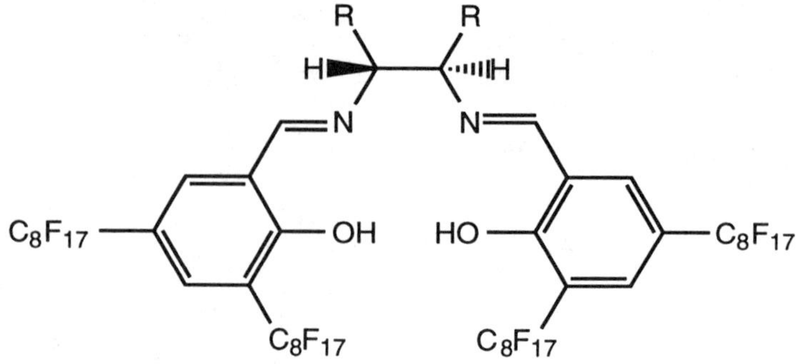

[Chemical structure of perfluorosoluble chiral SALEN ligand with C8F17 groups and imine/phenol moieties]

R-R = -(CH$_2$)$_4$-

R = Ph

Figure 5: Structures of the perfluorosoluble chiral SALEN ligands.

Using the above mentioned chiral SALEN-Mn complexes, epoxidation reactions were carried out under FBC conditions (perfluorocarbons/CH$_2$Cl$_2$) with O$_2$ at atmospheric pressure, and in the presence of a sacrificial aliphatic aldehyde, pivalaldehyde. High reaction yields (> 70%) of isolated epoxides were obtained in the oxidation of indene, 1,2-dihydronaphtalene, styrenes, and *cis*- and *trans*-stilbenes. Unfortunately, only indene was epoxidized with high enantioselectivity (> 90% ee), whereas other substrates gave low (1,2-dihydronaphtalene (13 % ee) or no selectivity at all. Interestingly, the catalyst was reused for a second run without loss of catalytic activity. Moreover, the epoxidation reactions were found to be very efficient, even at low temperature, with oxygen atom donors such as KHSO$_5$ or the couple, *m*-chloroperbenzoic acid/ N-methylmorpholineoxide, being commonly used with chiral SALEN complexes.

Conclusions

In this chapter, we wanted to introduce this novel FBC concept to the green chemistry community (*21*) and demonstrate the oxidation of alkanes and alkenes as pertinent examples of the powerful nature of this biphasic technique. While we have not developed the critical economic factors for using non-toxic and environmentally benign fluorcarbons in these alkane/alkene oxidation reactions on an industrial scale, it is clearly evident that if an industrial process needed to be biphasic, to separate the homogeneous catalyst from the products, then scrutinizing the FBC process would be a wise exercise for comparison to other plausible biphasic approaches. Finally, the FBC concept has caught the imagination of many colleagues globally, and thus, the outpouring of new fluorocarbon soluble catalysts, for the above-mentioned application and others, will continue unabated into the next millennium (*8, 9*).

Acknowledgments

We gratefully acknowledge Elf Aquitaine Inc. for support of our LBNL FBC program and the U. S. Department of Energy under contract No. DE-AC03-76SF00098.

Literature Cited

1. Horváth, I. T.; Rábai, J. *Science* **1994**, *266*, 72.
2. Horváth, I. T.; Rábai, J. US. Patent 5,463,082, 1995.
3. Cornils, B. *Angew. Chem. Int. Ed. Engl.* **1997**, *36*, 2057.
4. Vogt, M., Ph.D. Thesis, 1991; The Application of Perfluorinated Polyethers for Immobilization of Homogeneous Catalysts. Rheinisch-Westfälischen Technischen Hochschule, Aachen, Germany
5. Curran, D. P.; Hadida, S. *J. Am. Chem. Soc.* **1996**, *118*, 2531.
6. Curran, D. P.; Hoshino, M. *J. Org. Chem.* **1996**, *61*, 6480.
7. Curran, D. P.; Hadida, S.; He, M. *J. Org. Chem.* **1997**, *62*, 6714.
8. Horváth, I. T. *Acc. Chem. Res.* **1998**, *31*, 641 and references therein.
9. Fish, R. H. *Chem. Eur. J.* **1999**, *5*, 1677 and references therein.
10. Meunier, B. *Chem. Rev.* **1992**, 1411.
11. Lindsey, J. S. *Metalloporphyrins Catalyzed Oxidation*, ed. Montanari, F. and Casella, L., Kluwer, Dordrecht, 1994, p. 49 and references cited therein.
12. Pozzi, G.; Banfi, S.; Manfredi, A.; Montanari, F. Quici, S. *Tetrahedron* **1996**, *52*, 36, 11879.
13. Pozzi, G.; Montanari, F.; Quici, S. *Chem. Commun.* **1997**, 69.
14. Koola, J. D.; Kochi, J. K. *Inorg. Chem.* **1987**, *26*, 908.
15. Yoon, H.; Wagler, T. R.; OConnor, K. J.; Burrows, C. J. *J. Am. Chem. Soc.* **1990**, *112*, 4568.
16. De Vos, D.; Bein, T. *Chem. Commun.* **1996**, 917.
17. Vincent, J.-M.; Rabion, A.; Yachandra, V. K.; Fish, R. H. *Angew. Chem. Int. Ed. Engl.* **1997**, *36*, 2346.
18. Pozzi, G.; Cavazzinni, M.; Quici, S.; Fontana, S. *Tetrahedron letters* **1997**, *38*, 43, 7605.
19. Klement, I.; Lutjens, H.; Knochel, P. *Angew. Chem. Int. Ed. Engl.* **1997**, *36*, 1454.
20. Pozzi, G.; Cinato, F.; Montanari, F.; Quici, S. *Chem. Commun.* **1998**, 877.
21. For other aspects of Green Chemistry see for example: Anastas, P.T.; Williamson, T. C.(1998). *Green Chemistry: Frontiers in Benign Chemical Syntheses and Processes*. Oxford University Press, Oxford and references therein.

Chapter 15

Polymer-Facilitated Biphasic Catalysis

David E. Bergbreiter

Department of Chemistry, Texas A&M University, P.O. Box 30012, College Station, TX 77842-3012

Approaches using soluble polymers to facilitate recovery and reuse of homogeneous catalysts are discussed. Separation strategies based on soluble polymers that precipitate on cooling, soluble polymers that precipitate on heating and polymers that dissolve selectively in fluorous or aqueous phases are all described.

The focus of our work in catalysis chemistry is on homogeneous catalysis. While developing new catalysts is an appropriate and desirable way to new greener chemical processes, our work has focused on new ways to carry out homogeneous catalysis. This has involved two approaches. The first approach explores the use of polymer-supported catalysts in non-traditional media like water or fluorocarbon phases. The second approach expands on the use of polymers to recover reusable homogeneous catalysts after a reaction is complete. Essentially all of our work has been with soluble polymer supports.

The idea of using polymers as supports for catalysis chemistry is not new. Prior work with heterogeneous polymer bound transition metal catalysts extends back to the 1960s.(*1-3*) At that time there was considerable industrial interest in insoluble polymer supports. That interest continued through the 1970s but dwindled because homogeneous and 'heterogeneized' catalysts were not always as comparable to one another as anticipated and because catalysts could not always be recovered and re-used with the expected simplicity. Since that time there has been a rejuvenation of

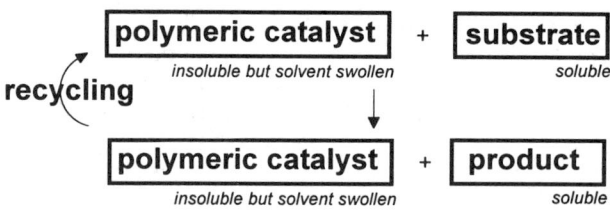

Figure 1. Recycling of insoluble, cross-linked polystyrene-supported catalyst. The product is separated from the catalyst by its solubility.

© 2000 American Chemical Society

interest in this area because of unrelated developments in medicinal chemistry, specifically combinatorial chemistry has developed into a practical method for drug discovery, materials research and catalyst development. (4) Figure 1 illustrates the general approach to catalyst recovery and reuse using insoluble cross-linked polymers.

The reasons why these original 'heterogeneized' catalysts were never widely adopted are complex. The original idea back in the '70s was that one could take an insoluble polymer like cross-linked polystyrene, functionalize it, attach a ligand like diphenylphosphine (Figure 1) and then attach a transition metal complex. This chemistry would then lead to a homogeneous catalyst that was heterogenized on an insoluble polymer support. Ideally, such a catalyst could be used in a continuous reactor. It would minimize loss of toxic and valuable transition metal salts and ligands since the catalysts would be easily recovered for reuse by a physical process like filtration. Product purification too would be simplified. This clever idea is the progenitor for much of the subsequent work in phase separation and polymer-supported catalysis.

However, although the ideas above were attractive, researchers applying and developing these catalysts encountered certain real problems. Most of these problems resulted from the use of cross-linked polystyrene supports. Indeed, the disadvantage and also the advantage of cross-linked polymers are that they are always insoluble. They are insoluble before the reaction starts. That means if you were going to do some synthesis that was a little bit elaborate, you now had insoluble organic materials to work with. Such materials are hard to analyze. Instead of using modern techniques like NMR spectroscopy, one was left with analytical tools like elemental analysis and infrared spectroscopy, techniques dating from the 1950s. Subsequent technological developments have mitigated but not eliminated these problems. For example, gel phase NMR and solid state NMR overcome these disadvantages to a certain extent.

In addition to analytical problems, the fact that cross-linked polymer-supported ligands and catalysts are always insoluble means that many of the advantages of homogeneous catalysis are also lost. For example, the dynamic ligand-ligand and ligand-metal interactions that occur in solution cannot readily occur in a cross-linked polymer matrix. Likewise, it is difficult to control ligand-metal ratios. Since these effects are, in part, the basis for the selectivity of the homogeneous catalyst, heterogenized catalysts often have different reactivity than their homogeneous counterparts. When the selectivity and reactivity of the homogeneous catalyst in question was the outgrowth of extensive optimization of reaction conditions, a heterogeneized catalyst was most often less useful and not practical.

Nearly everything my group has done in polymer-facilitated catalysis uses soluble polymers, a practice that has been followed by others as well.(5-7) Our group has emphasized chemistry using soluble polymers where the polymers are separable by some sort of precipitation or phase isolation technique. This paper presents examples of several catalysts and discusses several sorts of polymers. First, there is a discussion of some of our older work using functionalized polyolefins as the support. In this case, catalyst use, recovery and separation are possible because polyethylene is insoluble in all solvents at room temperature but soluble on heating. Next, the paper discusses catalysts that separate on heating. Two examples, poly(alkene oxide)-supported catalysts and poly(N-isopropylacrylamide)-supported catalysts that both regulate reactions and facilitate catalyst recovery and reuse are described. Finally, the paper describes strategies where polymers are used to facilitate separations on the basis of differential phase solubility. This section includes both work with fluorocarbon polymers and new work using a polymer like poly(*N*-isopropylacrylamide) (PNIPAM) to facilitate catalyst recovery in a so-called thermomorphic system.

Polyethylene-Bound Catalysts

Our work with soluble polymers did not begin with a goal of developing catalysts. Rather, we started off trying to do some polymer surface chemistry. We first prepared carboxyl-terminated polyethylene oligomers by the chemistry shown in Figure 2 below. We then found that oligomers prepared with a pendant nitroxyl spin label covalently attached via a carboxamide were soluble in toluene at 110 °C but insoluble in toluene at 25 °C based on the absence of an ESR signal in the toluene supernatant at 25 °C. Even though this material was prepared for a different purpose than catalysis (it was used in surface modification of polyethylene),(*8*) we immediately realized that the temperature controlled on/off solubility would be useful in the design of new sorts of polymer-supported catalysts.

Our efforts to make polyolefin-supported soluble catalysts require synthetic routes to terminally-functionalized ethylene oligomers. Such oligomers can be prepared by our approaches (Figure 2),(*9*) newer catalytic routes(*10*) or by polymethylenation.(*11*)

Strictly linear polyethylene oligomers are most suitable as catalyst supports. In the case of polyethylenes, a molecular weight of 2,000 is sufficient to get the desired on/off solubility behavior. Ethylene oligomers of this size are easily modified

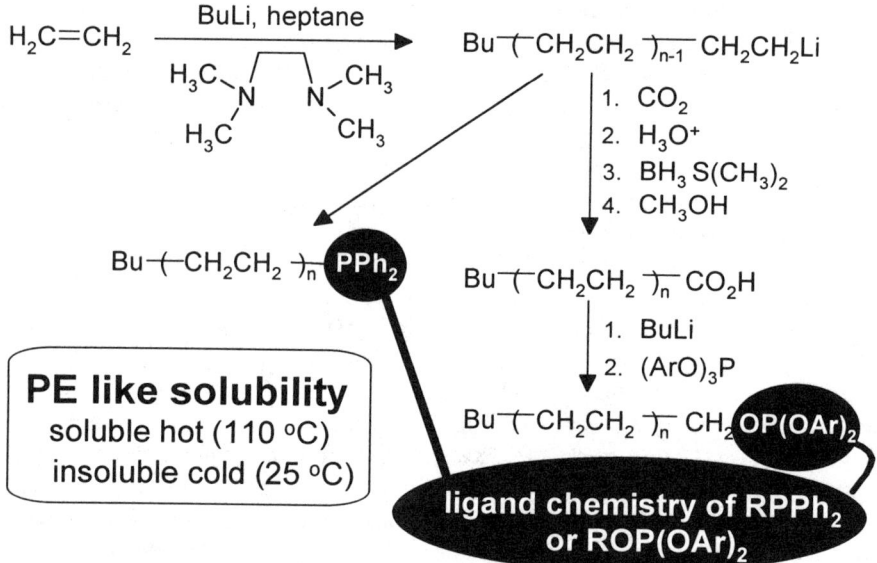

Figure 2. Synthesis of PE$_{Olig}$-bound ligands.

synthetically so long as the chemistry is carried out at 100 °C (conditions under which the oligomers are soluble). The terminal group can be an alcohol, a carboxylic acid, a phosphine ligand, a phosphite ligand or a chiral ligand. Regardless of the ligand used, the ligand has polyethylene-like solubility by virtue of the ligand being attached to ca. 200 carbons. These polymers and their metal complexes have the polyethylene-like property of being completely insoluble in all solvents at 25 °C but soluble on heating to 70-80 °C in solvents like toluene or dibutyl ether.

An attractive feature of catalysts like those derived from the ligands prepared in Figure 2 is that the PE_{Olig}-bound catalysts have the same chemistry as their low molecular weight analogs. That is one of the attractive features of this chemistry because one often wants exactly the same chemistry in a supported catalyst as in a homogeneous low molecular weight catalyst. Chemists go to a lot of trouble to optimize the ligand structure and to optimize ligand to metal ratios to get a particular product selectivity or particular stereoselectivity. They do not want to reinvent the whole process and reoptimize everything in order to use a polymer-bound catalyst. We demonstrated this feature using Ni(0) catalysts invented earlier by Wilke for diene cyclodi- and cyclotrimerization.(12) The plots shown in Figure 3 compare a PE_{Olig}-bound Ni(0) catalyst to its low molecular weight analog. A range of ligand:Ni ratios spanning several orders of magnitude were used and the product selectivity of the polymeric catalyst looks exactly the same as its low molecular weight cousin. The PE_{Olig}-bound catalyst selectivity is predictably dependent on ligand structure. The only difference is that the PE_{Olig}-bound Ni(0) catalyst can be recovered and reused with no detectable loss of metal to solution.

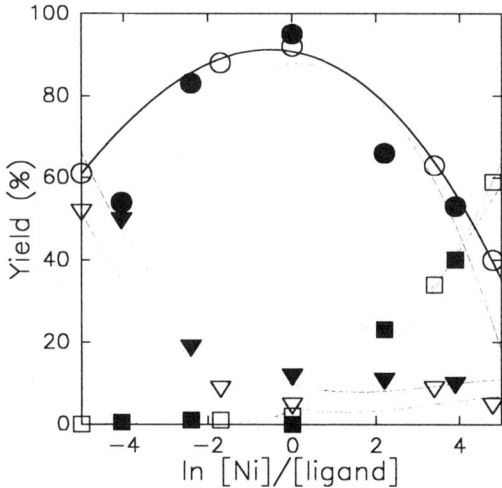

Figure 3. Ni(0) catalysts for butadiene cyclodimerization and cyclotrimerization. Filled circles, triangles and squares represent results for 1,5-cyclodiene, vinyl-cyclohexene and cyclododecatriene formation, respectively with a low molecular weight catalyst ($-C_6H_4OP(OC_6H_4)_2$). Open circles, triangles and squares are for PE_{Olig}-bound Ni(0) catalysts using $PE_{Olig}-C_6H_4OP(OC_6H_4)_2$ as a ligand.

Table I illustrates some of the catalysts we have prepared using polyethylene oligomers. Basically, we have carried out this chemistry with a wide variety of homogeneous catalysts, including various isomerization catalysts, hydrogenation catalysts, hydroformylation catalysts, etc. This approach to catalyst immobilization and recovery seems to be very general. In all cases, we recover 99.9 percent of the catalysts after the first or second cycle whenever we measure it by looking for metal in the filtrate.

Table I. PE_{Olig}-bound catalysts and levels of catalyst recovery based on analyses of the product-containing filtrate isolated from the first or second reaction cycle.

Type of Reaction	Catalyst	Recovery
Hydrogenation	L_3RhCl[a]	>99.9%
Hydroformylation	$L_2Rh(CO)Cl$[a]	>99.9%
Pd(0) chemistry	L_4Pd[a]	>99.98%
Cyclopropanation	$[(LCO_2)_2Rh]_2$[b]	>99.9%
Asymmetric cyclopropanation/insertion	$[LCO_2)_2Rh]_2$[c]	>99.98%
Pd(II) chemistry	L_2PdX_2[a]	[d]
Diene polymerization	$(LCO_2)_3Nd$[b]	[d]
Phase transfer catalysis	LPR_3^+[b]	[d]

[a]L in these cases was PE_{Olig}-$P(C_6H_5)_2$. [b]L was PE_{Olig}. [c]L in this case was PE_{Olig}-2-pyrrolidone-5S-carboxylate. [d]This value was not measured.

However, although this polyethylene chemistry is general, it has disadvantages. Its principle disadvantage is that it requires elevated temperatures to dissolve the polymer. Below about 70 °C, these catalysts have zero or minimal reactivity. If the PE_{Olig}-bound catalyst doesn't go into solution, the reaction does not work.

Water-soluble Polymer-Bound Catalysts

In addition to interests in making recoverable, reusable catalysts, there is interest in developing catalysts that work in nontraditional solvents. Water is an especially interesting solvent in this regard. Recently, we have developed some water-soluble polymers that work well either in an aqueous phase system or in biphasic systems. The polymeric systems we have developed serve two roles. They make the catalysts water- soluble and, the polymer portion of these catalysts allows us to recover the catalysts at the end of a reaction. This recovery is effected either in a biphasic manner or by means of the polymer's lower critical solution temperature (LCST)

(Figure 4). LCST behavior is a common and tunable property of many polymers in water.(*13*) This behavior is explained by considering the simple thermodynamic equation for ΔG (equation 1). Polymers dissolve because their free energy of solution is negative. This favorable free energy derives mainly from the exothermic ΔH term.

$$\Delta G = \Delta H - T\Delta S \qquad (1)$$

The ΔS term is typically unfavorable. In these cases, heating gradually increases the magnitude of the TΔS term until that term overwhelms the ΔH term and ΔG turns positive. At that temperature, the polymer phase separates from solution.

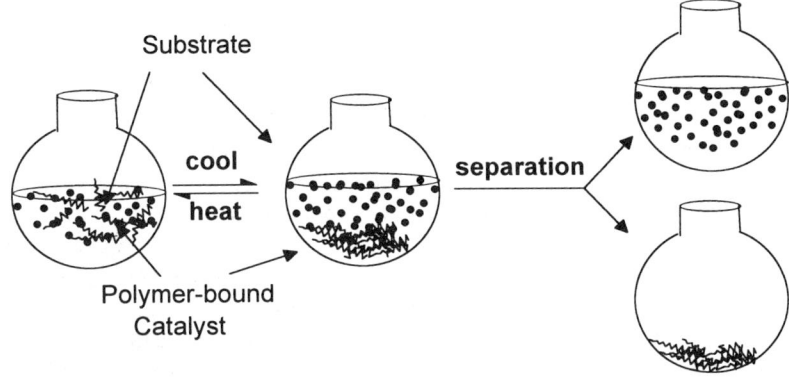

Figure 4. Separation of a polymer-bound catalyst whose polymer has LCST behavior from a solution by heating.

The water-soluble polymers we have examined that have LCST behavior include poly(alkene oxide)s and poly(*N*-alkyl acrylamide)s.(*14-16*) The poly(alkene oxide)s we used are mainly triblock copolymer of ethylene oxide, propylene oxide and ethylene oxide with the ratio of ethylene oxide/propylene oxide adjusted to create a more or less hydrophobic polymer with a lower or higher LCST, respectively. The poly(alkene oxide)s contain terminal hydroxyl groups that can be manipulated synthetically to form ligands for catalyst ligation. The reactions shown in equation 2 illustrate the typical chemistry used to prepare ligands and catalysts with these polymers.

$$\text{HO}(CH_2CH_2O)_n((CH_3)CHCH_2O)_m(CH_2CH_2O)_n\text{H} \xrightarrow[\text{Et}_3\text{N}]{\text{MsCl}} \text{MsO}(PEO)_n(PPO)_m(PEO)_n\text{OMs}$$

HO(PEO)$_n$(PPO)$_m$(PEO)$_n$OH   CH$_2$Cl$_2$

$$\xrightarrow[\text{THF}]{\text{LiP}(C_6H_5)_2} (C_6H_5)_2P(PEO)_n(PPO)_m(PEO)_nP(C_6H_5)_2 \xrightarrow[\text{toluene}]{\text{Rh}_2\text{Cl}_2[\text{C}_8\text{H}_{12}]_2} \qquad (2)$$

[(C$_6$H$_5$)$_2$P(PEO)$_n$(PPO)$_m$(PEO)$_n$P(C$_6$H$_5$)$_2$]$_{1.5}$RhCl
LCST; 0 - 50 °C

Poly(*N*-alkyl acrylamide)s and poly(*N*-isopropylacrylamide) in particular are the other type of LCST polymers our group has studied. Poly(*N*-isopropylacrylamide) is soluble below 31 °C in water but insoluble above that temperature. Our group has used this temperature induced phase change has been used as a way to isolate, recover and reuse water-soluble polymer-bound catalysts. It is also a way to make a 'smart' catalysts, catalysts that can turn off an exothermic reaction without external temperature control. Such on/off behavior is seen for both catalysts and substrates. An example of substrate behavior is nitroarene hydrogenation.(*17*) In this instance, a poly(*N*-isopropylacrylamide) copolymer containing a nitroarene substrate was hydrogenated using a heterogeneous Pt catalyst in water. This hydrogenation occurred readily below 33 °C but ceased above 39 °C (the LCST of this copolymer). In contrast, in ethanol, normal Arrhenius behavior was seen both below and above this temperature range. Similar 'smart' behavior has been seen and reported for other substrates and PNIPAM-bound hydrogenation catalysts.(*16,18,19*)

The ability to make co- or terpolymers from poly(*N*-isopropylacrylamide) by chemistry like that shown in equations 3-5 below is an attractive aspect of this support. Such syntheses allow us to tune the temperature dependent solubility of the catalyst by changing the hydrophilicity of the component monomers. It is also advantageous to use co- or terpolymers like **1** as substrates in these syntheses (equation 5). A copolymer like **1** allows one to prepare a wide variety of PNIPAM-bound catalysts because the reactive carboxylic acid derivative (*N*-acryloxysuccinimide) (NASI) reacts readily with amines. Thus, it is possible to attach a catalyst ligand or substrate to the resulting polymer by formation of a new amide bond from some of the NASI groups in **1**. It is then possible to tune the solubility and LCST behavior of the product by converting the rest of the NASI groups to amides of varying hydrophilicity with ammonia, *iso*-propylamine or *tert*-butylamine to balance the hydrophilicity of the catalyst ligand or substrate.

Polymer-Bound Catalysts in Biphasic Systems

In the examples above, polymers facilitate homogeneous catalysis by providing a practical way for solid/liquid separation and catalyst recovery after a reaction. A second conceptually different approach uses polymers to facilitate separation in a biphasic system. Indeed, during the time we were working with the soluble polymer-bound catalysts discussed above, a second commercially successful approach to catalyst recovery and recycling was developed. This chemistry used water-soluble phosphine ligands.(20) An apt example of this chemistry is the commercial process for hydroformylation of propene to form butyraldehyde. In this process, the reaction occurs in the aqueous phase.(21) However, if one uses a somewhat larger more hydrophobic olefin instead of propene, the reaction doesn't work quite so well. Thus, there is ongoing interest in other sorts of biphasic catalysis where there is an organic phase and where the substrate or catalyst either migrates from one phase to another or where the interfacial reaction is faster than in an aqueous biphasic system.(22)

Our work with biphasic systems was inspired by work first published by Horvath in 1994.(23) This seminal paper described a fluorous biphasic approach to catalysis, an idea that has subsequently mushroomed with applications in diverse areas of synthesis and catalysis.(24,25) The idea was to use fluorocarbon/organic phases as a substitute for the two phases in the aqueous biphasic catalysis system. Horvath showed that if there were a fluorocarbon phase represented by the black phase in Figure 5 below and a hydrocarbon phase represented by the clear phase, the two phases are normally immiscible. Shaking generates an emulsion. Heating, in certain cases, forms a single phase. When a homogeneous catalyst with a fluorocarbon tail is used, the catalyst will selectively dissolve in the fluorocarbon phase. When the substrate is organic, it dissolves in the organic phase with little or no solubility in the fluorocarbon or fluorous phase. Thus, one can carry out a homogeneous catalytic reaction either in an emulsion or in a homogeneous system. Simply letting the mixture stand or cool reproduces the initial biphasic system that is separated by a liquid/liquid method.

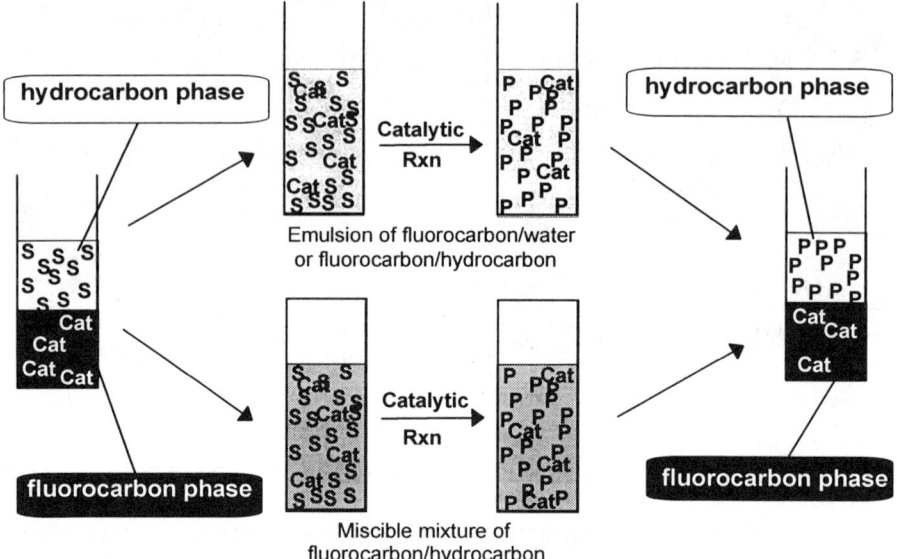

Figure 5. Fluorous phase approach for carrying out homogeneous catalysis in a system in which catalysts are recovered and separated from products by a liquid/liquid separation at the end of the reaction.

Our group's contribution to this area was the idea the preparation of a generic support substrate for doing fluorous phase chemistry using polymer chemistry. This avoids the use of catalysts or ligands with so-called Teflon-like ponytails attached to each catalyst.(*26, 27*) This was accomplished using fluoroacrylate monomers available from either DuPont or 3M by preparation of copolymers like **2** using the chemistry shown in equation 6 that were in turn used to prepare ligands like **3** or **4**.

(6)

This new type of soluble polymeric ligand derived from fluoroacrylate polymers that are soluble in fluorous solvents was used to prepare neutral and cationic rhodium(I) hydrogenation catalysts as well as palladium(0) catalysts. The rhodium catalysts were readily recycled and reused by liquid/liquid separation. The palladium catalysts were less readily recycled because their phosphine ligands were too susceptible to adventitious oxidation.

More recently we have also described another approach to using polymer in phase separation and catalyst recovery. This approach (equation 7) is similar to the fluorous phase chemistry shown in Figure 5 but uses less expensive solvents. For example, we showed that we can make a polymeric catalyst like **5** or **6** and that

these catalysts dissolve selectively in the lower phase of a biphasic mixture containing 90% aqueous ethanol and heptane. Under biphasic conditions, the polymeric hydrogenation catalyst is inactive with a heptane-soluble substrate like 1-octadecene or 1-dodecene. However, at 70 °C, the heptane/aqueous ethanol phases become miscible. Reaction between the catalyst and substrate readily occurs in this monophasic solution (Figure 6). In hydrogenation reactions, either of these two alkenes is quantitatively hydrogenated. In

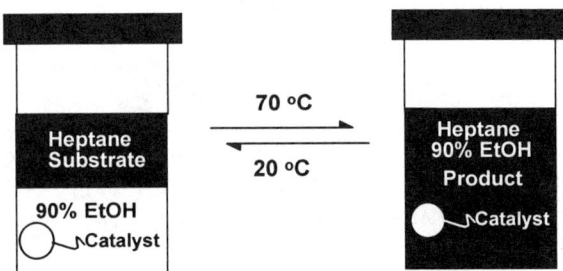

Figure 6. Thermomorphic system where the catalysis is carried out homogeneously at 70 °C in a monophasic system but where the separation is carried out at room temperature in a biphasic system with the soluble polymer-supported catalyst (e.g. **5** *or* **6**) *exclusively dissolved in the aqueous ethanol phase at 20 °C.*

allylic substitution reactions of cinnamyl acetate with dicyclohexylamine catalyzed by **6**, quantitative allylic substitution occurs. In either the hydrogenation or allylic substitution reactions, cooling reforms the original biphasic mixture and the product is selectively soluble in the heptane phase. Liquid/liquid separation and reuse of the polymeric catalyst (**5** or **6**) is straightforward.

Conclusions

In summary, new strategies for the use and recovery of homogeneous catalysts and for carrying out chemical processes are of increasing concern because there are problems associated with the use of organic solvents and because there are high costs associated with the purification and removal/disposal of by-products. The chemistry described here uses polymeric ligands and new separation strategies to facilitate homogeneous catalysis. By using the well-known properties of polymers to recover and separate catalysts and ligands for reuse and by employing relatively simple polymer chemistry, our group has been able to attach a wide variety of known homogeneous catalysts to soluble polymer without significant alteration of these catalysts' reactivity or selectivity. Separation/recovery strategies that use solid/liquid separation of precipitated polymers or liquid/liquid separations of polymer/product solutions have both been demonstrated. The utility of simple linear polymers in formation of aqueous and fluorous phase soluble catalysts has also been shown by our work. Finally, the work discussed above shows how polymer can facilitate catalysis by both regulating and controlling reactions when soluble polymer-bound "smart" ligands are chosen that precipitate from solution on heating.

Acknowledgments

Support of this chemistry through funding by the National Science Foundation (CHE-9707710) and the Robert A. Welch Foundation is gratefully acknowledged.

Literature cited

1. Holy, N. L. in *Homogeneous Catalysis with Metal Phosphine Complexes*, Pignolet, L. H., Ed.; Plenum Press: New York, NY 1983; p. 443.
2. Hartley, F. R. *Supported Metal Complexes. A New Generation of Catalysts*; D. Reidel: Dordrecht, 1985.
3. Bergbreiter, D. E. in *Chemically Modified Surfaces in Catalysis and Electrocatalysis*, Miller, J. S., Ed.; ACS Symposium Series No. 192; American Chemical Society: Washington, DC, 1982; pp 1-8.
4. Brown, R. *Contemporary Organic Synthesis* **1997**, *4*, 216-237.
5. Bergbreiter, D. E. *Catalysis Today*, **1998**, *42*, 389-397.
6. Geckeler, K. E. *Polym. Syn./Polym. Eng.* **1995**, *121*, 31-79.

7. Gravert, D. J.; Janda, K. D. *Chem. Rev.* **1997**, *97*, 489-509.
8. Bergbreiter, D. E. *Prog. Polym. Sci.* **1994**, *19*, 529-560.
9. D. E. Bergbreiter, D. E.; Blanton, J. R.; Chandran, R.; Hein, M. D.; Huang, K. J.; Treadwell, D. R.; Walker, S. A. *J. Polym. Sci., Part A: Polym. Chem.*, **1989**, *27*, 4205-4226.
10. Brookhart, M.; Desimone, J.M.; Grant, B. E.; Tanner, M. J. *Macromolecules*, **1995**, *28*, 5378-5380.
11. Shea, K. J.; Walker, J. W.; Zhu, H.; Paz, M.; Greaves, J. *J. Am. Chem. Soc.*, **1997**, *119*, 9049-9050.
12. Bergbreiter, D. E.; Chandran, R. *J. Org. Chem.*, **1986,** *51*, 4754-4760.
13. Schild, H. G. *Prog. Polym. Sci.* **1992**, *17*, 163-249.
14. Bergbreiter, D. E.; Zhang, L.; V. M. Mariagnanam, V. M. *J. Am. Chem. Soc.,* **1993**, *115*, 9295-9296.
15. Bergbreiter, D. E.; Mariagnanam, V. M.; Zhang, L. *Advanced Materials*, **1995**, *7*, 69-71.
16. Bergbreiter, D. E.; Case, B. L. Case, Liu, Y. S. Liu; Caraway, J. W. *Macromolecules*, **1998**, *31*, 6053-6062.
17. Bergbreiter, D. E.; Caraway, J. W. *J. Am. Chem. Soc.*, **1996**, *118*, 6092-6093.
18. Bergbreiter, D. E.; Liu, Y. S. *Tetrahedron Lett.*, **1997**, *38*, 7843-7846.
19. Bergbreiter, D. E.; Liu, Y. S; Furyk, S.; Case, B. L. *Tetrahedron Lett.*, **1998**, *39*, 8799-8802.
20. Joó, F.; Somsák, L.; Beck, M. T. *J. Mol. Catal.* **1984**, *24*, 71-5.
21. *Applied Homogeneous Catalysis with Organometallic Compounds*; Cornils, B.; Herrmann, W. A., Eds. VCH: Weinheim, 1996.
22. Cornils, B.; Herrmann, W. A.; Eckl. R. W. *J. Mol. Catal., A-Chem.*, **1997**, *116*, 27-33.
23. Horvath, I. T.; J. Rabai, J. *Science (Washington D. C.)*, **1994**, *266*, 72-4.
24. Horvath, I. T. *Acc. Chem. Research*, **1998**, *31*, 641-650.
25. de Wolf, E; van Koten, G.; Deelman, B. J. *Chem. Soc. Rev.*, **1999**, *28*, 37-41.
26. Bergbreiter, D. E.; Franchina, J. G. *Chem. Commun.*, **1997**, 1531-1532.
27. Bergbreiter, D. E.; Franchina, J. G.; Case, B. L.; Koshti, N.; Williams, L. K.; Frels, J. *J. Am. Chem. Soc.*, submitted for publication.
28. Bergbreiter, D. E.; Liu, Y. S.; Osburn, P. L. J. Am. Chem. Soc., 1998, 120, 4250-4251.

Chapter 16
Environmentally-Benign Liquid-Phase Acetone Condensation Process Using Novel Heterogeneous Catalysts

A. A. Nikolopoulos[1], B. W-L. Jang[1], R. Subramanian[1,4], J. J. Spivey[1,5], D. J. Olsen[2], T. J. Devon[3], and R. D. Culp[3]

[1]Research Triangle Institute, P.O. Box 12194, Research Triangle Park, NC 27709-2194
[2]Eastman Chemical Company, P.O. Box 1972, Kingsport, TN 37662-5150
[3]Eastman Chemical Company, P.O. Box 7444, Longview, TX 75607-7444

The condensation/hydrogenation of acetone to methyl isobutyl ketone (MIBK) was studied on a series of hydrotalcite-supported noble metal catalysts in a liquid-phase batch microreactor at 118°C and 400 psig. The 0.1wt.% Pd/HTC gave the highest acetone conversion (38%) and selectivity to MIBK (82%). The HTC catalyzes the condensation of acetone to diacetone alcohol (DAA) and its subsequent dehydration to mesityl oxide (MO), whereas the noble metal selectively hydrogenates MO to MIBK. A multi-functional catalyst is required for this acetone-to-MIBK reaction.

There is a clear economic incentive to develop heterogeneous catalysts to replace conventional homogeneous catalysts in many industrial processes. Efforts to formulate highly active and selective solid catalysts have been the subject of intense research. One such process of interest is the condensation of alcohols, aldehydes and ketones, known as aldol condensation, which is used industrially for the synthesis of a series of products used extensively in various industries (*1*).

The liquid-phase condensation of acetone is an important industrial process for the synthesis of a series of commodity chemicals like diacetone alcohol (DAA), mesityl oxide (MO) and methyl isobutyl ketone (MIBK) (*1*). MIBK is extensively

---

[4]Current address: Precious Metals Division, Johnson Matthey, 2001 Nolte Drive, West Deptford, NJ 08066
[5]Corresponding author

used as a solvent in paints, resins and coatings, as well as an extracting agent in the production of antibiotics and lubricating oils (2). Over 210 million lb/yr of MIBK are curently produced in the United States (3). A schematic of the major reaction pathways in the acetone condensation process is given in Figure 1.

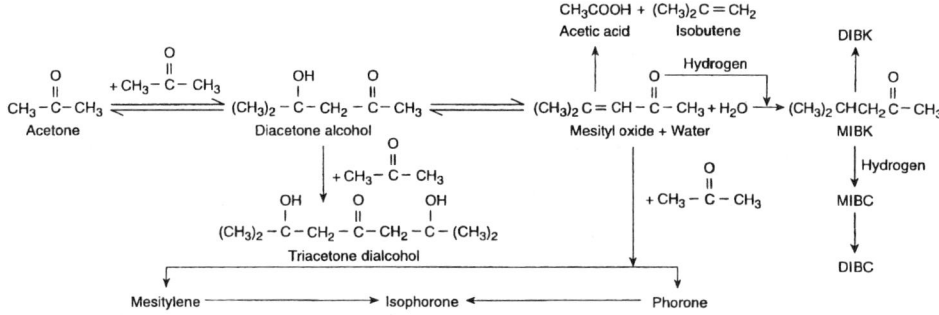

Figure 1. Major reaction pathways in the acetone condensation process (Ref. 1).

Aldol condensation reactions (and acetone condensation in particular) are practiced industrially using homogeneous liquid-based catalysts, primarily sodium and calcium hydroxide. Such processes generate significant wastewater streams that must be neutralized and properly disposed, resulting in substantial cost. The product separation and waste disposal steps are both energy- and cost-intensive, adding significantly to the selling price of the product MIBK. An active and selective heterogeneous catalyst would, in principle, eliminate this waste and reduce the cost significantly. Therefore, the development of suitable heterogeneous catalysts for aldol condensation would make this process more economically attractive and environmentally benign.

A number of solid base catalysts have been reported in the literature to be active for acetone condensation. These include alkali oxides (Na_2O, K_2O, Cs_2O) (1), alkaline earth oxides (MgO, CaO, BaO) (1,4-6), transition metal oxides (7) and phosphates (8-11), ion-exchange resins (12), zeolites (13) and clay minerals and hydrotalcites (HTC) (14,15). A suitable catalyst for the acetone-to-MIBK reaction must have several properties: the condensation of acetone to DAA is catalyzed by either basic or acidic sites, the dehydration of DAA to MO is acid-catalyzed, and the selective hydrogenation of MO to MIBK requires appropriate metal sites (7,8).

The reaction is further complicated by thermodynamic equilibrium limitations, as indicated in Table I. The condensation/dehydration of acetone to MO is limited to about 20% conversion at 120°C (16). However, there is no equilibrium limitation to the overall acetone-to-MIBK reaction. This, coupled with the possibility of numerous thermodynamically favorable side reactions that are also acid/base-catalyzed (Fig. 1), suggests the need to balance the acid/base and hydrogenation properties of the selected catalyst.

Table I. Thermodynamic Equilibrium Properties of the Acetone Condensation/Hydrogenation Reaction

Reaction	ΔH_{400K} (kcal/mol)	ΔG_{400K} (kcal/mol)
$2\ CH_3COCH_3 \Leftrightarrow (CH_3)_2{=}CHCOCH_3 + H_2O$ (Acetone) (Mesityl Oxide)	-5.2	+3.3
$(CH_3)_2{=}CHCOCH_3 + H_2 \Leftrightarrow (CH_3)_2CHCH_2COCH_3$ (Mesityl Oxide) (MIBK)	-14.5	-12.1
$2\ CH_3COCH_3 + H_2 \Leftrightarrow (CH_3)_2CHCH_2COCH_3 + H_2O$ (Acetone) (MIBK)	-19.7	-8.8

One promising catalyst formulation consists of a hydrotalcite (Mg-Al hydroxide) support, possessing both basic and acidic sites, loaded with a dispersed noble metal (like Pt or Pd), known to have good hydrogenation activity. Indeed, Pt and Pd catalysts have been reported to selectively hydrogenate the C=C bond of unsaturated ketones (as opposed to the C=O bond) to produce saturated ketones (6,17). Pd- and Pt-supported hydrotalcites are currently receiving increasing attention as promising catalysts for aldol condensation reactions (18-20).

The objective of this research is to develop a fundamental understanding of acetone condensation on noble-metal supported HTC catalysts and to examine the relationship between the physical and chemical properties of these catalysts and their activity/selectivity for the condensation of ketones. This study could result in demonstrating the feasibility of such an economically and environmentally sound process for the industrial liquid-phase aldol condensation.

Experimental

Catalyst Synthesis and Characterization

A hydrotalcite (HTC, Mg-Al hydroxide, Mg/Al = 1.8) powder (ID# 95-194-HT-001, sample 570, obtained from LaRoche Industries Inc.) was used as the support for this study. Aqueous solutions of $PdCl_2$ and H_2PtCl_6 of various concentrations (calculated to form catalysts with nominal metal loadings of 0.1, 0.7, and 1.5 wt.%) were used for the wet impregnation of the HTC support. The impregnated catalysts were dried at 120°C for 2 hours, calcined at 350°C for 2 hours, and then crushed and sieved to particle size of 20 to 40 mesh (370-840 μm). Prior to reaction, the samples were reduced *ex-situ* in flowing hydrogen (heating to 230°C at 5°C/min, holding for 1.5 hours, then heating to 450°C at 5°C/min and holding for 8 hours).

The surface area of the catalysts was measured by conventional BET methods (nitrogen physisorption at $-196°C$ using a Quanta Chrome-NOVA 1000 instrument). The actual metal loading was measured by inductively-coupled plasma / optical emission spectroscopy (ICP/OES). The acidity and basicity of the synthesized catalysts were measured by NH_3 and CO_2 thermoprogrammed desorption, respectively, using an AMI-100 (Zeton-Altamira, Pittsburgh, PA) characterization system. The catalyst samples were reduced in 10% H_2/Ar at 450°C for 8 hours, followed by treatment in 10% NH_3/He or 10% CO_2/He at 35°C and then by desorption up to 400°C with a heating rate of 10°C/min.

Reaction System and Procedure

The reaction system consisted of the reactor, the gas control panel, and an online data acquisition system. The reactor was a high-pressure 50-ml vessel equipped with a 900-W electric heater, and a high-temperature, high-pressure (HTHP) micro-Robinson catalyst basket (Autoclave Engineers Inc., Erie, PA), with a 50-mesh opening and 8.4-ml nominal internal volume (Figure 2). The liquid in the reactor was stirred using a magnedrive impeller, the speed of which was controlled by a motor (Saftronics non-regenerative DC Drive) and indicated by a digital tachometer. The reactor headspace pressure was monitored using an analog pressure gauge and a pressure transducer (OMEGA) providing input to a digital pressure meter. The reactor was rated for 5000 psig (340 atm) at 650°F (343°C). It was also equipped with a rupture disk (Autoclave Engineers), rated at 5375 psig at 72°F. The reactor and heater temperatures were continuously monitored by two thermocouples and were digitally displayed.

The gas control panel was used for controlling the flow rates of the feed gases to the reactor (by manual metering valves), and measurement by electronic mass flow meters (Brooks 5860). A flow totalizer measured the flow and also displayed the cumulative flow (at STP). The gas control panel and the reactor were located in a fume hood. The data from the flowmeters, pressure transducers, thermocouples, and the digital tachometer were collected by a data acquisition interface (OMEGA Multiscan 1200) and recorded on-line.

A typical experiment involved loading ca. 2.7 g of catalyst (reduced *ex-situ*) into the catalyst basket, which was then placed into the reactor vessel, and adding 20 ml of acetone. The reactor was purged with N_2 to remove the oxygen. It was then pressurized (by closing the reactor vent valve) in a 25% H_2 : 75% N_2 blend to 350 psig while heating to the required reaction temperature and stirring continuously at 200 rpm. This continuous stirring minimized diffusion limitations between the bulk liquid phase and the catalyst particles. After reaching the desired reaction conditions (350 psig and typically 118°C), the N_2 inlet was shut off, while the H_2 inlet was kept open and the reaction pressure was increased to 400 psig. Any loss in pressure, attributed to the consumption of H_2 by the MO hydrogenation to MIBK or the undesired hydrogenation of acetone to isopropanol (IPA), was compensated by feeding H_2 to maintain a constant reactant composition (3:1 N_2/H_2 ratio).

The experiment was typically carried out for 5 hours under these conditions. Upon completion, the H_2 was shut off, the reactor was cooled down, and the liquid sample was collected and analyzed. Analysis was performed in a GC (HP 6890) with FID and a capillary column (HP-1 methyl siloxane) for separation of compounds. Methanol was used as an internal standard.

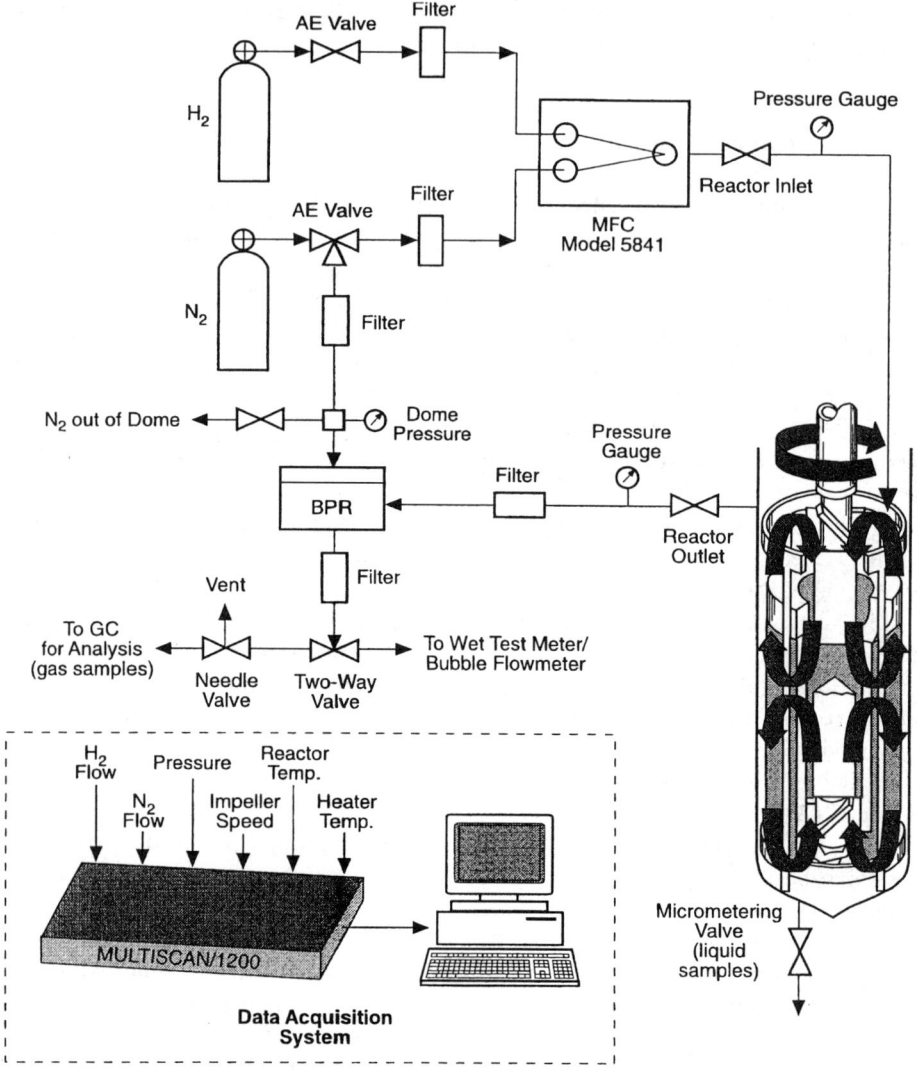

Figure 2. Schematic of the liquid-phase acetone condensation reaction system.

Results and Discussion

Catalyst Characterization

The physicochemical properties of the catalysts are reported in Table II. A quantitative estimate of the acidity and the basicity of the catalysts, expressed in terms of the amounts of desorbed NH_3 and CO_2 per g of catalyst, is also given in this table. The NH_3 desorption data showed no clear correlation between acidity and metal loading for either of the two types of samples. This observation suggests that the HTC support possesses a high concentration of acid sites which appeared to be unaffected by the metal impregnation process.

On the other hand, the basicity of the examined catalysts appeared to have a strong negative correlation with metal loading, as shown in Table II. The temperature profiles for the desorption of CO_2 from the parent HTC and selected Pd/HTC and Pt/HTC samples are shown in Figures 3a and 3b, respectively. The desorption profile of the non-impregnated, calcined HTC sample (included in both Figures 3a and 3b for the sake of convenience) showed two desorption peaks, one at the low temperature of ca. 100°C, and one at ca. 235°C, corresponding to weakly- and strongly-basic sites, respectively.

Table II. Physicochemical Properties of Acetone Condensation Catalysts

Catalyst	Metal loading[a] (wt.%)	Surface area[b] (m^2/g)	Acidity[c] ($\mu mol/g$)	Basicity[d] ($\mu mol/g$)
HTC	-	254	980	550
0.1% Pd/HTC	0.13	81	1350	312
0.7% Pd/HTC	0.69	92	NM	NM
1.5% Pd/HTC	1.53	59	1270	162
0.1% Pt/HTC	0.14	147	1020	234
0.7% Pt/HTC	0.67	146	NM	NM
1.5% Pt/HTC	1.47	150	1450	105

[a] Measured by ICP/OES on calcined samples
[b] Measured by N_2 physisorption at –196°C (BET method) on calcined samples
[c] Measured by NH_3 desorption (up to 400°C by 10°C/min), of samples reduced at 450°C for 8 h and treated with 10% NH_3/He at 35°C
[d] Measured by CO_2 desorption (up to 400°C by 10°C/min), of samples reduced at 450°C for 8 h and treated with 10% CO_2/He at 35°C
NM: Acidity/basicity properties of these intermediate-loaded catalysts were not measured

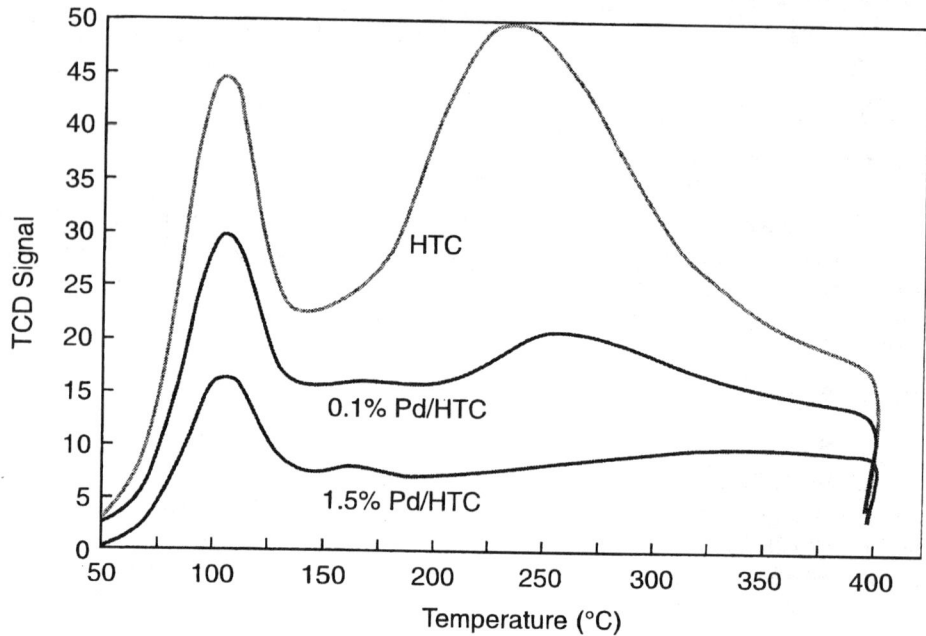

Figure 3a. CO_2 TPD profiles of HTC and Pd/HTC catalysts.

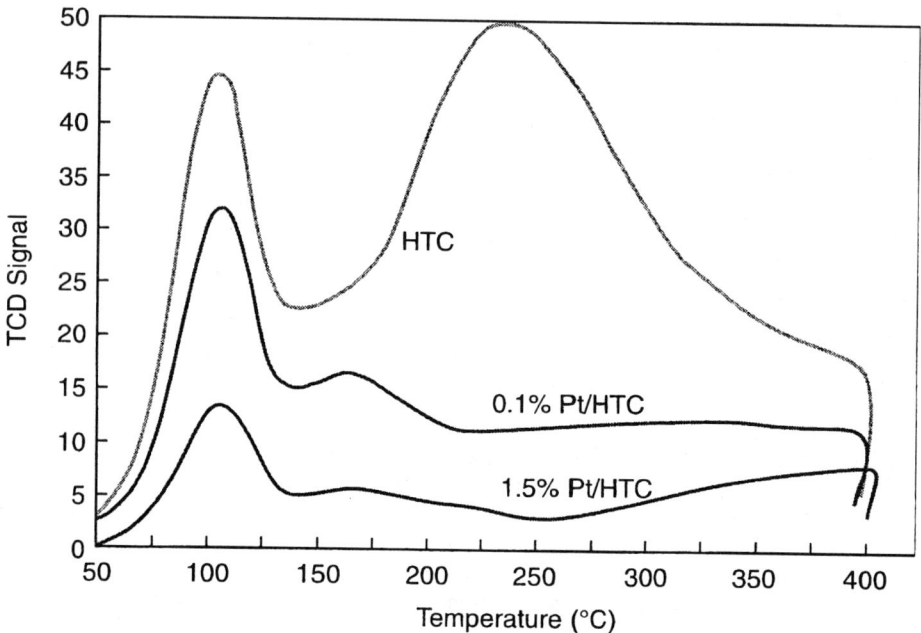

Figure 3b. CO_2 TPD profiles of HTC and Pt/HTC catalysts.

The CO_2 desorption profiles of the Pd/HTC and Pt/HTC samples showed moderately smaller low-temperature desorption peaks (ca. 100°C), which became progressively smaller at higher metal loadings. A more dramatic decrease was observed for the high-temperature desorption peaks of the impregnated samples as compared to the parent HTC sample. In fact, the high temperature peak of the high loading samples had essentially disappeared. It can therefore be concluded that the impregnation process resulted in a significant decrease in the basic character of the supported HTC catalysts.

Batch reactor experiments

The activity and selectivity of the HTC-supported Pd and Pt catalysts for the condensation of acetone, expressed in terms of reactant (acetone) conversion and selectivity to DAA, MO, MIBK, IPA, as well as other by-products is shown in Table III. The homogeneous reaction (i.e., no catalyst present) was also examined for comparison purposes. The results indicate that there was minimal acetone conversion in the absence of catalyst, with the direct hydrogenation of acetone to IPA being the only reaction. This result was not unexpected since, as indicated above, the MIBK formation requires multifunctional catalysis: acidity/basicity for condensation, acidity for dehydration, and metal sites for selective hydrogenation (*7,8*).

Table III. Catalyst Activity and Selectivity for Acetone Condensation (T=118°C, P=400 psig)

Catalyst	Conv. (%)	Selectivity (mol%)						
		DAA	*MO*	*MIBK*	*IPA*	*DMH*	*DIBK*	*HGL*
No catalyst	3.1	0	0	0	>99.9	0	0	0
HTC	19.6	14.4	84.9	0.6	0	0	0	0
0.1% Pd/HTC	38.0	14.3	0	82.2	0.6	1.2	1.8	0
0.7% Pd/HTC	29.7	11.8	0	71.7	14.7	0.6	1.0	0
1.5% Pd/HTC	28.2	9.0	0	61.8	27.7	0.5	0.8	0.2
0.1% Pt/HTC	34.0	12.0	0	52.4	32.6	0.4	0.8	0
0.7% Pt/HTC	32.7	6.0	0	38.1	52.5	0.3	0.5	0
1.5% Pt/HTC	30.8	6.6	0	33.6	56.8	0.2	0.3	0.3

DAA: diacetone alcohol (4-hydroxy-4-methyl-2-pentanone)
MO: mesityl oxide (4-methyl-3-penten-2-one)
MIBK: methyl isobutyl ketone (4-methyl-2-pentanone)
IPA: isopropanol (2-propanol)
DMH: dimethyl heptanone (4,6-dimethyl-2-heptanone)
DIBK: diisobutyl ketone (2,6-dimethyl-4-heptanone)
HGL: hexylene glycol (2-methyl-2,4-pentanediol)

The parent HTC sample (with no Pd or Pt) catalyzed the condensation of acetone to DAA (implying adequate basicity of the HTC catalyst), as well as the subsequent dehydration of DAA to MO (implying adequate acidity of this catalyst). Maximum (equilibrium) conversion of acetone was obtained under the given experimental conditions. The selectivity to MO was 85%, and no other higher condensation products were detected. The formation of MIBK was minimal, as expected, due to the absence of hydrogenation sites.

Both series of HTC-supported noble metal catalysts showed enhanced activity (expressed as acetone conversion) and improved selectivity to MIBK compared to the parent HTC sample. The acetone conversion was clearly greater than that in the absence of hydrogenation, due to a shift in the equilibrium-limited acetone-to-MO condensation reaction by direct removal of the product MO (via its hydrogenation to MIBK). No MO was detected at the end of the experiment, suggesting that these catalysts were very active for MO hydrogenation. The main reaction products were MIBK, IPA (formed by the undesirable direct hydrogenation of acetone) and the condensation intermediate DAA. Minor amounts (less than 2 mol%) of other higher condensation products (such as DMH and DIBK) were formed. Also, it is interesting to note that no measurable amounts of methyl isobutyl carbinol (MIBC, see Fig. 1), which is formed by the unselective hydrogenation of the C=O bond, were detected on any of the examined catalysts. This result is a clear indication of the selective C=C bond hydrogenation of these HTC catalysts.

A comparison of the activity and selectivity of the Pd- and Pt-supported HTC catalysts (Table III) indicated that both sets of catalysts exhibited similar overall acetone conversions, with a maximum conversion of 38% obtained on the 0.1% Pd/HTC sample. This catalyst also showed the highest selectivity to MIBK (82%), therefore resulting in maximum MIBK yield (31%).

The acetone conversion, as well as the DAA and MIBK selectivity, decreased slightly, but consistently, with higher metal loading for both Pd/HTC and Pt/HTC catalysts. A corresponding increase in IPA selectivity with metal loading showed that higher metal loadings favor the direct hydrogenation of acetone, due to the apparently slower rate of the acetone-to-DAA-to-MO-to-MIBK reaction. In addition, the hydrogenation of MO to MIBK was found to be complete in all cases since essentially no MO was detected in the products. These observations support the hypothesis that MO hydrogenation is not the rate-determining step in MIBK synthesis under the given experimental conditions.

The increasing hydrogenation activity of the catalysts with high metal loadings can be more clearly seen in the plot of IPA and MIBK selectivity vs. metal loading, shown in Figure 4. A higher metal loading favored the direct hydrogenation of acetone to IPA in comparison to its condensation/dehydration, which would lead to the formation of MIBK. Apparently a minimal metal loading (0.1wt.% or less) is required to accomplish complete MO hydrogenation to MIBK while minimizing IPA formation.

The acidity of the metal-containing catalysts was greater than that of the HTC, and their basicity was less (Table II). This is likely due to the applied preparation procedure (presence of chloride precursors). Despite the observed variation in acidity and basicity, no clear correlation between catalytic activity or selectivity and either of these pysicochemical properties of the examined catalysts was determined. Apparently the HTC-supported catalysts have at least the required basicity for the acetone condensation to DAA and the required acidity for the dehydration of DAA to MO.

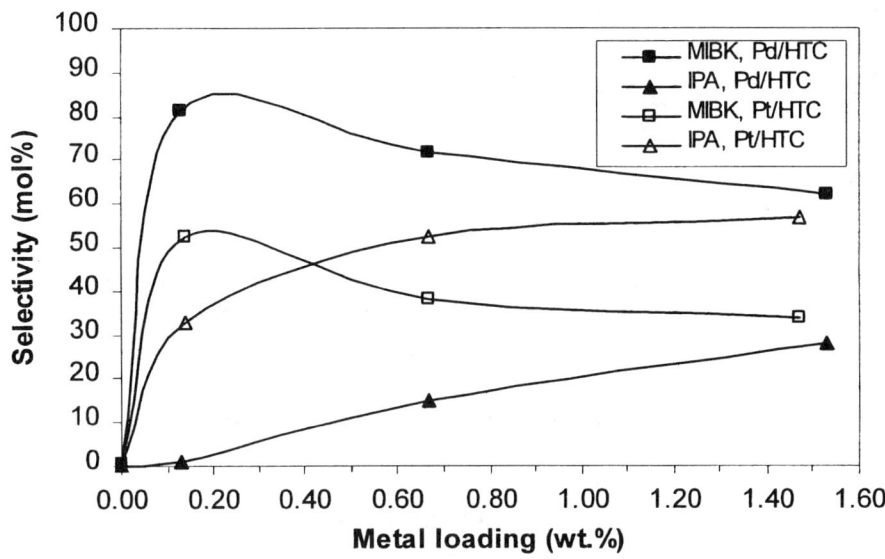

Figure 4. Effect of metal loading on the selectivity of Pd/HTC and Pt/HTC towards MIBK and IPA; T=118°C; P=400 psig.

The effect of reaction temperature on the formation of MIBK was also examined. An increase in temperature resulted in increased MIBK yield for the 0.1% Pd/HTC, as shown in Figure 5. The three data points at 118°C correspond to replicate experiments and are presented as an indication of the reproducibility of the activity measurements. Interestingly, this increase in reaction temperature resulted in decreasing selectivity for the major by-products DAA and IPA, and increasing selectivity to MIBK and to heavier by-products (DMH, DIBK). However, extensive formation of heavier (C_9+) by-products at elevated temperatures could result in blocking the active sites and thus decrease the activity for MIBK formation (*12*). Consequently, the choice of the appropriate reaction conditions may be a critical factor for optimizing the activity of a given catalytic system for the formation of MIBK.

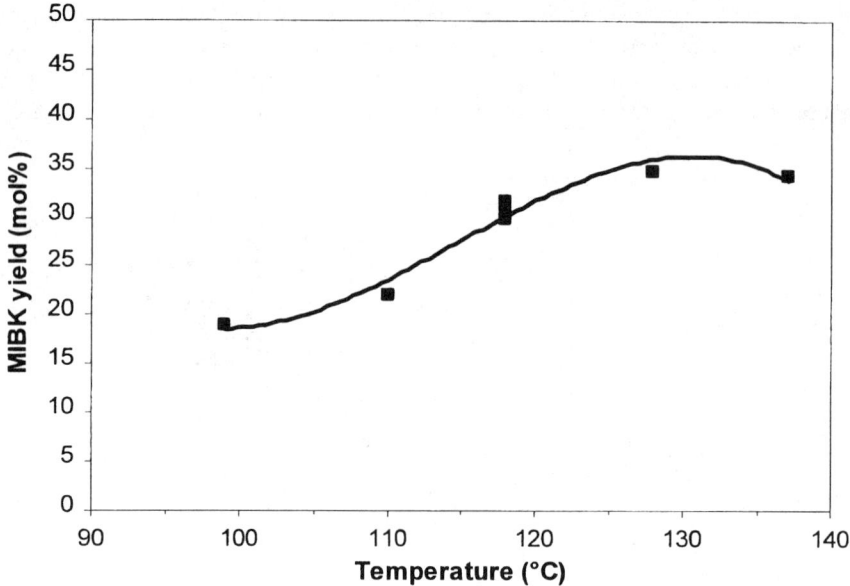

Figure 5. Effect of temperature on the MIBK yield of 0.1wt.% Pd/HTC; P=400 psig.

Conclusions

The results of this study clearly demonstrate that the single-step synthesis of MIBK from acetone condensation is feasible on noble-metal-supported hydrotalcites. This process offers significant advantages over the 3-step homogeneously-catalyzed acetone-to-MIBK process currently practiced. It is more efficient, giving higher acetone conversion (by overcoming reaction equilibrium limitations on acetone condensation) and high MIBK selectivity. Also, it is more environmentally benign, since it does not generate any significant amounts of strongly-basic waste streams. Finally, it is more economically attractive, by eliminating the cost-intensive product separation and waste disposal steps from the overall MIBK formation process.

The examined noble-metal supported HTC catalysts showed up to 35% yield to MIBK with minimal by-product formation (mainly IPA). Pd was found to be more selective to MIBK than Pt at every loading, despite similar acetone conversions for Pd/HTC and Pt/HTC catalysts. Activity is apparently dictated mainly by the acidity/basicity of the catalyst, which are properties of the support rather than the metal itself. The highest MIBK yield was obtained on the 0.1% Pd/HTC catalyst, indicating that minimal hydrogenation activity is required for the MO-to-MIBK reaction. Higher hydrogenation activity (higher metal loading, Pt instead of Pd) resulted in enhanced selectivity to IPA.

Although acidity is critical for the DAA dehydration to MO, the HTC support appeared to possess at least the required acidity for the MO formation. Also, basicity appears to be important for acetone condensation to DAA (the initial step in the reaction sequence), and the HTC support was observed to satisfy this requirement, thus giving maximum (equilibrium) conversion under the examined conditions. However, a direct effect of basicity on the activity and selectivity performance of the metal/HTC samples is rather ambiguous, due to its link with the varying metal loading of the examined HTC catalysts.

An increase in reaction temperature favored the formation of MIBK and moderately shifted the product selectivity towards heavier (C_9+) condensation products. Selection of the appropriate reaction conditions may be critical for maximizing the activity and selectivity for MIBK formation.

Disclaimer

Although the research described in this article has been funded wholly or in part by the United States Environmental Protection Agency through grant R-825331 to RTI, it has not been subjected to the Agency's required peer and policy review and therefore does not necessarily reflect the views of the Agency and no official endorsement should be inferred.

Literature Cited

1. Salvapati, G. S.; Ramanamurty, K. V.; Janardanarao, M. *J. Mol. Catal.* **1989**, *54*, p. 9.
2. Wilde, S.; Sommer, D. *Petroleum Technology Quarterly* **Spring 1998**, p. 127.
3. *Chemical Market Reporter* **1999**, *255(10)*, p. 37.
4. Dabbagh, H.; Davis, B. H. *J. Molec. Catal.* **1988**, *48*, p. 117.
5. Di Cosimo, J. I.; Diez, V. K.; Apesteguia, C. R. *Appl. Catal. A* **1996**, *137*, p. 149.
6. Lin, K.-H.; Ko, A.-N. *Appl. Catal. A* **1996**, *147*, p. L259.
7. Higashio, Y.; Nakayama, T. *Catal. Today* **1996**, *28*, p. 127.
8. Onoue, Y.; Mizutani, Y.; Akiyama, S.; Izumi, Y.; Watanabe, Y.; Maekawa, J. *Bull. Japan Petr. Inst.* **1974**, *16(1)*, p. 55.
9. Watanabe, Y.; Matsumura, Y.; Izumi, Y.; Mizutani, Y. *Bull. Chem. Soc. Japan* **1974**, *47*, p. 2922.
10. Watanabe, Y.; Matsumura, Y.; Izumi, Y.; Mizutani, Y. *J. Catal.* **1975**, *40*, p. 76.
11. Watanabe, Y.; Matsumura, Y.; Izumi, Y.; Mizutani, Y. *Bull. Chem. Soc. Japan* **1977**, *50*, p. 1539.
12. Pittman, C. U., Jr.; Liang, Y. F. *J. Org. Chem.* **1980**, *45*, p. 5048.
13. Hattori, H. *Chem. Rev.* **1995**, *95*, p. 537.
14. Reichle, W. T. *J. Catal.* **1985**, *94*, p. 547.
15. Rao, K. K.; Gravelle, M.; Valente, J. S.; Figueras, F. *J. Catal.* **1998**, *173*, p. 115.
16. Klein, F. G.; Banchero, J. T. *Ind. Eng. Chem.* **1956**, *48(8)*, p. 1278.
17. Kizling, M. B.; Bigey, C.; Touroude, R. *Appl. Catal. A* **1996**, *135*, p. L13.
18. Chen, Y. Z.; Hwang, C. M.; Liaw, C. W. *Appl. Catal. A* **1998**, *169*, p. 207.
19. Narayanan, S.; Krishna, K. *Appl. Catal. A* **1998**, *174*, p. 221.
20. Chen, Y. Z.; Liaw, C. W.; Lee, L. I. *Appl. Catal. A* **1999**, *177*, p. 1.

Chapter 17

Photooxidation of Toluene in Cation-Exchanged Zeolites

A. G. Panov, K. B. Myli, Y. Xiang, V. H. Grassian, and S. C. Larsen

Department of Chemistry, University of Iowa, Iowa City, IA 52242

Photochemical oxidation of hydrocarbons with molecular oxygen is potentially an environmentally benign method for the selective oxidation of hydrocarbons. In this study, the photooxidation of toluene in zeolites, BaX, BaY, CaY, BaZSM-5, and NaZSM-5 was investigated using in-situ Fourier Transform Infrared (FT-IR) spectroscopy and ex-situ gas chromatography (GC) for product analysis. In BaX and BaY, the primary product formed from the photooxidation of toluene and oxygen was benzaldehyde (~87%). GC analysis revealed the presence of small amounts of other products. In-situ FT-IR product spectra following the photooxidation of toluene in CaY, BaZSM-5, and NaZSM-5 showed similarities between the three zeolites and revealed a loss of product selectivity compared to that found in BaX and BaY. A large number of products (aromatic alcohols, aldehydes, and condensation products) were detected using GC analysis The loss of product selectivity can be correlated with the presence of residual Brønsted acid sites in CaY, BaZSM-5, and NaZSM-5.

Introduction

The selective photooxidation of hydrocarbons in zeolites is an example of new technology that is directed toward the development of environmentally benign processes for the production of chemicals as an alternative to conventional liquid phase oxidation reactions. Green methodology for the catalytic syntheses of industrially important chemicals is necessary if the production of hazardous waste is to be reduced *(1)*. The selective photooxidation of hydrocarbons combines several key aspects of eliminating waste. First, the reactions are done in the gas phase thereby eliminating the use of organic solvents. Second, the process has the potential to be energy efficient because sunlight can be used to initiate the reaction. Third, zeolite materials, long exploited in industry are used as reaction vessels to control the selective oxidation reaction.

In this study, a mercury arc lamp was used as the light source, but visible wavelengths were selected for irradiation using long pass filters. Products were extracted from the zeolite using organic solvents in order to facilitate product analysis. The focus of future work will be to eliminate the extraction step by addressing the issue of product desorption using zeolite thin films or membranes.

If it can be shown that the photooxidation of hydrocarbons in zeolites is a general method, then the shape and size-selective properties of zeolites may potentially be used to control the selectivity of specific oxidation reactions *(2,3)*. For example, ZSM-5 is an important shape-selective catalyst in many reactions, such as the disproportionation of toluene *(4)*. Para-xylene is the dominant product because the transport of the other isomers, ortho- and meta-xylene, is restricted due to the pore size of ZSM-5. Thus, stereochemical aspects of selective photooxidation reactions may also be influenced by the zeolite and may be used to design environmentally benign processes for the synthesis of industrially useful molecules.

The objective of this study is to further develop a methodology for the selective photooxidation of hydrocarbons in zeolites by exploring the effects of the zeolite host on the photooxidation of an aromatic molecule, toluene. Frei and coworkers observed the photooxidation of toluene to benzaldehyde in zeolite Y using visible light *(5-7)*. The use of visible rather than ultraviolet light allows access to a low energy pathway and eliminates many secondary photoprocesses. They attributed the high selectivity for benzaldehyde to the formation of a toluene-oxygen charge transfer complex that is stabilized by the zeolite framework *(5-7)*. They also cited cage effects as being important in the photooxidation reaction.

The zeolites used in this study were chosen because they differ in chemical composition (Si/Al and exchanged cation) and framework topology. The framework of ZSM-5 (shown in Figure 1a) is composed of *straight* 10-ring, elliptical channels (pore dimension: 5.3 by 5.6 Å) running along the (010) direction and *sinusoidal* 10-ring, elliptical channels (pore dimension: 5.1 by 5.5 Å) along the (100) direction *(8)*. Zeolite Y has a faujasite structure composed of sodalite units as shown in Figure 1b. Ten sodalite units form a supercage large enough to accommodate a sphere with a diameter of ~12 Å *(4)*. Apertures leading into the supercages are composed of 12 membered oxygen rings and are approximately 7.4 Å in diameter. Zeolite X has the same structure as Y but with a different Si/Al ratio.

We have investigated the photooxidation of toluene in several different zeolite hosts (BaX, BaY, CaY, BaZSM-5, and NaZSM-5) using in-situ Fourier Transform Infrared (FT-IR) spectroscopy and ex-situ Gas Chromatography (GC) to analyze product formation and product yields. This combined approach allows for a more detailed analysis of the product distribution. The product selectivity in these reactions appears to be governed by the presence of a small number of acid sites rather than by the framework composition or topology of the zeolite host.

Experimental Section

In-situ FT-IR spectra were recorded with a Mattson RS-10000 infrared spectrometer equipped with a narrowband MCT detector. Each spectrum was taken by averaging 500 scans at an instrument resolution of 4 cm^{-1}. The infrared sample cell used in this study has been described previously *(2,9)*. Briefly, 50-70 mg zeolite was sprayed from a water slurry onto a photo-etched tungsten (3x2 cm^2) grid held at 40 °C. The tungsten grid coated with the zeolite was mounted onto nickel jaws that are attached to a copper feedthrough. The sample can be heated and cooled from –123 to 927 °C. The temperature of the sample was measured with a thermocouple wire spotwelded to the center of the grid. The entire assembly was mounted inside of the IR cell, a 2.75" stainless steel cube with BaF$_2$ windows. The IR cell was then evacuated by a turbomolecular pump to a pressure of 1 x 10^{-7} Torr. Zeolites were heated under vacuum to 300 °C for BaX, BaY, and CaY, 350 °C for BaZSM-5 and NaZSM-5 overnight to remove adsorbed water.

Toluene was loaded into the zeolite by adsorption under an equilibrium vapor pressure of the liquid at room temperature. The excess toluene was then pumped out for 5 minutes and oxygen was added to the IR cell at a pressure of approximately 600 Torr. A 500 Watt mercury lamp (Oriel Corp.) with a water filter was used as the light source for photolysis. Broadband long pass filters were placed in front of the lamp (Oriel Corp. -filter 59492, %T = 0 at 495 nm; filter 59472, %T = 0 at 400 nm). For room temperature experiments, even with the water filter in place, the zeolite sample warmed to between 35-45°C after several hours of irradiation.

Products were extracted from the zeolite in an organic solvent, such as acetonitrile. GC (gas chromatography) with an FID detector and a 5% phenyl/95% methylpolysiloxane capillary column (Alltech, SE-54) was used for product identification. If available, standards of the products were injected separately to determine retention times and response factors.

Zeolites BaX, BaY, CaY, BaZSM-5, and NaZSM-5 were prepared from NaX (Aldrich), NaY (Aldrich), NaZSM-5 (Zeolyst) by standard ion-exchange procedures at 90 °C using aqueous 0.5 M BaCl$_2$ (Aldrich) and CaCl$_2$ (Aldrich) solutions. The Si/Al ratios for BaX, BaY (CaY), and BaZSM-5 (NaZSM-5) were determined by ICP-AES (inductively coupled plasma atomic emission spectroscopy) to be 2.4, 1.4, and 19, respectively. The Ba/Al ratios for BaX, BaY, and BaZSM-5 were determined to be 0.39, 0.33, and 0.26, respectively. The Ca/Al ratio was determined to be 0.49.

To qualitatively test for Brønsted acidity using the colorimetric method, 50 mg of the zeolite sample was pretreated at 300°C under vacuum for approximately 12 hrs. 1 mL of a dilute solution of retinol (Aldrich) or retinyl acetate (Aldrich) in dry hexane was injected onto the activated zeolite *(10,11)*. A color change to blue indicated the presence of Brønsted acid sites.

Toluene (Aldrich, 99% purity), toluene-d$_6$ (Aldrich, 99%) and toluene-(methyl-^{13}C), (Cambridge Isotopes, 99%) were subjected to several freeze-pump-thaw cycles prior to use. Standards of benzaldehyde and benzyl alcohol were purchased from Aldrich and subjected to several freeze-pump-thaw cycles prior to use.

Results

Toluene Photooxidation in BaX and BaY

Toluene photooxidation was investigated in BaX and BaY. Toluene was introduced into the infrared cell at a pressure of approximately 10 Torr and allowed to equilibrate for one hour. Gas-phase toluene was subsequently pumped out of the sample cell leaving strongly adsorbed toluene. Quantitative measurement of toluene adsorption indicated that the loading was approximately 2 toluene molecules per supercage. Molecular oxygen was then added to the infrared cell at a pressure of approximately 600 Torr. The reaction mixture was left for approximately 1 hour to equilibrate the system and to ensure that a dark reaction did not occur. Subsequently, the sample was irradiated with a broadband lamp. A 400 nm broadband filter (Oriel filter 59472) was placed at the output of the lamp so that only wavelengths above 400 nm were incident on the sample. Figure 2 shows the difference spectra in the region extending from 1850 to 1300 cm^{-1} following broadband excitation of the toluene-O_2 complexes in zeolites BaX and BaY at room temperature. The difference spectra were made by spectral subtraction of the spectrum recorded prior to irradiation from the spectrum recorded after irradiation. The spectra show that new product bands have appeared in the spectrum after irradiation. For BaX, these peaks are at 1681, 1650, 1592, 1577, 1394 and 1317 cm^{-1}, for BaY, these peaks are at 1677, 1654, 1592, 1578, 1391 and 1313 cm^{-1}. These bands agree well with the IR spectrum of an authentic sample of benzaldehyde in BaY which is shown for comparison. The major benzaldehyde peaks appear at 1687, 1652, 1594, 1580, 1456, 1393, and 1313cm^{-1}. Therefore, it is concluded that the primary product for this reaction is benzaldehyde. Water is also formed in the reaction, as indicated by a shoulder at 1645 cm^{-1} in the spectrum, due to the water bending mode. No other products were observed by FTIR in agreement with previous work by Frei and coworkers *(5, 6)*.

The presence of additional photoproducts was determined in the ex-situ product analysis using GC. The results of the GC analysis are given in Table I.

Table I: Photoproduct Distributions from Ex-situ GC Analysis

Zeolite	BZ^a	BA^a	Phenol, Cresols	Condensation Products	Other
BaY[b]	87%	4%	3%	2%	4%[c]
BaZSM-5[b]	31%	1%	31%	15%	22%[d]

[a] BZ = benzaldehyde and BA = benzyl alcohol
[b] Extracted acetonitrile
[c] Other = p-xylene (3%) and unidentified (1%)
[d] Other = doubly oxygenated products (17%), p-tolualdehyde (4%) and p-xylene (1%)

The product distribution from GC analysis of the extracted products in acetonitrile from BaY was approximately 87% benzaldehyde, 4% benzyl alcohol and small amounts of cresol and phenol (3%) and condensation products (2%). Benzyl alcohol and the other products were not readily detected in the FT-IR experiments.

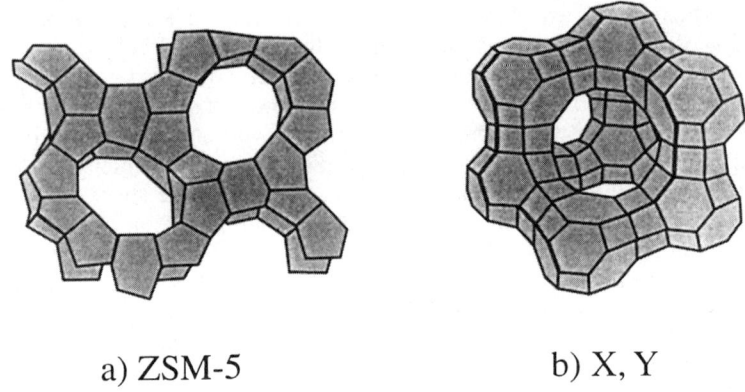

a) ZSM-5   b) X, Y

Figure 1. Structures of zeolites, a) ZSM-5 and b) zeolite Y, used in this study.

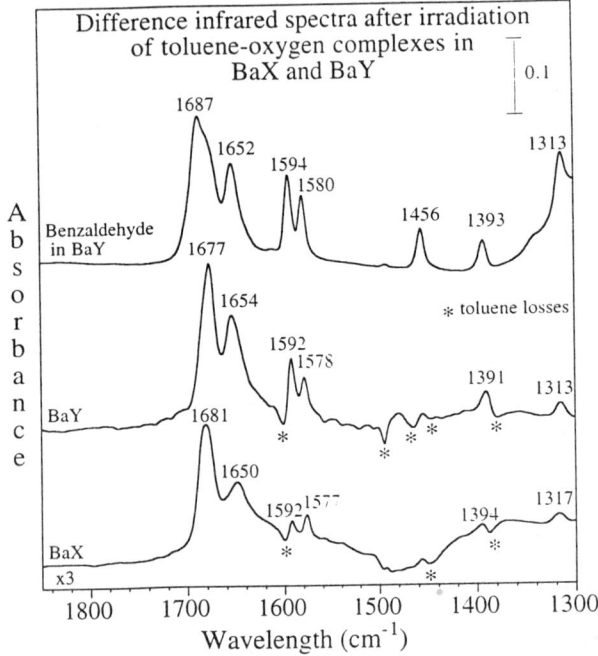

Figure 2. Difference infrared spectra before and after visible light photolysis of toluene and oxygen in zeolites BaX and BaY near room temperature with $\lambda > 400$ nm. Toluene losses are indicated by asterisks. Also shown is an authentic spectrum of benzaldehyde adsorbed in BaY for comparison with product spectra.

Toluene Photooxidation in CaY, BaZSM-5, and NaZSM-5

The photooxidation of toluene and oxygen in other zeolites, such as CaY, BaZSM-5, and NaZSM-5 was investigated next. The in-situ FT-IR difference spectra obtained after irradiation of toluene and oxygen in CaY, BaZSM-5, and NaZSM-5 are shown in Figure 3. The spectral features in all of the FT-IR spectra in Figure 3 are similar. The spectral features are broad with the most prominent peaks present near 1630-1650, 1585-1595, 1511, and ~1700 cm^{-1}. The FT-IR spectrum of benzaldehyde adsorbed on NaZSM-5 is included in Figure 3 for comparison. The agreement between the FT-IR spectrum of benzaldehyde and the photoproducts of toluene oxidation is not very good, suggesting that a mixture of products with overlapping IR bands is formed Most notably, the peaks at ~1700 cm^{-1} and 1511 cm^{-1} are not observed in the benzaldehyde spectrum and these bands are attributed to photoproducts other than benzaldehyde. There are a number of possible assignments for these peaks. The peak at ~1700 cm^{-1} is in the carbonyl stretching region of the FT-IR spectrum. The ~1700 cm^{-1} band does not shift when toluene-(methyl-^{13}C) is photolyzed on BaZSM-5 suggesting that the carbon atom in the observed carbonyl band does not originate from the toluene methyl group. This suggests that toluene may undergo disproportionation or cleavage followed by subsequent photooxidation. The 1511 cm^{-1} band observed in Figure 3 is due to a ring mode of an aromatic molecule. When toluene-d_6 was photolyzed on BaZSM-5 the band at 1511 cm^{-1} shifted to 1435 cm^{-1} supporting this assignment.

For the BaZSM-5 sample, the photoproducts were extracted in acetonitrile and analyzed by GC. The results are given in Table I. It can be clearly seen that the product selectivity to benzaldehyde observed for photooxidation of toluene in BaY is lost when the photoreaction is conducted in BaZSM-5. A number of different products have been identified. The three dominant products or product groups are benzaldehyde (31%), combined phenol and cresols (31%), condensation products (15%) and other oxygenated products (22%). Given that the FT-IR spectra are very similar for CaY, BaZSM-5, and NaZSM-5, a similar product distribution would be expected in each case.

Colorimetric Detection of Brønsted Acid Sites

In this study, the product selectivity and product distribution following photooxidation was found to depend on the zeolite used. All of the zeolites that exhibited a loss of product selectivity for toluene photooxidation are susceptible to acid site formation during cation exchange and subsequent pretreatment or calcination. The colorimetric test described by Ramamurthy and coworkers was utilized to detect Brønsted acid sites in the zeolites used in this study (10, 11). The colorimetric test is based on differences in the electronic absorption properties of protonated and unprotonated forms of probe molecules, such as retinol and retinol acetate. Both retinol and retinol acetate form a blue retinyl cation in an acid solution. The zeolites were activated by heating to 300°C under vacuum for approximately 12 hrs. Samples of BaZSM-5 (12) and CaY both turned dark blue when a dilute solution of retinol was added to the activated zeolite indicating the presence of Brønsted acid sites. Activated NaZSM-5 turned blue and activated BaY zeolite turned light blue

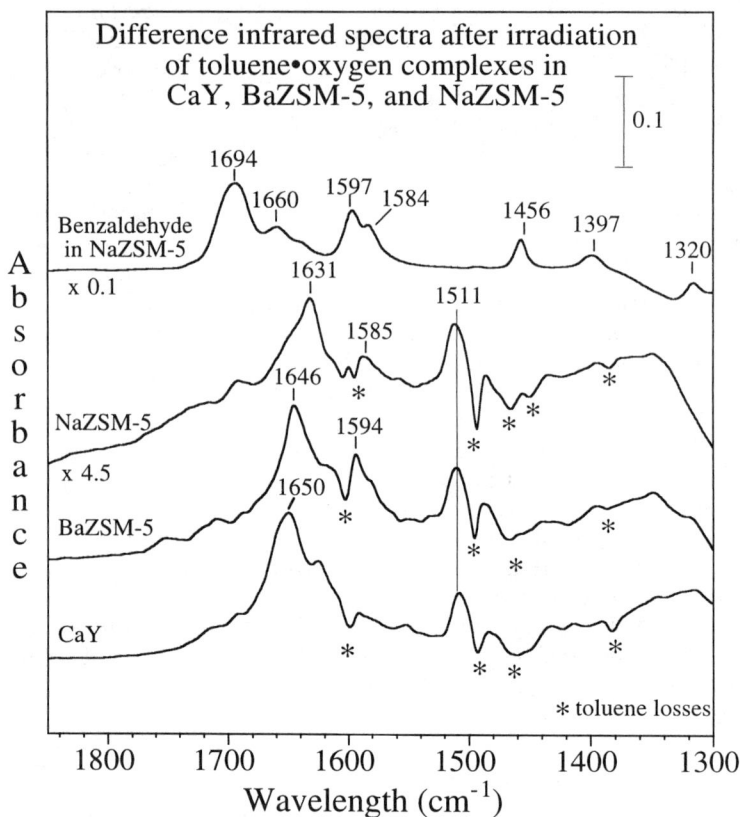

Figure 3. Difference infrared spectra before and after visible light photolysis of toluene and oxygen in zeolites CaY (λ> 455 nm), BaZSM-5 (λ> 495 nm) and NaZSM-5 (λ> 455 nm) near room temperature. Toluene losses are indicated by asterisks. Also shown is an authentic spectrum of benzaldehyde adsorbed in NaZSM-5 for comparison with product spectra.

when retinol in hexane was added to the samples. Activated samples of BaX, NaX, and NaY exhibited no color change when a dilute solution of retinol was added to the zeolite. These results suggest that for the zeolites used in this study, the number and strength of Brønsted acid sites trend in the following way: BaZSM-5, CaY > NaZSM-5 > BaY > BaX. This is qualitatively consistent with the observed trends in product distributions for toluene photooxidation and will be discussed further in the next section.

Discussion

Mechanism for the Formation of Benzaldehyde on BaX and BaY

Frei and coworkers previously suggested that the photoxidation of hydrocarbons in cation-exchanged zeolite Y proceeds through an intermolecular hydrocarbon-O_2 charge transfer state *(5, 13-15)*. The stabilization of this charge transfer state by the cation-exchanged zeolite leads to the remarkable product selectivity. The second step in the proposed mechanism involves transfer of a proton to O_2^- to form a hydrocarbon radical / $HO_2^{\bullet}$ radical pair that reacts further to yield a hydroperoxide intermediate. The hydroperoxide intermediate fragments to form the corresponding aldehyde and/or ketone products and water. The mechanism for the photooxidation of toluene to yield benzaldehyde is shown in Scheme 1. The mechanism was proposed by Frei and coworkers and accounts for the formation of the predominant photoproduct, benzaldehyde, that is formed in both BaX and BaY *(5)*.

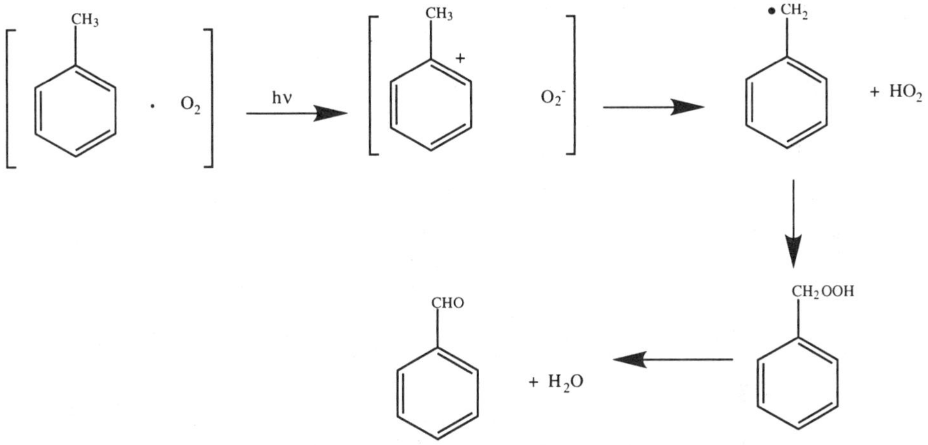

Scheme 1: Mechanism for the Photooxidation of Toluene to Benzaldehyde in BaX and BaY

Using GC product analysis, a small amount of benzyl alcohol (4% of products) is also observed after the photooxidation of toluene on BaX and BaY. Previous photochemical oxidation studies of toluene in solution provide conflicting results with respect to the formation of benzyl alcohol *(16, 17)*. Wei and coworkers observed a mixture of benzaldehyde and benzyl alcohol when a neat toluene solution saturated with oxygen was photolyzed with UV light *(16)*. The ratio of benzaldehyde/benzyl alcohol decreased as the photolysis time and the temperature were increased. In contrast to those results, Hashimoto and Akimoto concluded that photochemical excitation of toluene lead to the oxidative cleavage of the aromatic ring to form CO_2 and a dicarbonyl *(17)*. In the present case, it is most likely that the benzyl alcohol orginates from secondary photolysis or from bimolecular reactions involving toluene and the benzyl hydroperoxide intermediate. Further studies are in progress to investigate the mechanism for the formation of benzyl alcohol.

Correlation of Loss of Product Selectivity and Brønsted Acidity in CaY, BaZSM-5, and NaZSM-5

A dramatic loss of selectivity to benzaldehyde was observed when toluene and oxygen were photolyzed in CaY, BaZSM-5, and NaZSM-5. The FT-IR spectra shown in Figure 3 all exhibit similar spectral features suggesting that the same products are formed in each case. Extraction of the photoproducts from BaZSM-5 revealed that many different products were formed. The product distribution was benzaldehyde (31%), combined phenol and cresols (31%). various condensation products (15%) and other oxygenated and doubly oxygenated products (22%). The data shown in Table 1 and the results of the colorimetric test for acid sites indicate that the formation of phenol, cresol, and condensation products correlates with the presence of Brønsted acid sites on BaZSM-5. This suggests that acid chemistry is responsible for the loss of selectivity in this system. Specifically, it is well-known that alcohols and phenols can be produced from hydroperoxides with acid catalysts *(18)*. Further experiments involving thermal reactions of benzyl alcohol and toluene in BaZSM-5 produced isomers of monomethyl substituted diphenyl methane, which is the major condensation product in these reactions. In order to eliminate the nonselective products and to improve the product selectivity, future work will focus on preparing higher quality ZSM-5 samples that are free of acid sites.

The source of Brønsted acid sites in BaZSM-5 is most likely due to the following reaction that occurs during pretreatment or activation of a divalent cation-exchanged zeolite at elevated temperature:

$$M^{2+} + H_2O + Si-O-Al \rightarrow M(OH)^+ + Si-O(H^+)-Al$$

where Si-O-Al represents a part of the zeolite framework *(10)*. It has also been shown that Brønsted acid sites can be present in low concentrations in alkali-metal zeolites. For example, Ramamurthy and coworkers have established the presence of low levels of Brønsted acidity in NaY and NaX zeolites using the color change of a base indicator *(10, 11, 19)*. The acidic forms of retinyl acetate, retinol and retinyl Schiff bases are colored blue and are easily identified by both visual inspection and spectrophotometrically. Using these bases, they were able to determine whether or not Brønsted acids were present in low concentrations in various zeolite samples. Importantly, these studies showed that the presence of small quantities of acid sites can alter the reactivity of various alkenic substrates. In another study using solid-state NMR spectroscopy, the presence of Brønsted acid sites, as well as other types of sites, were identified in the alkaline earth zeolite CaY *(19)*. The presence and concentration of sites was found to depend on whether the zeolite was activated in air or under vacuum. These studies indicate that the reactivity and properties of cation-exchanged zeolites can change from zeolite source to zeolite source and from the use of different pretreatment conditions *(11)*.

The zeolite samples used in this study contained various quantities of Brønsted acid sites. Using the colorimetric method with retinol as a probe molecule, the activated forms BaY, CaY, BaZSM-5, and NaZSM-5 were all found to contain Brønsted acid sites. BaY contained fewer Brønsted acid sites than NaZSM-5 and BaZSM-5 and CaY and no Brønsted acid sites were detected colorimetrically in BaX. The zeolites with similar quantities of Brønsted acid sites exhibited similar reactivity and selectivity patterns. Work is in progress to prepare zeolites that are free of acid sites so that the selectivity can be improved.

Conclusions

The photooxidation of toluene and oxygen in different zeolite hosts was investigated using in-situ FT-IR and ex-situ GC product analysis. The goal was to study the effect of zeolite chemical composition and framework topology on the selectivity of toluene photooxidation reactions as an environmentally benign alternative to conventional liquid phase oxidation reactions. Toluene and oxygen photooxidation in BaX and BaY yielded benzaldehyde as the primary product with high selectivity. A small amount of benzyl alcohol was also observed using GC analysis. The toluene photooxidation reaction in zeolites CaY, BaZSM-5 and NaZSM-5 exhibited a loss of selectivity to benzaldehyde. The product distribution resulting from photooxidation of toluene in BaZSM-5 was benzaldehyde (31%), combined phenol and cresols (31%), various condensation products (15%) and other oxygenated products (22%). This lack of selectivity is attributed to the presence of Brønsted acid sites in CaY, BaZSM-5, and NaZSM-5. These acid sites were subsequently detected colorimetrically. This work demonstrates that the presence of even small amounts of acid sites can alter the product selectivity in the photooxidation of hydrocarbons. Current work is in progress to prepare zeolites that are free of acid sites so that high product selectivity can be retained, an important requirement for environmentally benign synthesis.

Acknowledgments

Although the research described in this article has been funded wholly or in part by the Environmental Protection Agency through grant number R825304-01-0 to SCL and VHG, it has not been subjected to the Agency's required peer and policy review and therefore does not necessarily reflect the views of the Agency and no official endorsement should be inferred.

Literature Cited

1. Anastas, P. T.; Williamson, T. C. in Green Chemistry: Designing Chemistry for the Environment; Anastas, P. T.; Williamson, T. C., Eds.; ACS Symposium Series # 626; American Chemical Society: Washington, D.C., **1996**, pp 1-17.
2. Myli, K. B.; Larsen, S. C.; Grassian, V. H. *Catal. Lett.* **1997**, *48*, 199-202.
3. Xiang, Y.; Larsen, S. C.; Grassian, V. H. *J. Am. Chem. Soc.* **1999**, *121*, 5063-5072.
4. Gates, B. C. *Catalytic Chemistry*; Wiley: New York, 1992.
5. Sun, H.; Blatter, F.; Frei, H. *J. Am. Chem. Soc.* **1994**, *116*, 7951 - 7952.
6. Frei, H.; Blatter, F.; Sun, H. *CHEMTECH* **1996**, 24-30.
7. Blatter, F.; Sun, H.; Vasenkov; Frei, H. *Catalysis Today* **1998**, *41*, 297-309.
8. Kokotailo, G. T.; Lawton, S. L.; Olson, D. H.; Meier, W. M. *Nature* **1978**, *272*, 437-438.
9. Miller, T. M.; Grassian, V. H. *J. Am. Chem. Soc.* **1995**, *117*, 10969-10975.
10. Rao, V. J.; Perlstein, D. L.; Robbins, R. J.; Lakshminarasimhan, P. H.; Kao, H.-M.; Grey, C. P.; Ramamurthy, V. *Chem. Comm.* **1998**, 269-270.
11. Thomas, K. J.; Ramamurthy, V. *Langmuir* **1998**, *14*, 6687-6692.
12. It is possible that the retinol is only able to detect surface acidity in the ZSM-5 zeolites due to the small pore size relative to the size of the probe molecule.
13. Blatter, F.; Frei, H. *J. Am. Chem. Soc.* **1994**, *116*, 1812 - 1820.
14. Blatter, F.; Moreau, F.; Frei, H. *J. Phys. Chem.* **1994**, *98*, 13403 - 13407.
15. Blatter, F.; Sun, H.; Frei, H. *Catal. Lett.* **1995**, *35*, 1-12.
16. Wei, K. S.; Adelman, A. H. *Tetrahedron Lett.* **1969**, 3297-3300.
17. Hashimoto, S.; Akimoto, H. *J. Phys. Chem.* **1989**, *93*, 571-577.
18. Sheldon, R. *The Chemistry of Peroxides*; Wiley: New York, 1983.
19. Kao, H.-M.; Grey, C. P.; Pitchumani, K.; Lakshminarasimhan, P. H.; Ramamurthy, V. *J. Phys. Chem. A* **1998**, *102*, 5627-5638.

Chapter 18

Oxygenation of Hydrocarbons Using Nanostructured TiO_2 as a Photocatalyst: A Green Alternative

Endalkachew Sahle-Demessie and Michael A. Gonzalez

ORD, National Risk Management Research Laboratory,
U.S. Environmental Protection Agency, Cincinnati, OH 45268

High-value organic compounds have been synthesized successfully from linear and cyclic hydrocarbons by photocatalytic oxidation using the semiconductor material, titanium dioxide (TiO_2). Various hydrocarbons were partially oxygenated in both aqueous and gaseous phase reactors using ultraviolet light and titanium dioxide under mild conditions. Gas phase reaction conditions eliminate the need for a separation step involving with liquid solvents and minimizes the adsorption of products to the catalyst. The conversions and selectivities obtained for partial oxidation of hydrocarbons have been comparable to those achieved with conventional methods. Initial life-cycle analysis showed that the technology has the potential to reduce water contaminants and eliminate the use of toxic catalysts. Light-induced catalysis opens up possibilities of the use of oxygen in partial oxidation reactions now being conducted with far more expensive polluting oxidants. The high selectivity of the mild photochemical routes are especially attractive for the manufacture of fine chemicals. This chapter describes the chemistry of TiO_2 catalyzed reactions for environmentally beneficial chemistry based on the work done in our laboratory and others.

The chemical industry is a significant component of the domestic economy, generating over $250 billion in sales and maintaining a trade surplus of more than $15 billion for each of the last 5 years. The industry is also a major source of industrial waste and is the dominant source of hazardous waste in the United States. The costs of handling, treating and disposing of wastes generated annually in the United States has climbed to 2.2% of the gross domestic product, and continues to rise (1). As a consequence, the chemical industry spends billions of dollars annually on managing pollutants and has hundreds of billions of dollars invested in pollution control equipment. Costs to clean up existing toxic sites are estimated to approach $750 billion over a 30 year period. Cleaning up our water and air and preventing future deterioration of the environment is estimated to cost even more. The average environmental cost per firm in various industries is broken down as follows: high-tech firms, $2 million (6.1% of revenues); utilities, $340 million (6.1% of revenues); steel and metals, $50 million (2.9% of revenues); and oil companies, $430 million (1.9% of revenues). Given that the average pre-tax profits of the 500 largest U.S. manufacturing companies were 7.7 percent of sales in 1996, these figures are staggering. These costs will only grow when one considers "superfund legislation", which provides government support and mandate to clean up existing sites.

The chemical manufacturing industry generates more than 1.5 billion tons of hazardous waste and 9 billion tons of non-hazardous waste annually. Roughly half of the releases and transfers of chemicals reported through the Toxic Release Inventory and 80-90% of hazardous waste generation reported through the Resource Conservation and Recovery Act (RCRA) are due to chemical manufacturing (2,3,4). Organic chemicals constitute the largest source of the toxic releases (2). Many of these releases can be minimized by improving conventional house keeping methods. Other strategies for pollution prevention include better management of material and energy, more efficient process control, optimized process conditions, and recycle and reuse of waste and by-products (5). However, cleaner production methods can be achieved by adopting "green synthesis" methods. Cleaner chemical processes, involving both evolutionary and revolutionary technologies, could generate less waste and emission of toxic substances, use materials that are less toxic, and require less energy than their predecessors. These new technologies, which can emerge from advances in basic research, will help the chemical industry address the dual goals of global competitiveness and environmental stewardship.

Partial Oxidation Reactions. Oxidation is used in industry for producing aliphatic and aromatic alcohols, aldehydes, ketones and acids. Generally, oxidation involves splitting of C-C or C-H bonds and formation of C-O bonds. For example, the partial oxidation of hydrocarbons by molecular oxygen, to form oxygenates that are used as building blocks in the manufacturing of plastics and synthetic fibers, is an important process in the chemical industry.

Oxidation of alkanes is especially important both industrially and in organic synthesis. Among the products, aldehydes and ketones are useful synthetic intermediates both industrially and in organic synthesis. Oxidation reactions are usually catalyzed and carried out in liquid or gas phase. Stoichiometric oxidation is widely used and large quantities of by-products are formed. Replacement of stoichiometric oxidation processes with new catalytic, low or no salt technologies could avoid the creation of such by-products. The current processes are energy intensive, have low conversion efficiencies and generate environmentally hazardous waste and by-products. In addition, the desired oxygenated product is subject to further oxidation, which must be prevented or minimized. In particular, autooxidation of small alkanes, alkenes, or aromatics is inherently unselective, whether conducted in the gas or liquid phase or catalyzed by transition metals. The major reason why selectivities are low is that the desired products (such as aldehydes or alcohols) are more easily oxidized by O_2 than the parent hydrocarbon. Most oxidations are highly exothermic and may generate high localized temperatures such that the catalyst surface is degraded. Careful control of partial oxidation reactions to prevent further oxidation of desired products is difficult. Overoxidation can be minimized only by keeping conversions low, a serious disadvantage from a chemical processing standpoint. Therefore, it is a major challenge to find a reaction pathway that affords the primary product with high selectivity and high conversion of the hydrocarbon. In addition, several commonly used catalysts for oxidation employ toxic heavy metals (like chromium, vanadium) or strong acids (like H_2SO_4, HNO_3) increasing the environmental hazards of these reactions. Pollution is inevitable in loading, recovering, and regeneration of these catalysts. Thus, a cleaner alternative is needed.

We describe here our investigations of an alternative method for the activation and oxidation of hydrocarbons. The substrate is reacted using air under ambient conditions in a photochemical reactor that uses ultraviolet (preferably solar) light and a specially prepared semiconductor catalyst. The use of photocatalysis has many advantages because of its environmentally friendly nature, its high oxygen atom efficiency, and its versatility. This research incorporates the principles of Green Chemistry and Engineering (6) by allowing the production of oxygenates in a selective manner and producing less by-products and pollutants than comparable conventional techniques.

Photocatalysis

Principles. A photocatalyst is a substance that is activated by a photon. Activation of the semiconductor photocatalyst for reaction is achieved through the absorption of a photon of ultra-band energy, which results in the promotion of electrons from the filled level (i.e. the valence band edge) to the vacant level (i.e. the conduction band edge), generating an electron-hole pair (Figure 1). This electron-hole pair is the primary photoproduct formed upon photoexitation of a semiconductor and has a finite lifetime to allow a separate site for the oxidative and reductive half-reactions to occur. Semiconductor photocatalysis can be more attractive than the more conventional chemical oxidation methods because semiconductors are inexpensive, nontoxic, and capable of repeated use without loss of photoactivity. Therefore, the development of efficient photocatalysis for selective oxidation of alkanes with TiO_2 is very appealing.

Mechanism of Photocatalyzed Oxidations. Heterogeneous photocatalysis is a technology based on the irradiation of a semiconductor (SC) photocatalyst. Upon irradiation, a semiconductor generates electron/hole pairs with free electrons produced in the nearly empty conduction band (CB) and positive holes remaining in the valency band (VB) (7). The holes migrate to the semiconductor surface and react with organic compounds, acting as strong oxidizing agents. Figure 1 shows the typical reaction scheme of a n-type semiconductor such as TiO_2. Depending on the ambient conditions, the lifetime of an electron/hole separation process can be from a few nano-seconds to a few hours (8).

Absorption of light energy greater than or equal to the band gap (E_{bg}) of the semiconductor results in a shift of electrons from the valence band(VB) to the conduction band (CB) and the creation of holes (h+) in the valence band:

$$SC + h\nu \rightarrow h^+{}_{VB} + e^-{}_{CB} \qquad (1)$$

These charge carriers recombine, radiatively and/or nonradiatively, in competition with rapid diffusion to the surface where the resulting nonequilibrium distribution of electrons and holes gives rise to reduction or oxidation processes of adsorbed species, surface groups, and the semiconductor component. If the irradiated semiconductor is in contact with a suitable redox system, a redox reaction may take place. The process occurring on the irradiated semiconductor is dependent on pH, temperature, the concentration of reactants, and a redox potential that lies between E_{cb} and E_{vb}. From the thermodynamic point of view, possible oxidative processes are those characterized by potentials lower than the valence band energy. A semiconductor is chosen based on the potentials of its valence and conduction band edges as measured against the reference electrode. These positions govern the oxidizability and reducibility of the half-reactions to be conducted.

The band positions of several commonly used semiconductor photocatalysts are listed in Table I (9). Chemical transformation can occur, only if the process competes kinetically with electron-hole recombination

The recombination of electron/hole pairs can take place either between energy bands or on the surface. As a result the photocatalytic efficiency is reduced. To impede the recombination process, conducting materials such as noble metals can be incorporated into the semiconductor to facilitate the electron transfer and prolong the lifetime of the electron/hole separation process (10). Although, there has been considerable efforts in using photocatalysis for complete oxidation of organic compounds in air and water streams, incomplete or partial oxidation has been reported (11, 12).

The valence band positions of most of these semiconductors lie higher than the oxidation potentials of many organic compounds, indicating that oxidation of many functional groups can be facilitated. The redox potential of hydrogen evolution (H^+/H_2) and oxygen evolution (O_2/H_2O) are -0.3 eV and 1eV, respectively (13). Many conjugated groups or functional groups bearing non-bonded electrons should participate in photoinduced single-electron oxidation (14). The band gap energy for TiO_2 varies from 3.0 to 3.2 eV depending on the environment in which it is found (15). Titanium dioxide is the most commonly used semiconductor since it has many advantages: it is inexpensive, widely available, insoluble in water and many other solvents, is stable over a wide range of pH and has fewer problems with photocorrosivity (16, 17).

Photocatalytic oxidation processes involving the use of semiconductors have gained increased attention for innovative treatment of hazardous wastes as well as photoreduction of toxic compounds and purification and disinfection of drinking water.

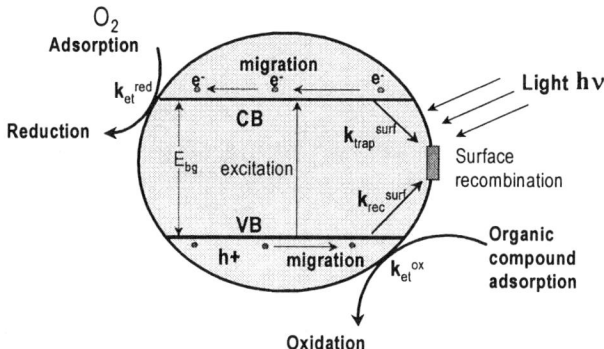

Figure 1. Illustration of major processes occurring on semiconductor particle following electronic excitation

Table I. Band Position* of Semiconductor Photocatalyst (9)

Semiconductor	Valence Band	Conduction Band
TiO_2	+3.1	-0.1
SnO_2	+4.1	+0.3
ZnO	+3.0	-0.2
WO_3	+3.0	+0.2
CdS	+2.1	-0.4
CdSe	+1.6	-0.1
GaAs	+1.0	-0.4
GaP	+2.2	-1.0
SiC	+1.6	-1.4

* Band position in water at pH 1

There is also great interest at present in photocatalytic degradation of organic molecules in air and water streams. The focus of the research described here, however, is the photocatalyzed partial and selective oxidation of a hydrocarbon to alcohol and ketone, derived by kinetic control of the reaction in gas phase.

Semiconductor Catalysis for Chemical Synthesis. In recent years, the use of semiconductor photocatalysts have been investigated in a variety of applications. Anatase phase titania has applicability as a photocatalyst for several problems of environmental interest (*18,19*), including use as catalyst for sulfur removal (*20*), for toxic metal capture (*20,21*) and as an additive in cosmetics due to its efficient sunscreen properties. Theoretically, photocatalysts function as a pool of electron holes, providing efficient charge separation at the interface between the semiconductor and a gas or liquid. Two critical features that a photocatalyst must posses in order to be effective are sufficiently long excited-state lifetime to allow for interaction with the organic substrate of interest and sufficient oxidizing and reducing power.

We found that aqueous phase photooxidaton of hydrocarbons (such as cyclohexane) using a batch reactor for two hours resulted in 3 to 5 % initial conversion to their respective alcohols and aldehydes (*23*). Similar work was done on the photoinduced conversion of acetic acid to methane and carbon dioxide as an alternative to Kolbe reaction (*24*). When water was used as a solvent, deep oxidation, or the complete oxidative mineralization of the organic substrate, usually predominated due to the generation of highly oxidizing hydroxyl radicals. Because aqueous phase reactions encounter separation problems, it is better to carry out organic chemical synthesis using photocatalysis either in an inert solvent, such as acetonitrile, in the neat organic solvent for liquid substrates or in gas phase.

We performed gas phase photocatalytic oxidation reactions of hydrocarbons by flowing a known mixture of heated humid air with organic vapor through an annular reactor (*25*). We used nanostructured TiO_2 coated using flame aerosol technique at optimized process conditions, which enabled us to achieve higher conversions and selectivities than obtained by coating prepared by other methods.

Photoxidations. The photocatalytic oxidation of many organic molecules, including saturated hydrocarbons, by optically-excited semiconductor oxides is thermodynamically allowed in the presence of oxygen at room temperature. UV light assisted reactions have been promising for oxidation and epoxidation of small olefins (*26, 27*). Selectivities for ketones and aldehydes were higher than those obtained by other oxidation means have been reported (*9, 28*). Some of the substrates investigated, the products formed, and the conversion and selectivities are shown in Table II.

Distinction between photocatalytic oxidations and dehydrogenations, both of which give oxidized organic products, have been made (*28*). A photocatalytic oxidation involves a primary electron transfer to the photogenerated hole, producing, at least transiently, a single electron oxidized intermediate. Incontrast, a photocatalytic dehydrogenation, may involve loss of hydrogen atoms from the molecule without forming intermediate radicals. Some interesting examples of oxidative cleavage of photoreactions sensitized by semiconductors include the photoxidation of cyclohexane and toluene. Photoxidation of toluene in water resulted in the formation of cresols (*11, 30*). In acetonitrile toluene oxidation resulted in benzaldehyde and then benzoic acid. In our laboratory, gas-phase photocatalyzed oxidation of toluene resulted primarily in benzaldehyde and secondarily in benzyl alcohol (Table II). The high selectivity of toluene-to-benzaldehyde oxidation by O_2 without overoxidation or side reaction is remarkable. The selectivity and the efficiency with which the photoreactor operates were influenced by the oxygen concentration, the illumination, the properties of the photocatalytic coating, and the

Table II. Results of photoxidation of hydrocarbons.

Reactants	Products			Conversion perpass %	Selectivity
toluene	benzyl alcohol	benzaldehyde		6-12	96-98% benzaldehyde
cyclohexane	cyclohexanol	cyclohexanone		2-6	65-84% cyclohexanone
n-pentane	1-pentanol, 2-pentanol, 3-pentanol, 2-pentanone, 3-pentanone			2-4	80% pentanones
methylcyclohexane	hydroxy- and keto- methylcyclohexanes			2-5	86-89% methylcyclohexanones
CH_4	CH_3OH	HCHO		< 1	70% H_2CO
propene	allyl alcohol, acrolein, isopropanol, acetone			1-2	75% ketone
CH_3-CH_2-OH	CH_3CHO	CH_3COOH		3-5	96% acetaldehyde

conditions within the fluid phase affecting contact of the substrate to the TiO_2 surface. Current industrial processes lack this selectivity, due to overoxidation to benzoic acid which occurs unless conversions are kept to a few percent. The same is true for other hydrocarbon substrates. The conventional method of vapor-phase oxidation of cyclohexane using a nickel or boric acid catalyst also gives low conversions per pass (usually below 5%) in order to allow yields to reach 70+% (*31*).

Photocatalytic oxidation is a promising technology. We have obtained higher conversions and better selectivities for partial oxygenates with minimal side reactions or overoxidation. The active radical might be derived from the reduction of oxygen (*11*) or from the water vapor. Oxidative cleavage is attained by interception of the surface bound radical cation with superoxide or adsorbed oxygen. With alkanes or simply substituted alkenes, however, the capture of a photogenerated hole is often thermodynamically forbidden (*32*). Therefore, radicals formed by hydrogen abstraction by activated oxygenation represent the critical intermediates.

Photoreductions. Reduction reactions have received less attentions than oxidation reactions. These reactions must be carried out in an oxygen free environment since oxygen is a ready acceptor in the reduction reaction. Reduction reactions of organic substrates can be one of two cases. Either the organic substrate acts as an electron acceptor and is reduced by the conduction band electron, or the solvent or additive is sacrificially oxidized (*33*). An example of an organic reduction reaction in which water/HNO_3 was used as additive is given in Scheme 1 (*34*). Reductive hydrogenation of acetylene on colloidal suspension of TiO_2, loaded with MoS_4^{2-}, to ethane has also been reported (*35*). Reductive dehalogenation of chlorofluorocarbons (CFCs) takes place upon illuminated air free TiO_2 particles (*36*). Chlorofluorocarbons can interfere with the natural stratospheric cycle of ozone and are currently being replaced by their hydrogen-substituted analogs (HCFCs).

Process Variables. Work by our group and others have shown that product distribution can be markedly affected by a wide variety of parameters, including the nature of the semiconductor, crystal phase, particle size, surface morphology, light intensity and the presence of catalysts (*14, 38*). Immobilized titania film system were prepared using nanosized titania particles prepared by the flame aerosol coating method. The flame aerosol coating enabled the formation of high surface area aggregates with minimal pore area coatings that were much more active than the TiO_2 prepared by other methods and did not appear to create inter particle diffusion control (*38*). The rate of photoxidation is affected by the solvent type, colloidal preparation, oxygen flow, colloidal concentration, and light intensity. Aromatic aldehydes have been synthesized by the oxidation of the methyl groups in toluene and derivatives by routes which can also yield the corresponding alcohols and benzoic acids. Pavlik and Tatayanon (*40*) have demonstrated that lactams can be oxidized to imides following steps similar to those of the Fenton reaction. Various factors controlling the yield and specificity of reaction products have been investigated, such as the photophysics of the various excited-states

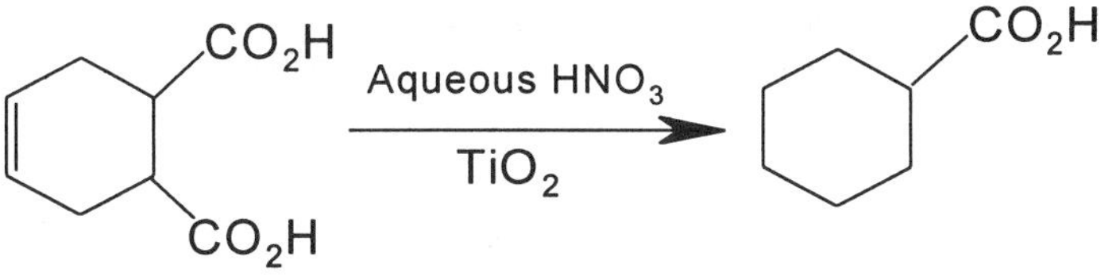

Scheme I. Photoreduction initiated by TiO_2 particles using aqueous HNO_3 (*34*)

and reactions occurring at the titanium dioxide/non-aqueous solution interface. The photooxidation of m-phenoxytoluene with colloidal TiO_2 catalyst is very selective and produce almost exclusively m-phenoxybenxaldehyde (*14*). Moreover, m-phenoxybenzaldehyde is not further oxidized to m-phenoxybenzoic acid. Selectivity in oxidation can be influenced by the environment, reflecting kinetic influences on reaction partitioning by the catalyst, adsorption effects, and trapping of surface-bound reactive intermediates (*41*). Such extra thermodynamic control of redox chemistry for systems containing semiconductor catalysts was previously ascribed to differences in the adsorption of the various oxidizable species adsorbed on semiconductor surface.

Safety Considerations

The two major safety concerns are explosions from the gas phase mixing of hydrocarbons and oxygen (air) at high temperatures, and the use of UV light The violent oxidations of many low boiling point hydrocarbons simply produce CO_2 and H_2O. However, when oxidation is gentle and carefully controlled, they can also produce useful quantities of partial oxygenates. Because of the danger of explosion, one needs to operate either below the lower explosion (LEL) or above the upper explosion limits (UEL). The safe range of operation for cyclohexane oxidation at 60°C is outside the shaded region as shown in Figure 2. The other safety concern is the danger of ultraviolet light to the eyes and skin. Cataracts and skin cancer are caused by UV-B and even UV-A. That is not to say that UV light cannot be used safely. The main precaution to avoid looking directly at a UV light source and to prevent UV light from shining directly on one's skin. Since most substances, including paper efficiently absorb UV radiation and do not efficiently reflect it, there is usually little danger in looking at an object that is being illuminated with UV light. Proper design of the system and judicious selection of materials can easily prevent UV light exposures.

Environmental Advantages of Photocatalyzed Chemical Synthesis

Synthesis of oxygenates using photocatalytic oxidation on TiO_2 surfaces creates the potential of using a low cost and non-toxic catalyst at much lower temperatures and

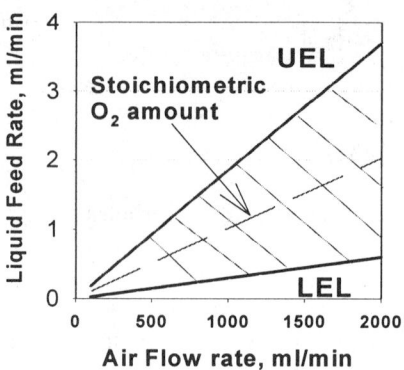

Figure 2. Safe range of operation for photocatalytic operation of cyclohexane

pressures than the current industrial methods. Release from a typical oxidation process occurs as fugitive gaseous emissions, vent gases, wastewater contaminats, and as catalyst residue. A comparative waste generation study was conducted for the conventional and photocatalyzed production of cyclohexanol and cyclohexanone by oxidation of cyclohexane. Cyclohexanol and cyclohexanone (AK oil) are primarily used, either isolated or in a mixture, in the production of nylon intermediates (adipic acid and caprolactam). If the desired product is cyclohexanol, boric acid is used as a catalyst; however, if cyclohexanone is desired, nickel is commonly used as a catalyst (*42*). The Toxic Release Inventory (TRI) was used as a source to inventory the waste generated and reduced for an ideal plant with a 100,000 kg/year capacity. About 5 kg of boric acid is used per ton of cyclohexane oxidation. The aqueous phase containing carboxylic acid by-products must be neutralized and separated from the AK oil. This creates about 200 kg of sodium salt and 10 kg of sodium hydroxide per ton of AK oil produced (*31*). Therefore, a successful application of photocatalyzed synthesis of AK oil from cyclohexane would annually eliminate 2×10^7 kg of potential water pollutants. This synthetic alternative, if developed for other products could prevent a large amount of pollution at the source.

Conclusions

The increasing local and global concern for environmental pollution offers incentive to explore new catalytic materials for safeguarding the environment. Catalysts have been applied in the design of benign synthetic processes to eliminate or minimize pollution at the source, as well as in the remediation of a variety of existing environmental pollution problems. The photoxidation technology developed and tested at US EPA, National Risk Management Research Laboratory, is directed towards the use of ultraviolet light with a specially prepared catalyst, titanium dioxide, to synthesize high-value organic compounds from cheap raw materials which include linear, cyclic and aromatic hydrocarbons. The photocatalytic oxidation of many organic molecules, including saturated hydrocarbons, by optically-excited semiconductor oxides like TiO_2 is thermodynamically feasible in the presence of oxygen. Selectivities different from those obtained by other oxidation means have been obtained with a sufficient amount of quantum yields showing the potential of this method for clean syntheses. Photocatalyzed oxidation of hydrocarbons has the potential to prevent pollution at the source by replacing the conventional oxidation catalysts, heavy metals, and strong acids with an 'environmentally friendly' catalyst, titanium dioxide. TiO_2 prepared using the flame aerosol coating method has shown high activities (*38*). This new technology has the ability to reduce water pollutants such as dissolved organic salts, generated by the neutralization of homogeneous acid catalysts in conventional processes. The success of this technology could potentially lead to the use of sun light and molecular oxygen for partial oxidation of hydrocarbons for synthesis of commercially important chemicals.

Acknowledgments

The authors would like to thank Tom Deinlein for collecting the pilot plant data and Diem Tran for helping in the chemical analysis.

Literature Cited

1. Allen and Jain, *U.S. EPA*, 199, **1992**.
2. *Fortune*, 19, September **1994**.
3. Pollution Prevention, *A National Progress Report*, EPA 742-R-97-00, p.49, **1997**
4. Shen T.T. *Industrial Pollution Prevention*, Berlin: Springer, P. 5, **1995**.
5. Sikdar, K. S., Howell, G. S., *J. Clean Production*, 6, 253-259, **1998**.
6. Anastas, P.T. and Warner, J.C., Green Chemistry: Theory and Practice. Oxford University Press, **1998**.
7. Bard, A. J., Photoelectrochemistry and Heterogeneous photocatalysis at Semiconductors, *J. Photochemistry,* 10, pp59-75, **1979**.
8. Tseng, J., and Huang, C. P., *American Chemical Society Symposium Series*, Edited Tedder, D. W., Pohland, F. G., ACS 422, pg 12., **1990**.
9. Fox, M. A., *Acc. Chem. Res.*, 16, 314-321, **1989**.
10. Izumi, I., Dunn W. W., Wilbourn, K. O., and Fan, F. R. F. and Bard, A. J., *J. Phys. Chem.*, 84, pp3207-3210, **1980**.
11. Fujihira et al., *Nature,* 293, 206, **1981**.
12. Fujihira, M., *Chem. Letter*, Chemical Society of Japan, 1053, **1981**.

13. Sakata, T., In *Photocatalysis - Fundamentals and Applications,* Ed. N. Serpone and E. Pelizzetti, publisher Jon Wiley & Sons, Inc., New York, **1989.**
14. Fox., M .A., and Dulay, M. T., *Chem. Rev.,* 93, 341-357, **1993.**
15. Schiavello, M. and Sclafani, A., In *Photocatalysis- Fundamentals and Applications,* Ed. N. Serpone and E. Pelizzetti, pub. Jon Wiley & Sons, Inc., N.Y., **1989.**
16. Pourbaix, M., *Atlas of Electrochemical Equilibria in Aqueous Solutions,* Ed. J. A. Franklin, Published by Pergamon Press, New York, **1966.**
17. Munuera, G., Espimos, J. P., Fernandez, A., Malet, P., Gonzalez-Elipe, A. R., *J. Chem. Soc. Faraday Trans.,* 86, 20, 3441, **1990.**
18. D'Olivera J.C, Al-Sayyed G., Pichat P., *Env. Sci. Technol.,* 24, 990-996, **1990.**
19. Hoffman, M. R., Martin, S. T., Choi W, and Bahnemann D. W., *Chem. Rev,* 95, 69-96,**1992.**
20. Beck, D. D., Siegel, R.W. *J. Electroanal. Chem.,* 386, 229-233,**1995.**
21. Owens, T. M., Wu S. Y., and Biswas, P., *Chem. Eng. Comm.,* 133, 31-52,**1995.**
22. Biswas, P., and Zachariah, M. R., *Envion. Sci. Technol.,*31, 2455-2463, **1997.**
23. Gonzalez, A. M., Howell, G. S., and Sikdar, S., *J. Catalysis,* 183, 159-162, **1999** .
24. Kraeutler, B., and Bard, A. J., *J. Am.. Chem. Soc.,* 100, 2239, **1978** .
25. Sahle-Demessie, E., Gonzalez, M. A., Sikdar, S., Harten T. presented at the 2nd Annual Green Chemistry and Engineering Conf., Washington DC., June 30, **1998.**
26. Maldotii, A.; Bartocci, C.; Amadelli, R., Polo, E.; Battioni, P.; Mansuy, D. J. Chem. Soc., Chem. Commun. 1487, **1991.**
27. Groves, J. T.; Quinn, R. *J. Am. Chem. Soc.,* 107, 5790, **1985.**
28. Ollis, D. F.; Hsiao, C. Y.; Budiman, L.; Lee, C. L., *J. Catalysis,* 88, 89, **1984.**
29. Fox, M. A. *Top. Curr. Chem.,* 142, 172, **1991.**
30. Fujihara, M., Satoh, Y., and Osa, T., Bull. Chem. Soc. Japan, 55, 666, **1982.**
31. Versar, *Analysis and Characterization of Chemical Manufact. Ind.,* Vol II, **1982**
32. Pichat, P. and Fox, M. A., In *Photocatalysis - Fundamentals and Applications,* Ed. N. Serpone and E. Pelizzetti, publisher Jon Wiley & Sons, Inc., New York, **1989.**
33. Weaver, S., and Mills, G., *J. Phys. Chem., B,* 101, 3769-3775, **1997.**
34. Fox, M. A., Chen, C. C., Park, K. H., Younathan, *Am. Chem. Soc. Symp. Ser.,* 278, 69, **1985.**
35. Lin, L.; Kuntz, R. R. *Langmuir,* 8, 870, **1992.**
36. Ravishankara, A. R., Lovejoy, E. R., *J. Chem. Soc., Faraday Trans.,* 90,2159,**1994.**
37. Mills, A., Le Hunte, S., *J. Photochem. and Photobiol. A: Chemistry,* 108, 1-35, **1997.**
38. Sahle-Demessie, E., Gonzalez, M., Wang, Z., Biswas, P., *Ind. Eng. & Chem. Research,* vol. 38, No. 9, 3276-3284, **1999.**
39. Pavlik, J. W. and Tantayanon, S., *J. Am. Chem. Soc.,* 103, p6755, **1981.**
40. Fox, M. A., and Abdel-Wahab, A. A., *Tetrahedron Letters,* 31, 32, 4533-4536, **1990.**
41. Stahl, F., Chemical Economics Handbook, 68.7000, SRI, **1998.**

Green Solvent Systems

Chapter 19

The Design of Technologically Effective and Environmentally Benign Solvent Substitutes

Renhong Zhao[1,3], Heriberto Cabezas[1,4], and Subba R. Nishtala[2]

[1]National Risk Management Research Laboratory, Sustainable Technology Division, Systems Analysis Branch, U.S. Environmental Protection Agency, 26 West Martin Luther King Drive, Cincinnati, OH 45268
[2]Center for Environmental Analysis, Research Triangle Institute, 3040 Cornwallis Road, Research Triangle Park, NC 27709-2194

There is presently considerable interest in finding environmentally benign replacement solvents that can perform in many different applications as solvents normally do. The most efficient means of doing this is to design chemical mixtures to replace undesirable solvents. This requires chemical mixtures with desirable properties such as the ability to dissolve certain compounds but without other undesirable properties such as toxicity. This complex problem can be simplified because the behavior of solvents in all applications is governed by universal mathematical expressions. The identity of the solvent in these expressions is represented by coefficients such as viscosity, density, activity coefficients, etc. Thus, the problem of designing environmentally benign solvents has been addressed by using property prediction and phase equilibrium analysis methods in a computer algorithm to design mixtures with specific properties. A classification of the necessary properties for solvent design is presented. The property categories include general, dynamic, and equilibrium properties, environmental requirements (i.e., a VOC index and an environmental impact index), and performance requirements. These concepts form the basis of the PARIS II solvent design software, which is near commercialization.

There are two principal approaches for dealing with environmentally undesirable solvents: (1) finding alternative technologies that do not use solvents and (2) finding alternative solvents. Finding alternative solvents is often preferable because it uses existing processes and equipment with minor or no modifications. Unfortunately, the

---

[3]Research Associate, National Research Council.
[4]Corresponding author.

solvent alternative approach presents serious technical challenges because one must design substitute solvents that are as effective as current solvents but have better environmental properties, such as reduced toxicity. This is the challenge presented by the environmentally benign solvent design problem. Research efforts in the design of environmentally benign solvents can be roughly classified into three categories: (1) screening available solvent databases for single chemical substitutes, which includes the Solvent Substitution for Pollution Prevention (*1*), the Solvent Database Software Program (*2*), the Tool for Systematic Solvent Screening for Batch-Process Development and Revamping (*3*); (2) designing new chemicals that meet specified requirements, which includes Molecular Design of Solvents for Liquid Extraction based on UNIFAC (*4*), Designing Molecules Possessing Desired Physical Property Values (*5*), Computer-Aided Molecular Design by Group Contribution (*6*), Group Contribution Approach to Computer Aided Molecular Design (*7*), Design of Optimal Solvents for Liquid-Liquid Extraction and Gas Absorption Processes (*8*), Computer-Aided Molecular Design of Solvents for Separation Processes (*9*), and Design of Environmentally Safe Refrigerants Using Mathematical Programming (*10*); and (3) designing mixture replacements if no single chemical replacements are found or available, which includes Solvent Selection by Computers (*11*), Computer Aided Mixture Design with Specified Property Constrains (*12*), Computer Program for Assisting the Design and Replacement of Environmentally Objectionable Solvents (*13*), Computer Aided Solvent Substitution for Pollution Prevention (*14*), and Molecular Thermodynamics in the Design of Substitute Solvents (*15*). For the first category, the main advantage is relative simplicity, and the disadvantage is that the screening cannot always find substitutes because the information provided by the database is limited or because no existing chemicals meet the requirements. For the second category, the main advantage is that the computer program can design entirely new chemicals when no single chemical replacements are found or available, and the disadvantage is the increased complexity of molecular design. For the third category, the main advantage is the ability to tailor solvent properties precisely by using mixtures, and the disadvantage is the complexity of calculating and manipulating mixture properties. The focus of this paper is the third category, designing mixture replacements, and the objective is to discuss the concepts and ideas that are important for the design of substitute solvents, including physical properties and environmental effects.

Substitute Solvent Design Theory

Technological Performance: Theory

In designing substitute solvents, one needs to match or improve all aspects of solvent behavior, including equilibrium, dynamic, and phase behavior, while meeting or exceeding all technical and environmental performance requirements. This is basically a mapping of the behavior of one solvent into that of another in a space where the coordinates are physical properties and performance requirements. To illustrate the significance of this point, consider, for example, the isothermal equations of motion given below in Cartesian coordinates for the x, y, and z directions respectively. These expressions are,

$$\rho\left[\frac{\partial v_x}{\partial t}+v_x\frac{\partial v_x}{\partial x}+v_y\frac{\partial v_x}{\partial y}+v_z\frac{\partial v_x}{\partial z}\right]=-\frac{\partial P}{\partial x}+\frac{\partial}{\partial x}\left[2\mu\frac{\partial v_x}{\partial x}-\frac{2}{3}\mu(\nabla\cdot\vec{v})\right]$$
$$+\frac{\partial}{\partial y}\left[\mu\left(\frac{\partial v_x}{\partial y}+\frac{\partial v_y}{\partial x}\right)\right]+\frac{\partial}{\partial z}\left[\mu\left(\frac{\partial v_z}{\partial x}+\frac{\partial v_x}{\partial z}\right)\right]+\rho g_x \quad (1)$$

$$\rho\left[\frac{\partial v_y}{\partial t}+v_x\frac{\partial v_y}{\partial x}+v_y\frac{\partial v_y}{\partial y}+v_z\frac{\partial v_y}{\partial z}\right]=-\frac{\partial P}{\partial y}+\frac{\partial}{\partial y}\left[2\mu\frac{\partial v_y}{\partial y}-\frac{2}{3}\mu(\nabla\cdot\vec{v})\right]$$
$$+\frac{\partial}{\partial x}\left[\mu\left(\frac{\partial v_y}{\partial x}+\frac{\partial v_x}{\partial y}\right)\right]+\frac{\partial}{\partial z}\left[\mu\left(\frac{\partial v_z}{\partial y}+\frac{\partial v_y}{\partial z}\right)\right]+\rho g_y \quad (2)$$

$$\rho\left[\frac{\partial v_z}{\partial t}+v_x\frac{\partial v_z}{\partial x}+v_y\frac{\partial v_z}{\partial y}+v_z\frac{\partial v_z}{\partial z}\right]=-\frac{\partial P}{\partial z}+\frac{\partial}{\partial z}\left[2\mu\frac{\partial v_z}{\partial z}-\frac{2}{3}\mu(\nabla\cdot\vec{v})\right]$$
$$+\frac{\partial}{\partial x}\left[\mu\left(\frac{\partial v_z}{\partial x}+\frac{\partial v_x}{\partial z}\right)\right]+\frac{\partial}{\partial y}\left[\mu\left(\frac{\partial v_z}{\partial y}+\frac{\partial v_y}{\partial z}\right)\right]+\rho g_z \quad (3)$$

where ρ is the solvent density, and v_x, v_y, v_z are the fluid velocity in the x, y, z direction, P is the pressure, μ is the solvent viscosity, $\vec{v}$ is the velocity vector, and where g_x, g_y, and g_z represent the acceleration of gravity in the x, y, and z directions respectively. eqs 1, 2, and 3 together with the appropriate initial and boundary conditions embody all the motion that a fluid could possibly exhibit in all its applications; these mathematical expressions are generic for all fluids. The identity of the fluid in eqs 1, 2, and 3 appears only in the density ρ and the viscosity μ. Therefore, if two solvents have the same value for density and viscosity, their motion as represented by these expressions will be the same under similar conditions. However, there is no need to solve these expressions to design a solvent.

Under non-isothermal conditions, more complex expressions apply but the same principle for matching the behavior of one fluid to that of another still applies. To further illustrate this point, consider the simplest form of the transport expressions, which are Newton's law for momentum transport, Fourier's law for energy transport, and Fick's law for mass transport given by eqs 4, 5, and 6, respectively:

$$\tau_{yx} = -\mu \frac{dv_x}{dy} \qquad (4)$$

$$q_x = -\lambda \frac{dT}{dx} \qquad (5)$$

$$J_{Ax} = -D_{AB} \frac{dC_A}{dx} \qquad (6)$$

These three simple expressions plus momentum, energy, and mass balances can be used to derive general transport equations that can, in principle, represent solvent dynamic behavior in all possible uses or applications. The difference between one application and another depends on the initial and boundary conditions and other assumptions that are imposed on the equations. There is extensive literature (16) in this area. The important point is that, in these three simple equations as well as in the more general transport expressions, the fluxes and driving forces do not depend on the individual components present: the form of these equations is independent of the particular chemical components present. The identity of the components appears only in the proportionality factors μ, λ, and D_{AB}. In fact, because all the possible dynamic behavior of a solvent is embodied in the three modes of transport, if two fluids have the same values for these three quantities, then all of the dynamic behavior as represented by these expressions will be the same for both fluids.

This type of reasoning also applies to equilibrium behavior and to other solvent properties and requirements. For example, the central expression in computing the phase equilibrium behavior of solvents is the fugacity expression of eq 7 below,

$$f_i^{(l)} = x_i \gamma_i f_i^o \qquad i = 1,2\ldots n \qquad (7)$$

where $f_i^{(l)}$ is the fugacity of component i in a liquid mixture, x_i is the mole fraction of a component i, γ_i is the activity coefficient of component i, and f_i^o is the reference fugacity of component i that can often be set equal to the vapor pressure, P_i^s. Again, the form of eq 7 is the same for all solvents, and the identity of the solvents appears only in γ_i and f_i^o. Reid et al. have written a general review of fluid physical properties and property prediction methods (17).

Technological Performance: Fluid Properties

The following set of properties generally suffices to match the bulk behavior of one solvent to that of another: density, vapor pressure, boiling temperature, surface tension, thermal conductivity, viscosity, and diffusivity. For example, matching the density insures that fluid handling equipment would not be unduly overloaded by the substitution; matching the vapor pressure insures that evaporation rates and design pressures are matched; matching the boiling temperature insures that the replacement remains a liquid at operating conditions; matching the surface tension insures proper wetting of surfaces; matching the thermal conductivity insures proper heat transfer; matching the viscosity insures that mixing and fluid motion occurs as before; and matching the diffusivity helps to match performance in separations, chemicals reactions, and other operations. One could, of course, think of other properties that should be considered, but these have been adequate in the authors' experience.

In addition to matching bulk physical properties as already mentioned, it is also necessary to consider the activity coefficients to insure that the molecular interactions between the solutes and the solvent in the original and the substitute are generally similar. This insures that proposed substitute solvents will likely dissolve the same solutes and have similar effects to those of the original solvent. However, it is important to match only the activity coefficients of the solutes in the solvents at infinite dilution (zero solute concentration), so as not to include solute-solute interactions. The authors matched the activity coefficients at infinite dilution of a representative from six chemical families: alcohols, ethers, ketones, water, normal alkanes, and aromatics, i.e., they have matched these activity coefficients in the solvent to be replaced to those in the replacement solvent. The particular components used are ethanol, diethyl ether, acetone, water, normal octane, and benzene. However, one could conceivably use different compounds successfully. Activity coefficients can be estimated from group contribution methods (17).

Last, there are also other fluid properties such as molecular mass and flash point that need to be considered. Molecular mass is used to capture any other properties or behavior not already covered. The flash point is actually a safety-related property that helps to insure that the replacement solvent will not present more of a fire hazard than the original solvent did. This is not a simple question for the case of liquid mixtures because the volatility of the different components can vary due to the non-idealities.

Environmental Indexes

The environmental impacts of both the solvent to be replaced and the replacement solvent are considered using two indexes: an air index ψ_s^{AIR} and an overall environmental index ψ_s. Since the object here is to formulate substitute solvents that have better environmental performance, the indexes for the solvent to be replaced are not matched but are rather treated as an upper bound on the indexes of the acceptable replacement. This insures that the replacement solvent is environmentally better than the original solvent as measured by the indexes. The inherent toxic effects of the solvent and the toxic effects due to volatile organic emissions are considered separately because, when chemicals are mixed, their volatility changes due to the non-idealities in the mixture. Therefore, a chemical that has low risk by inhalation due to low volatility in pure form can have a much higher volatility and a much higher risk when mixed with other chemicals. The air index, ψ_s^{AIR}, which represents this effect, is defined by,

$$\psi_s^{AIR} = \frac{\sum_i x_i \gamma_i P_i^{vap} \psi_i M_i}{P \sum_i x_i M_i} \tag{8}$$

where ψ_s^{AIR} has units of potential environmental impact (*PEI*) per kilogram of solvent s, x_i is the mole fraction of chemical i, γ_i is the activity coefficient of chemical i, P_i^{vap} is the vapor pressure of chemical i, ψ_i is the environmental impact index for chemical i, P is the pressure, M_i is the molecular mass of i, and the summation is taken over all the chemicals making up the solvent. The index, ψ_i, which is used by itself and in the calculation of ψ_s^{AIR} is estimated from,

$$\psi_i = \sum_j \alpha_j \psi_{ij}^s \tag{9}$$

where ψ_i has units of *PEI* per kilogram of i, α_j is the dimensionless relative weighting factor for environmental impact category j, ψ_{ij}^s is the normalized specific environmental impact score of chemical i with units of *PEI* per kilogram of i for environmental impact category j, and the summation is taken over all the categories of environmental impacts (e.g., human toxicity, ozone depletion, etc.). ψ_{ij}^s is calculated from,

$$\psi_{ij}^s = \beta_j \frac{(Score)_{ij}}{\langle (Score)_{ij} \rangle_j} \tag{10}$$

where $\beta_j = 1$ with units of *PEI* per kilogram of i per units of $(Score)_{ij}$ per $\langle (Score)_{ij} \rangle_j$ that are specific to each impact category j, $(Score)_{ij}$, is a measure of chemical potential environmental impact such as $1/LD_{50}$, $1/LC_{50}$, etc. for chemical i in impact category j and has units that change according to impact category j, and $\langle (Score)_{ij} \rangle_j$ is the arithmetic average of all the $(Score)_{ij}$ in impact category j and has units that again change according to j. It should be noted that β_j converts the units of $(Score)_{ij}$ per $\langle (Score)_{ij} \rangle_j$ to units of *PEI* per kilogram of chemical i. Setting $\beta_j = 1$ implies that the value of ψ_{ij}^s for the average chemical in the database is 1, and that chemicals with values above and below 1 have respectively above and below average potential environmental impact for impact category j. The normalization procedure of eq 10 gives all of ψ^s_{ij} the same units and places them on the same scale where the value of ψ^s_{ij} for the average chemical in the database is 1. This makes it possible to add the different ψ^s_{ij}'s across impact categories according to eq 9.

For pure chemicals, eq 9 is used to calculate the overall environmental index of chemicals so that $\psi_s = \psi_i$. For mixtures, however, a mass fraction weighted average of the environmental index of individual chemicals is used as shown by,

$$\psi_m = \sum_i W_i \psi_i = \sum_i W_i \sum_j \alpha_j \psi^s_{ij} \qquad (11)$$

where ψ_m is the environmental index of the mixture m and W_i is the mass fraction of chemical i. It is critical to note, however, that eq 11 does not account for the synergistic toxic effects that occur when different chemicals are mixed.

Chemical Environmental Impacts

The chemical environmental impacts used here are based on the work of Young (18, 19). The $(Score)_{ij}$ measures of chemical environmental impacts used in eq 10 above fall into four general categories: (1) human toxicological impacts (human toxicity potential by ingestion or *HTPI* and human toxicity potential by dermal or inhalation exposure or *HTPE*), (2) ecological toxicological impacts (aquatic toxicity potential or *ATP* and terrestrial toxicity potential or *TTP*), (3) regional environmental impacts (acidification potential or *AP* and photochemical oxidation potential or *POP*), and (4) global environmental impacts (global warming potential or *GWP* and ozone depletion potential or *ODP*).

The $(Score)_{ij}$ for the two human toxicological and the two ecological toxicological impact categories were obtained from measures of toxicity as discussed below. For human toxicity potential by ingestion (*HTPI*) and terrestrial toxicity potential (*TTP*) for a chemical i were determined by,

$$(Score)_{i,HTPI} = (Score)_{i,TTP} = \frac{1}{(LD_{50})_i} \qquad (12)$$

where LD_{50} is the oral dose of chemical i that produced a 50% death rate in rats. The LD_{50} for rats is used for both *HTPI* and *TTP* out of simple necessity since the results of toxicological tests on rats are normally extrapolated to humans. For most chemicals there is no toxicity data directly measured on humans. However, *HTPI* and *TTP* do represent toxic effects on two different types of species, humans and non-human terrestrial animals, and one can not, therefore, combine the two into one. The

$(Score)_{ij}$ for aquatic toxicity potential (*ATP*) for a chemical *i* were determined by,

$$(Score)_{i,ATP} = \frac{1}{(LC_{50})_i} \qquad (13)$$

where LC_{50} is the concentration that produced a 50% death rate in *Pimephales promelas* (fathead minnows). The scores for human toxicity by dermal/inhalation exposure (*HTPE*) were obtained from time-weighted averages of threshold limit values.

The $(Score)_{ij}$ for the two regional environmental and the two global environmental impact categories were obtained from Heijungs et al. (20) and directly inserted into eq 10. The acidification potential (*AP*) was obtained by dividing the rate of release of H^+ promoted by chemical *i* by the rate of release of H^+ promoted by SO_2. The photochemical oxidation potential (*POP*) was obtained by dividing reaction rate of a unit mass of chemical *i* with hydroxyl radical ($OH\bullet$) by the rate for ethylene. The global warming potential (*GWP*) was obtained by dividing the absorption of infrared radiation by a unit mass of a chemical *i* by the absorption by a unit mass of CO_2 over a span of 100 years. The ozone depletion potential (*ODP*) was obtained by dividing the rate at which a unit mass of chemical *i* reacts with ozone to form molecular oxygen by the rate of the same reaction for CFC-11 (trichlorofluoromethane). Although the scores from Heijungs et al. (20) are dimensionless, it is necessary to normalize them according to eq 10 to place them on the same scale where the value of ψ_{ij}^s for the average chemical in the database is 1.

The PARIS II Algorithm

The theory discussed above has been implemented in the form of a computer program entitled PARIS II. This is an acronym for Program for Assisting the Replacement of Industrial Solvents, Version 2. This software embodies a sophisticated solvent design algorithm, a suite of fluid property prediction routines used by the algorithm, the AIChE/NIST DIPPR chemical physical property database, a US EPA developed health and environmental impact database (18), and the UNIFAC library/program from the Technical University of Denmark (17) for predicting activity coefficients. The PARIS II solvent design algorithm consists of nine steps, represented in the program interface by nine different screens. There are, in addition, a number of auxiliary screens associated with these nine major ones. The major steps of the PARIS II algorithm are:

Step 1. Define the composition and operating conditions of the current solvent.

The algorithm starts by defining the composition, the operating temperature, and the operating pressure of the solvent to be replaced. The solvent to be replaced may consist of up to nine chemicals chosen by name, family, or CAS number. PARIS II uses the operating temperature and pressure to insure that only chemicals that can exist as liquids under these conditions are used in the replacement. It also uses the temperature and pressure to compute the various properties of the solvent to be replaced and the replacement.

Step 2: Adjust relative weighting of the toxicity and environmental impact categories.
The different categories of potential environmental impact are weighted relative to each other on scale of 0 to 10 using the α_j values defined in eqs 9 and 11. The default value for α_j is 5. Adjusting the α_j values allows the user to construct appropriate environmental profiles for different regions and applications. For instance, the user may want to over-weight photochemical oxidation in a region suffering from smog and under-weight acidification in an area with few lakes and rivers, such as a desert.

Step 3. Compute the static, dynamic, performance, and environmental properties of the solvent to be replaced.
PARIS II estimates the relevant static, dynamic, performance, and environmental properties of the solvent to be replaced. It does this using the solvent composition, the operating temperature and pressure, and a suite of built-in estimation routines.

Step 4. Design the criteria to be met by the replacement solvent.
Starting with the properties of the solvent to be replaced, the user assigns the desired value and tolerance bounds for each of the solvent properties used by PARIS II. PARIS II essentially looks for replacement solvents that have properties as close as possible to the desired values within the tolerance bounds. The default desired values are those of the solvent to be replaced and the default tolerance limit is ±20%. However, the user may change both the desired values and the tolerance as needed for all the properties except the air index and the environmental index. For latter two, the lower bounds are zero, the desired values are zero, and the upper bounds are the index values for the solvent to be replaced. This insures that the replacement solvent will be environmentally better than the solvent to be replaced.

Step 5. Search for a single chemical replacement.
The PARIS II algorithm initially searches its database of pure chemicals, along with the information provided by the environmental/toxicological impact database and the UNIFAC library, for a single chemical that can meet all of the requirements. If one or more adequate replacements are found, PARIS II displays them in ranked order and compares the properties of each to those of the solvent to be replaced.

Step 6. View Single Chemical Properties.
If no adequate single chemical replacement is found, PARIS II displays the chemicals in its solvent database in ranked order and compares their property values to the desired values. This way, it is possible to introduce human judgment and experience to ascertain the adequacy of the computer-generated recommendations.

The ranking of a any particular chemical is based on how close its properties come to the desired values.

Step 7. Select chemicals for a solvent mixture.

If PARIS II has not found an adequate single chemical even after the analysis of Step 6, then the user adds chemicals from the ranked list one at time to make up a mixture. Only one chemical at a time is added so that the resulting replacement solvent will have the minimum number of components. The user may select any of the chemicals in the database with the ranking given by PARIS II used as a guide.

Step 8. Design q mixture replacement solvent

When at least two chemicals are selected for a proposed mixture, PARIS II calculates the properties of the mixture in small increments of composition. PARIS II then compares the mixture properties to see if there are any compositions whose properties lie within the tolerance ranges set for the solvent design requirements. If one or more compositions satisfy the requirements, PARIS II ranks them, shows their properties, and compares the properties to the solvent design requirements. If there are no compositions that satisfy the requirements, PARIS II again ranks the compositions, shows their properties, and compares them to the solvent design requirements. This allows human knowledge of chemistry to be brought to bear on the problem. For example, the user could further consider whether some of the unmet requirements are or are not truly important for a particular solvent application.

Step 9. Change the chemical candidates of the mixture.

If, in fact, none of the compositions considered meet the requirements of a particular application, then it is necessary to return to Step 7 to select either additional chemicals for the mixture or to change the candidate chemicals. Afterwards, one goes through Step 8 again as described above.

Case Study: Methyl Ethyl Ketone Replacement

To illustrate the use of the solvent replacement design theory and the PARIS II algorithm already discussed, we consider the design of a mixture replacement for methyl ethyl ketone (MEK). This is a widely-used solvent that is known to have serious effects on both human health and the environment; as a result, MEK is included in the regulatory listings of the U.S. Environmental Protection Agency under the Toxics Release Inventory (21) and the Clean Air Act (22). As already discussed, the design of an environmentally benign replacement solvent involves technological effectiveness and reduction of adverse environmental impacts.

Technological Effectiveness

The technological effectiveness of the replacement solvent is determined by considering the physical and performance properties of the solvent to be replaced and the replacement. Here we consider a replacement mixture to match the properties of MEK as closely as possible within a tolerance of 20%. It is also possible and often desirable not to match the exact property values of the solvent to be replaced, but to look for other values that would improve the performance of the replacement in a particular application. For example, if the solvent needs to wet surfaces as thoroughly as possible, one could consider lowering the surface tension below that of MEK. Table I shows the physical and performance properties of MEK and of the mixture replacement, calculated using the PARIS II software (15). Note that the properties of the replacement mixture are quite close to those of MEK.

TABLE 1. Physical Properties of Methyl Ethyl Ketone (MEK) and the Replacement Mixture of 90 mole % Ethyl Acetate and 10 mole % Ethanol.

Properties	MEK	Replacement Mixture
Molecular Mass (kg/kmol)	72.1	83.9
Mass Density (kg/m^3)	800	887
Boiling Temperature (K)	353	348
Vapor Pressure (kPa)	12.3	13.3
Surface Tension (N/m)	0.0240	0.0231
Viscosity x10^4 (N-s/m^2)	3.96	4.66
Thermal Conductivity (W/m-K)	0.145	0.145
Flash Point (K)	267	268

TABLE II. Infinite Dilution Activity Coefficients of Representative Solutes in Methyl Ethyl Ketone (MEK) and the Replacement Mixture of 90 mole % Ethyl Acetate and 10 mole % Ethanol.

Representative Solutes	MEK	Replacement Mixture
Ethanol	2.24	N/A
Diethyl Ether	1.46	1.25
Acetone	1.01	1.03
Water	7.38	8.40
Octane	5.26	4.61
Benzene	1.32	1.25

Table II shows the activity coefficients of six representative solutes at infinite dilution in MEK and in the replacement mixture as calculated by PARIS II. The six solutes are representatives of six different chemical families: alcohols, ethers, ketones, polar inorganics, normal alkanes, and aromatics. The activity coefficients at infinite dilution represent the characteristic molecular interactions between members of each of the six chemical families and MEK, and members of each of the six chemical families and the replacement mixture. Therefore, matching the two sets of activity coefficients, representative solutes in MEK and representative solutes in replacement mixture, matches the molecular interactions, and the consequent non-idealities in the liquid mixture. This is important for many applications including liquid extraction, cleaning, and chemical reactions. Note that there is a reasonably good match between the activity coefficients of the six chemicals in MEK and the replacement mixture.

Environmental Indexes

As already discussed, the potential environmental impact of the solvent is measured by the values of the Air Index, ψ_s^{AIR}, and the Environmental Index, ψ_m. The values for both of these indexes for MEK and the replacement mixture are shown in Table III below. The environmental indexes are treated differently than other properties: (1) the desired value for each index is not the value for MEK but zero and (2) the value of each index for MEK is treated as the upper bound on the respective index for the replacement solvent. Setting the desired value to zero implies that, all other considerations being equal, the lower the values of the indexes, the more desirable the solvent. Setting the value of the indexes for MEK as an upper bound insures that the replacement solvent will be environmentally better than MEK as measured by the two indexes.

TABLE III. Environmental Indexes for Methyl Ethyl Ketone (MEK) and the Replacement Mixture of 90% Mole Ethyl Acetate and 10% Mole Ethanol.

Indexes	MEK	Replacement Mixture
ψ_s^{AIR}, Air Index (Impact/kg)	0.7	0.4
ψ_m, Environmental Index (Impact/kg)	6	3

Summary

This paper introduces a systematic method of computer-aided solvent substitute design for pollution prevention. The method is based on the theoretical observation that the mathematical expressions representing solvent behavior, equilibrium and non-equilibrium, are generic for all solvents. These expressions contain the identity of the solvent only in certain coefficients representing physical properties such as viscosity, density, activity coefficients, etc. Matching these coefficients, therefore, matches the behavior of the original solvent to that of the replacement regardless of application. This is an important conclusion because it indicates that, in principle, it is possible to find true "drop-in" replacement solvents. One should note that the methodology is applicable to the design of solvent replacement mixtures as well as screening for pure component replacement solvents. This is particularly important because one can not completely replace one pure component solvent with another pure component solvent. The reason is that no two chemicals have exactly similar properties and behavior. It is possible, however, to design a mixture that matches another mixture or a pure component solvent. These considerations also give rise to a set of criteria and a set of properties necessary to design replacement solvents. Thus, the infinite dilution activity coefficients insure solvent effectiveness, i.e., its ability to perform its intended technical function such as dissolving solutes, extracting specific compounds, etc. Bulk physical properties such as density, viscosity, etc. insure that the replacement solvent will not require major changes to current equipment and processes. Two environmental indexes insure that the replacement solvent has lower potential environmental and human health impacts than the solvent it is replacing. The flash point is used to insure that replacement solvent does not present more of a safety hazard than the solvent it is replacing. Another important conclusion is that a successful solvent substitution for pollution prevention must incorporate information from different fields in a multiple-step decision and selection process. For example, a replacement solvent designed using only the activity coefficients may perform its intended function adequately, but it may well present serious practical operational difficulties. Lastly, it should emphasized that this methodology reduces the time required to find replacement solvents from months to hours. It should, therefore, lead to the creation of many new technically effective solvents that have minimal potential impact on human health and the environment.

Acknowledgments

Renhong Zhao acknowledges the support of the National Research Council Associateship Program. Renhong Zhao and Heriberto Cabezas also wish to acknowledge the encouragement and support of S.K. Sikdar, the Director of the Sustainable Technology Division at the National Risk Management Research Laboratory.

Literature Cited

1. Joback, K. G. In *Pollution Prevention via Process and Product Modifications*; Gaden, E. L., Ed.; AIChE Symposium Series; American Institute of Chemical Engineers: New York, 1994; Vol. 90, pp 98.
2. Hermansen, R. D. *Development of a Solvent Database Software Program;* Pollution Technology Review; U.S. Department of Energy & U.S. Air Force; Noyes Data Corporation: Park Ridge, NJ, 1993; Vol.212, p 159.
3. Modi, A. K.; Stephanopoulos G. *American Institute of Chemical Engineers Spring Meeting;* New Orleans, LA, 1996; Paper 90c.
4. Gani, R.; Brignole, E. *Fluid Phase Equil.* **1983**, *13*, 331.
5. Joback, K.; Stephanopoulos, G. In *Proceedings of the Third International Conference on Foundations of Computer-Aided Process Design*; Siirola, J .J., Grossmann, I. E., Stephanopoulos, G., Eds.; Elsevier Science: New York, 1990; pp 363.
6. Nielsen, B.; Gani, R.; Fredenslund, Aa. In *Computer Applications in Chemical*

Engineering; Bussemaker, H. Th., Iedema, P. D., Eds.; Elsevier Science: Amsterdam, The Netherlands, 1990; pp 227.
7. Gani, R.; Nielsen B.; Fredenslund, Aa. *AIChE J.* **1991**, *37*, 1318.
8. Macchitto, S.; Odele, O.; Omatsone, O. *Trans. Inst. Chem. Eng.* **1990**, *68*, 429.
9. Pretel, J.; Lopez, A.; Bottini, B.; Brignole, A. *AIChE J.* **1994**, *40*, 1349.
10. Duvedi, P.; Achenie, K. *Chem. Eng. Sci.* **1996**, *51*, 3727.
11. Hansen, C. M. In *Solvents Theory and Practice*; Tess, R. W., Ed.; Advances in Chemistry Series; American Chemical Society: Washington, D.C., 1973; Vol. 124, p 48.
12. Klein, A.; Wu, T.; Gani, R. *Comp. Chem. Eng.* **1992**, *16*, 229.
13. Zhao, R.; Cabezas, H.; Govind, R.; Fang, Y. *American Institute of Chemical Engineers Spring Meeting;* New Orleans, LA, 1996, Paper 40c.
14. Cabezas, H.; Zhao, R.; Bare, J. C.; Nishtala, S. R. *American Institute of Chemical Engineers Annual Meeting*, Chicago, IL, 1996; Paper 46h.
15. Zhao, R.; Cabezas, H. *Ind. Eng. Chem. Res.*, **1998**, *37*, 3268.
16. Bird, R. B.; Steward, W. E.; Lightfoot, E. N. *Transport Phenomena*; John Wiley & Sons: New York, 1960.
17. Reid, R.; Prausnitz, J. M.; Poling, B. *The Properties of Gases & Liquids*; McGraw-Hill: New York, 1987.
18. Young, D. M., U.S. Environmental Protection Agency, Office of Research and Development, National Risk Management Research Laboratory, *unpublished*, 1998.
19. Young, D. M.; Cabezas, H. *Comp. Chem. Eng. 1999, in press.*
20. *Environmental Life Cycle Assessment of Products: Guide-October 1992;* Heijungs, R., Ed.; Centre of Environmental Science: Leiden, Netherlands, 1992.
21. U.S. Environmental Protection Agency, Office of Pollution Prevention and Toxics. *1997 Toxics Release Inventory*; EPA 745-R-99-003; U.S. Environmental Protection Agency, Office of Pollution Prevention and Toxics: Washington, D.C., 1999.
22. The Clean Air Act, 42 U.S.C. s/s 7401 et seq. (1970).

Chapter 20

Volatile Methyl Siloxanes: Environmentally Sound Solvent Systems

Dwight E. Williams

Dow Corning Corporation, Midland, MI 48686

Linear volatile methyl siloxanes are a class of mild solvents having an unusual combination of environmentally benign qualities. They are low in toxicity, contribute little to global warming, do not contribute to urban ozone pollution, and do not attack the stratospheric ozone layer. They do not accumulate in the atmosphere, but rather are rapidly transformed to naturally occuring chemical species. They have received approval of the significant new alternatives program (SNAP) and VOC exemption from the U.S. EPA. We have shown that their solvency can be tailored to specific applications by using azeotropes, cosolvents and surfactants. We review here the underlying toxicological, physical, and chemical properties and phenomena related to their use in replacing less benign solvents in applications such as coating formulations or to remove particulates, oils, fluxes, and aqueous contaminants. Their mild but selectable solvency, environmental benignancy, and odorless character commend them for many uses.

Introduction

The linear volatile methyl siloxanes (VMS) are low molecular weight species which contain only silicon, carbon, hydrogen, and oxygen. Highly purified grades of the linear dimer, trimer and tetramer have recently been made commercially available as *Dow Corning®* OS-10, OS-20, and OS-30. These are the permethylated hexamethyldisiloxane, octamethyltrisiloxane, and decamethyltetrasiloxane, respectively, and are abbreviated as MM, MDM, and MDDM. A general purpose cleaning grade is also available as *Dow Corning* OS-2, and an azeotrope of 91% MM and 9% propyleneglycol-monomethyl ether is available as *Dow Corning* OS-120.

Linear VMS are a class of mild solvents which have an unusual combination of environmentally benign qualities. They are low in toxicity, contribute little to global warming, do not contribute to urban ozone pollution, and do not attack the stratospheric ozone layer. Dow Corning discovered that the VMS have these desirable properties during the course of several years of in-house research and supported extramural research at academic centers of excellence in atmospheric chemistry.

The VMS have received SNAP approval as a replacement for Ozone Depleting Substances (ODS) and exemption from federal regulation of Volatile Organic Compounds (VOC) by the U.S. EPA. Their solvency can be tailored to specific applications by use of cosolvents and surfactants. In addition, VMS fluids do not strongly chemisorb to surfaces. We have found that highly purified grades leave no surface residue. Application and customer requirements drove the acquisition by Dow Corning of new manufacturing capability to satisfy the need to produce these compounds at very high purity.

We review here the underlying toxicological, physical, and chemical properties and phenomena that determines the ability of VMS fluids to replace less benign solvents in selected coating and adhesive formulations as well as to act as cleaning agents for removal of particulates, oils, fluxes, and aqueous contaminants. Their mild but adjustable solvency, environmental benignancy, and odorless character commend them for many uses. We have shown, for example, that VMS fluids have potential value for precision water displacement drying during the many aqueous processing steps of Flat Panel Display and semiconductor manufacturing.

We have also found that the mildness of these solvents confers compatibility with a wide variety of plastics and some rubbers. Their solvent strength is weaker than that of saturated hydrocarbons but stronger than that of the commercially available saturated hydrofluorocarbons. Their solvency may be enhanced by adding moderate amounts of stronger solvents. The addition of very small amounts of soluble surfactants confers the ability to displace water from aqueous-cleaned parts while retaining their favorable environmental profile. In some applications, the VMS may be blended with brominated or highly fluorinated solvents to impart non-flammability to the mixture while retaining the performance advantages of the VMS.

Health and Environmental Profile

Toxicological Profile

Dow Corning OS-10, OS-20, and OS-30 are low in toxicity as demonstrated in acute oral, inhalation, and dermal exposure testing; subchronic oral and inhalation studies; and tissue culture biocompatibility and genotoxicity evaluations. In the acute studies, no adverse effects were observed, even at the maximum achievable doses in laboratory animals; however, it was noted that mild, temporary discomfort might result from eye contact similar to windburn. Subchronic oral and inhalation studies revealed no toxicological responses significant to human health. Tissue culture biocompatibility studies conducted to assess the ability of these fluids to damage or destroy cells revealed no effects for any of the materials tested. Further, the fluids were found to be non-genotoxic in short-term tests for DNA damage or mutation.

Threshold Limit Values (TLVs), which are set by the American Conference of Governmental Industrial Hygienists (ACGIH), have not been established for VMS fluids. However, based on a 90-day study, an Industrial Hygiene Guideline (IHG), defined as the maximum average exposure level to which workers may be exposed over an 8-hour work day, was set at 200 ppm for *Dow Corning* OS fluids.

Exposure to hexamethyldisiloxane, the major ingredient of *Dow Corning* OS-120, was monitored at two typical electronics assembly facilities. The results of these studies demonstrate that the use of *Dow Corning* OS-120 for manual cleaning from pressure-dispensed containers can be handled safely by operators in terms of long-term inhalation exposure. For the 14 different operators tested, vapor exposures never reached even 10 percent of the 200 ppm guideline (*1*). In addition to hexamethyldisiloxane (91 percent), *Dow Corning* OS-120 contains propylene glycol monomethyl ether (9 percent) whose TLV is 100 ppm.

Environmental Profile

In 1994, the U.S. EPA issued a formal ruling on VMS materials, declaring them exempt from federal VOC regulations (*2*). This action was taken in response to a petition and supporting data submitted by Dow Corning to demonstrate that VMS fluids do not contribute to air quality problems. Individual states have also been petitioned to request that similar action be taken to coincide with the federal government's exemption. As of February 1998, 47 states and major districts of California had granted VOC exemption to VMS fluids. The South Coast Air Quality Management District of California has designated OS-10, OS-20, and OS-30 as Certified Clean Air Solvents which are exempt from normal tracking and reporting requirements in that District. A similar designation for OS-2 is pending.

Further, VMS materials are not regulated as "hazardous air pollutants" (HAPs) and are not controlled under National Emission Standards for HAPs (NESHAP) requirements or the Clean Air Act National Ambient Air Quality Standards. In 1994, VMS materials were included on a list of acceptable precision and electronics cleaning substances under SNAP. In addition, VMS fluids are not regulated under U.S. EPA Title III Air Toxics and are not EPA Criteria Pollutants.

Due to the volatility of the linear VMS fluids, the most likely route for environmental exposure is through evaporation. These fluids degrade quickly via ongoing natural photo-oxidation and have an atmospheric half-life of just 4 - 9 days (*3-5*). Ultimate degradation products include water-dissolved silica, water, and carbon dioxide, all abundant, naturally occurring, benign compounds. Because VMS fluids are rapidly and effectively oxidized before reaching the stratosphere, they are believed to have no impact on the earth's protective ozone layer and negligible contribution to global warming. A complete discussion of their atmospheric chemistry has been published (6).

Applications

Cleaning of Non-Polar Soils

VMS fluids are typically used to remove oils, light greases, fingerprints, and particulates while cleaning optics, gyroscopes, fiber optic assemblies, plastics, and metals. Unmodified VMS fluids provide excellent solvency for non-polar oils and viscous fluids (7). The ability of VMS fluids to remove a number of industrial liquids is illustrated in Figure 1. It is ironic that these valuable liquids themselves become redefined as dirt or soil when they must be removed before further manufacturing steps or during maintenance and repair operations. It is noteworthy that high molecular weight silicones are easily removed by VMS fluids: This class of materials is used in many industries and can otherwise be difficult to remove.

Note that there is little difference in efficacy between MM, MDM, and MDDM for removal of non-polar soils. However, drying times after cleaning vary greatly: The drying-off time for MM is similar to that of acetone, MDM is similar to that of isopropyl alcohol (IPA), and MDDM is similar to that of water. Selection of the appropriate VMS fluid depends on both cleaning and drying properties and on the use of the proper equipment. Several equipment manufacturers offer equipment designed to use flammable or combustible cleaning fluids safely. Note that MM and MDM are classified as flammable liquids, while MDDM is classified as a combustible liquid.

Azeotropes

The use of VMS fluids to remove semi-polar soils requires the use of additives. We have explored many possible applications of such solvents modified using additives, including the use of reflux and of room temperature azeotropes to provide consistent cleaning performance (8-13). Azeotropy is the property of certain mixtures to volatilize with unchanged composition. In general, the temperature at which a particular composition displays azeotropy is temperature dependent. Room temperature (RT) azeotropes have the advantage that evaporation of the more volatile component does not cause a change in the composition during use in an open container or during benchtop cleaning. Furthermore, RT azeotropes tend to dry more rapidly from cleaned articles than similar non-azeotropic compositions because of their slightly higher than expected vapor pressures and because less volatile components are not left behind. Reflux azeotropes have the advantage that they can be recovered to the same composition by distillation. This is beneficial in

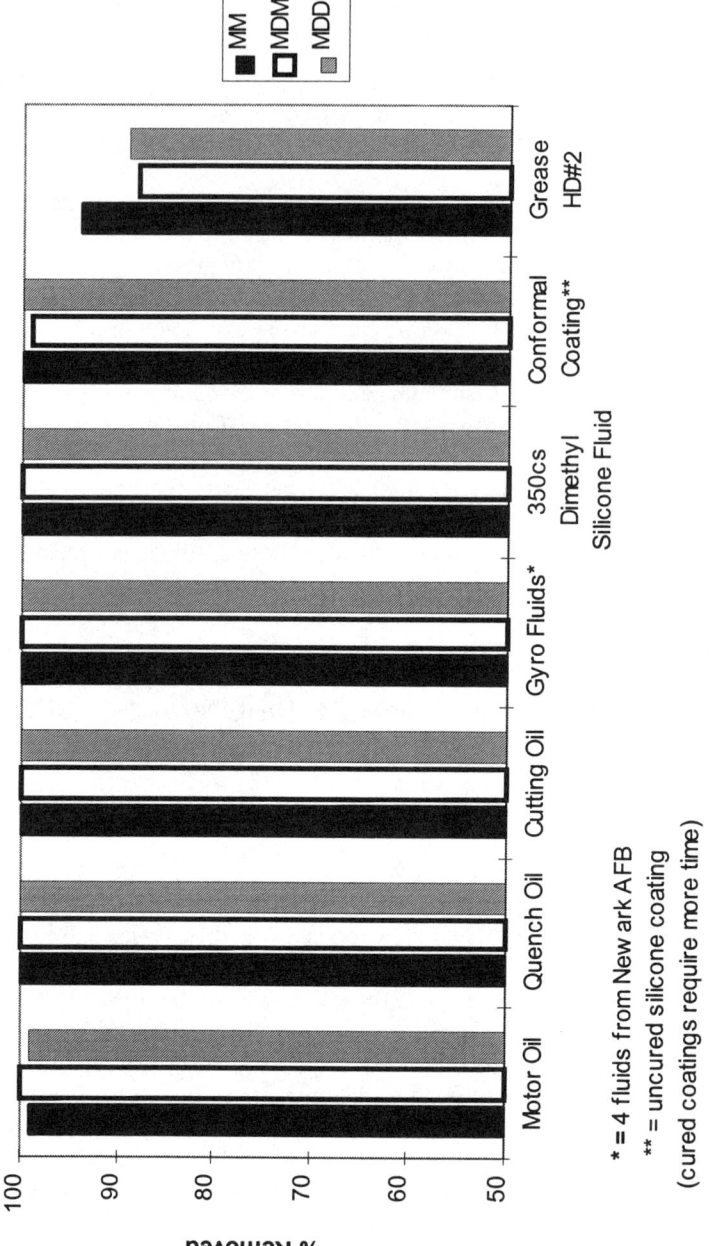

Figure 1. Cleaning Performance for Non-Polar Soils

closed cycle cleaning operations and in so-called vapor phase cleaning. We have discovered and patented 26 binary azeotropes of VMS fluids, including 8 that display RT azeotropy (*8-13*). The azeotropic and azeotrope-like composition of each of these binary systems typically spans a range of 60 to 95 wt% VMS.

Cleaning of Semi-Polar Soils

The use of VMS fluids to remove semi-polar soils requires the use of additives to tailor the solubility. We chose a rosin-based solder flux as a model for such soils. Rosin is a natural product which is a complex mixture rich in tricyclic, partially unsaturated carboxylic acids that contain twenty carbon atoms, and related derivatives. Figure 2 illustrates the impact that addition of a polar solvent (designated a polar solvency enhancer) has on the ability of the VMS fluids to remove solder flux. Although none of the VMS fluids were able to remove much solder flux unassisted, addition of 18 wt% 1-methoxy-2-propanol greatly accelerated removal rate and resulted in thorough removal in 5-10 minutes at RT. Fortunately, the VMS fluids are miscible with many polar solvents, so it is easy to tailor solvency to particular soils.

Table I presents comparisons of cleaning power for a number of RT azeotropes of VMS. For purposes of comparison, power is defined as completeness of removal of solder flux, a semi-polar soil, after 5 minutes of cleaning. The unassisted VMS fluids have little ability to remove solder flux. The RT azeotropes are considerably more effective. The cleaning power of the reflux azeotropes is greater than that of the RT azeotropes, even when both are used at RT. This is due to the fact that these reflux azeotropes happen to have a greater amount of the more polar organic component than the corresponding RT azeotropes.

The reflux azeotropes are especially useful when distillative regeneration of spent solvent is desired, or when the azeotrope is to be used at its boiling point to remove solder flux. High cleaning power is essential in the latter case. On the other hand, benchtop cleaning is conducted at RT, often by manually applying the fluid to the article to be cleaned. Benchtop cleaning typically employs a scrubbing tool. This accelerates soil removal and is a requirement when using these RT azeotropes to remove solder flux.

Note that the azeotropes of higher molecular weight VMS fluids are somewhat less effective at similar organic content. The higher oligomers have higher viscosities and somewhat lower Hildebrand solubility parameters: Both factors can reduce cleaning power.

Precision Water Removal

Currently, the water removal step at the end of precision aqueous cleaning processes requires high capital and operating costs for the large volumes of ultra-high purity (UHP) water needed for rinsing to a spot-free finish upon drying. This

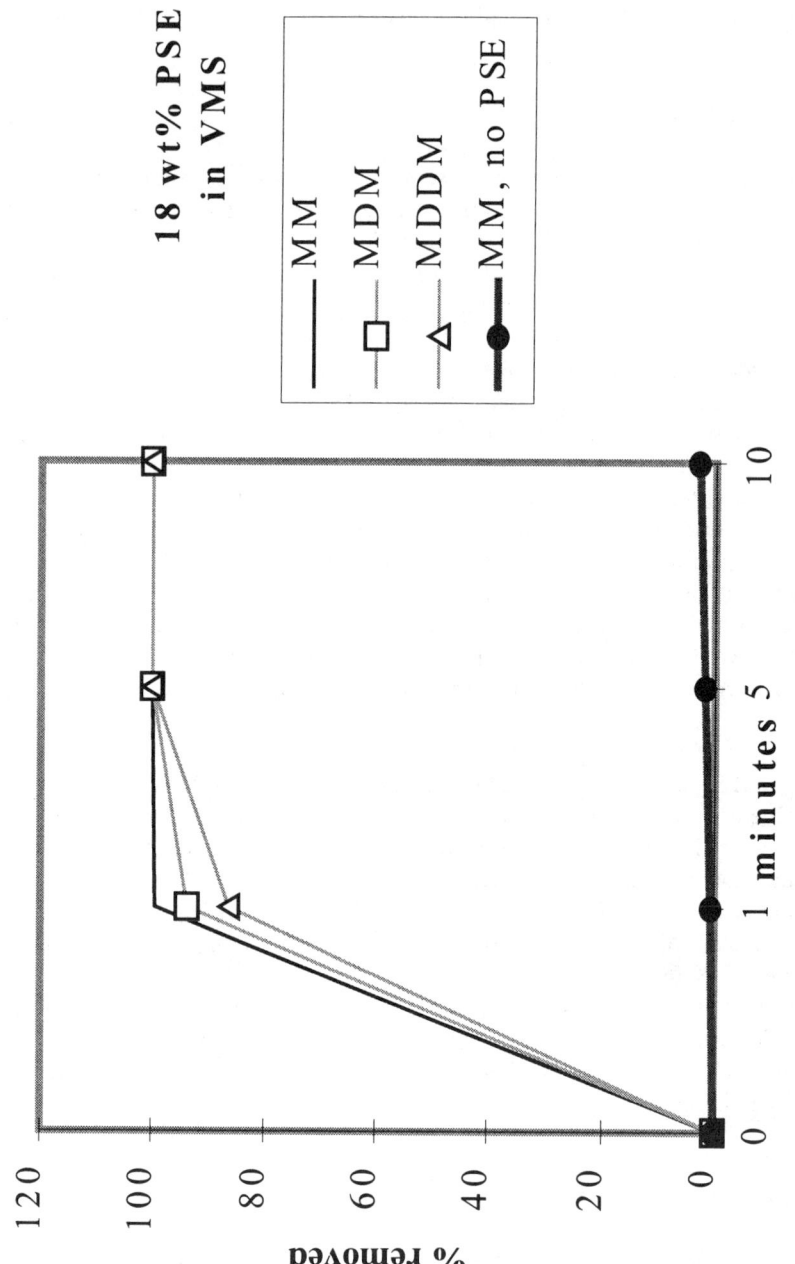

Figure 2. Effect of Polar Solvency Enhancer on Flux Removal.

Table I. Rosin Flux Cleaning Power of Selected VMS Azeotropes

VMS	%WT VMS	Organic Partner[a]	Rosin Cleaning Power, % at RT	Azeotropic Type[b]	Source
MM	100	None	4.1	NA[d]	8
MM	91	MPP	96	RT	8
MM	82	MPP	100	Reflux	8
MM	76	NPA	97	RT	9
MM	61	NPA	100	Reflux	9
MDM	100	None	1.5	NA	10
MDM	59	ELT	ND[c]	RT	10
MDM	63	ELT	100	Reflux	10
MDM	87	MPL	97	RT	10
MDM	61	MPL	100	Reflux	10
MDM	86	PNP	97	RT	11
MDM	60	PNP	100	Reflux	11
MDM	79	IPL	98	RT	12
MDM	62	IPL	100	Reflux	12
MDDM	100	None	0	NA	13
MDDM	60	DPM	99	RT	13
MDDM	61	DPM	100	Reflux	13
MDDM	84	DPE	89	RT	13
MDDM	66	DPE	100	Reflux	13

[a] MPP = 1-methoxy-2-propanol, NPA = n-propylacetate, ELT = ethyl lactate, MPL = 2-methyl-1-pentanol, PNP = 1-n-propyloxy-2-propanol, IPL = isopropyl lactate, DPM = dipropyleneglycolmonomethyl ether, DPE = dipropyleneglycolmonoethylethyl ether.
[b] Temperatures at reflux were 96-97 °C for the MM azeotropes, 139-143 for the MDM azeotropes, and 180-187 for the MDDM azeotropes.
[c] ND = not determined.
[d] NA = not applicable.

is an important consideration for the manufacture of flat panel displays, memory disks, and microelectronics. In these industries, current methods include hot air-assisted evaporation, capillary pull-out drying, spin-rinse drying, Marangoni drying, and IPA solubilization. Immiscible fluid displacement of water was a very popular method during the peak of the CFC-113 era. Immiscible fluid water displacement drying (WDD) offers many advantages over the existing methods, including high throughput, completeness of removal of water *and its dissolved impurities*, and rapid regeneration of the displacement fluid.

WDD requires a non-aqueous solution of an enabling surfactant. The surfactant is retained in the non-polar fluid and the dissolved aqueous salts and colloids are removed with the water. The basis of the phenomenon is illustrated by the cartoon shown in Figure 3. A water-covered hydrophilic solid is immersed in a non-polar solution containing dissolved surfactant molecules (designated as V). The water initially is a film, but increasing adsorption of surfactant at the oil/water and oil/solid interfaces rapidly causes the water to bead up. Then the water rolls off as the contact angle approaches 180°. This phenomenon is controlled by the venerable Young Equation:

$$\cos\theta = [\gamma(OS) - \gamma(WS)]/\gamma[OW], \tag{1}$$

where $\gamma(OS)$, $\gamma(WS)$, and $\gamma(OW)$ are the interfacial tensions (IFT) of the non-polar Fluid/Solid, Water/Solid, and Fluid/Water interfaces, respectively. The quantity, $[\gamma(OS) - \gamma(WS)]$, is sometimes called the differential IFT. Upon immersion, the surfactant must be adsorbed at both new interfaces for the change in IFTs to result in water beading up and rolling off. An advantage of VMS over halogenated fluids for WDD is that the water drops to the bottom of the displacement tank rather than floats to the top of the fluid, facilitating removal of the dewatered article. In addition to the required interfacial effects, two other important criteria for the surfactants are that they must be soluble in VMS fluids and must prefer to remain in the VMS rather than transfer into the displaced water. Various non-ionic and zwitterionic surfactants meet these criteria.

We have found a number of compositions of surfactants in VMS fluids which provide a high degree of water removal (*14*), as shown in Figure 4. Note that some surfactants provide thorough removal of water regardless of which VMS base fluid is used. However, there is a tendency for MM to be a more potent base fluid, probably due to more rapid sedimentation of water in MM than in the denser and more viscous MDM or MDDM.

Several experimental WDD formulations based upon MM greatly ameliorate the need for UHP water rinsing since dissolved solids and colloids are removed simultaneously with the water. Some VMS formulations can even displace unrinsed soap solutions from hydrophilic surfaces such as clean glass, resulting in a dewatered and spot-free surface with little or no aqueous rinsing required.

253

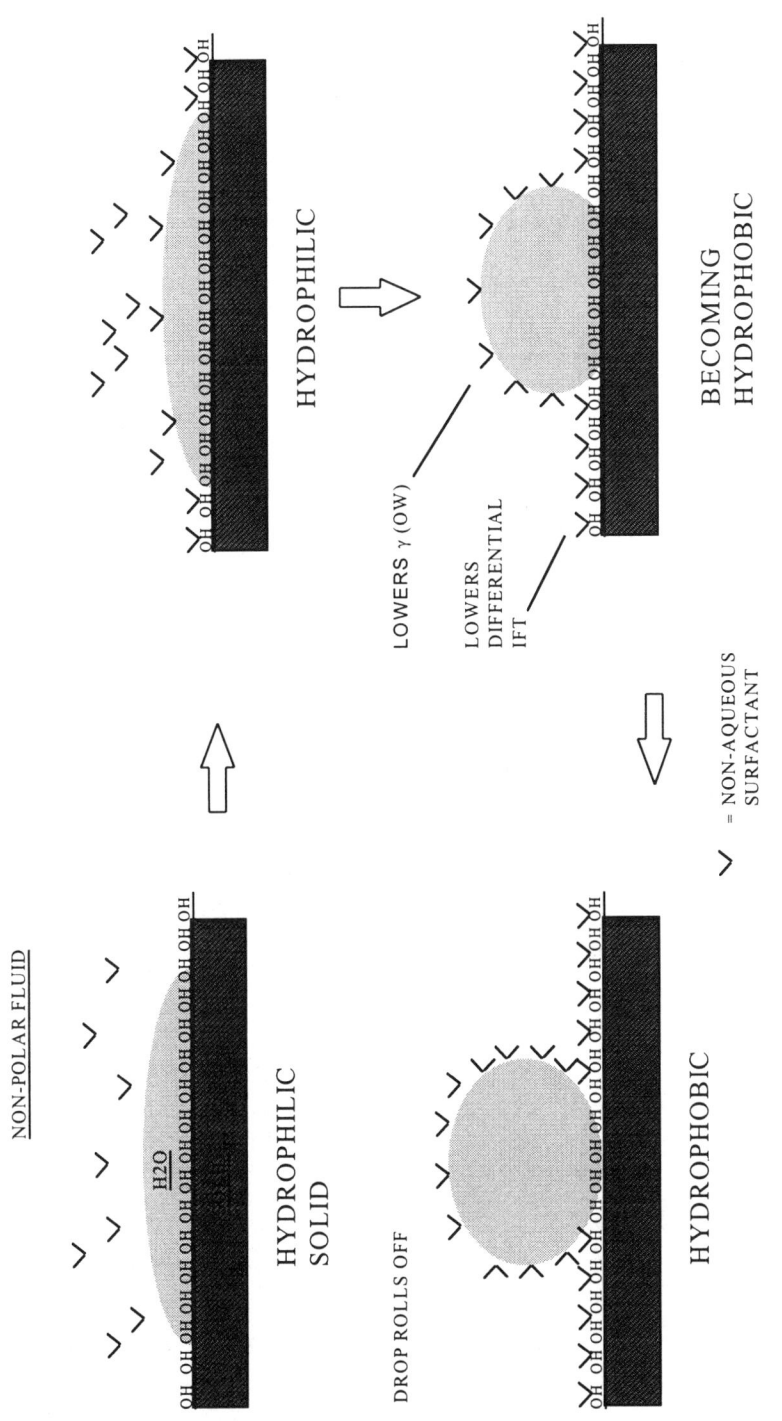

Figure 3. Dynamics of Water Displacement. Reproduced with permission from reference 14.

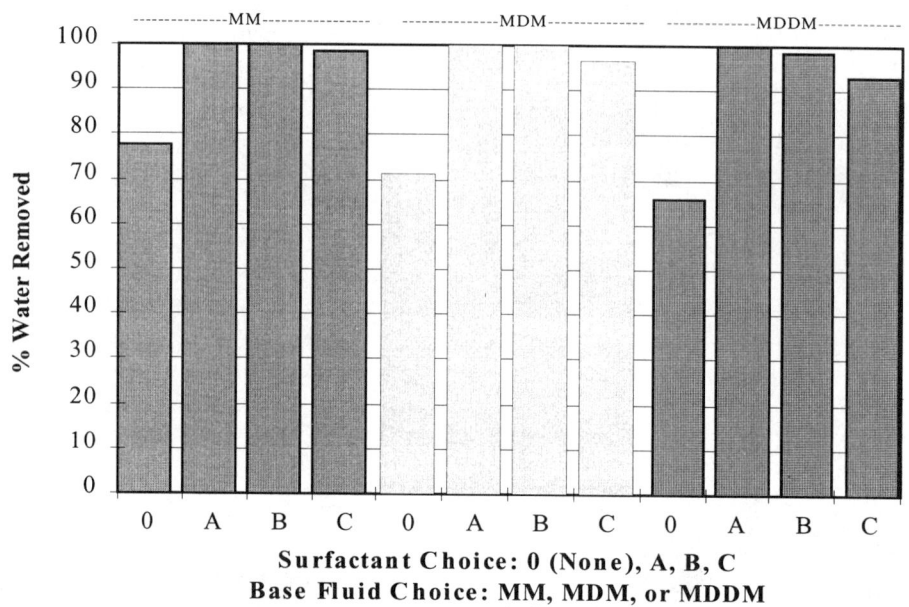

Figure 4. WDD Performance of Experimental Formulations

Carriers

Dow Corning Corporation and other companies have incorporated VMS fluids into a variety of formulated products to replace a variety of solvents. These include solvents classified as ODS, VOCs, or HAPs such as CFC-113, saturated hydrocarbons like hexane and stoddard solvents, and in some cases aromatic solvents like toluene and xylene. All of the replaced solvents are VOCs except CFC-113 which is an ODS. Furthermore, hexane is a neurotoxin, and many of the others are thought to cause liver damage. These characteristics are in contrast to the environmental friendliness and low order of toxicity of the linear VMS fluids. Many applications depend on the excellent solvency of VMS fluids for other silicone materials. The impact of VMS is multiplied by the versatility of silicone products which serve a broad spectrum of US companies, including those in the transportation, construction, electronics, health care, textiles, paper, personal care, coatings, and process industries.

Experimental

Non-Polar Soil Cleaning

Tared steel plates were immersed to a constant depth in the non-polar liquid soil, removed, and held several minutes until dripping stopped. They were weighed while hanging to measure soil loading. Then the soiled region was immersed in the cleaning liquid at room temperature in a beaker for 3 minutes with rapid stirring using a magnetic stirring bar. The plates were removed, allowed to air dry at RT, and reweighed. The weight percent removed was reported.

Rosin Flux Cleaning

Kester #1544 rosin flux (Kester Solder Division, Litton Industries, Des Plaines, IL) was applied as a thin layer to a tared 2"x3" aluminum panel and spread out across a constant portion of the panel surface with a spatula. The rosin flux contains 50 wt% modified rosin and 1 wt% proprietary activator in mixed alcohols. The coating was allowed to dry in a hood and then cured at 100 °C for ten minutes in an explosion-proof air oven. The weight of the panel with the cured flux was measured. The amount of flux applied was adjusted so as to provide about the same weight of cured flux on each panel. The coated portion of the panel was immersed in the cleaning liquid in a beaker. Cleaning was conducted at room temperature, even when cleaning with reflux azeotropes, with rapid stirring using a magnetic stirring bar. At timed intervals, the panel was removed, dried at 80 °C for ten minutes, and weighed. The weight percent removed was recorded.

Azeotropic Determinations

Azeotropes were discovered and characterized using a single plate distillation apparatus. The liquid mixture was boiled and the vapor condensed in a small receiver which contained an overflow path to return condensate at the boiling point to the boiling liquid. The composition of the boiling liquid and of condensed vapor phases were quantitated by gas chromatography after equilibrium was attained. The temperature, ambient pressure, and equilibrium compositions were measured at several different overall compositions. The azeotropic composition and vapor pressure at different temperatures was determined by analyzing the data with ASPENPLUS®, a commercial software program.

Water Displacement Drying

Hydrophilic substrates, either glass slides or scrolls of wound Cr/Ni wire mesh trapped in an open Teflon® tube, were used to test for the ability to remove water by WDD. The glass slides were cleaned just before use by immersion in a saturated solution of KOH in ethanol, followed by rinsing in deionized (DI) water. This produced a water contact angle below 5°. The wire mesh was cleaned in dilute nitric acid and rinsed in DI water. These hydrophilic substrates were immersed in water containing controlled amounts of dissolved impurities, then immersed in and gently agitated in the water displacement fluid, removed, and rinsed in pure VMS. Displacement was quantitated by placing the wire mesh in anhydrous methanol. Water transfer to the methanol was measured by Karl-Fisher titration. Percent removal was determined by comparing the amount of transferred water to that transferred without immersing in VMS. Glass slides were removed and allowed to dry following the VMS rinse step. Visual examinations were done for spots or streaks following displacement from the glass slides.

Summary

The linear VMS fluids are environmentally benign base fluids used for precision cleaning, drying, and carrier/coating uses. Their use in precision cleaning requires highly purified grades that are completely volatile. Their solvency may be tailored with additives to broaden their uses to include removal or deposition of semipolar oils and viscous fluids as well as to the removal of water by displacement. Their outstanding environmental and toxicological profiles and their effectiveness help to ensure that they will remain in industry's repertoire of cleaning and carrier fluids for a long time to come.

Acknowlegements

Many people have made significant contributions to the work reviewed here. I would like to acknowledge Grish Chandra, John Moore, Cecil Frye, Frank Lewis, Mike Thelen, Ray Cull, Jeff Baker and Steve Kamin, Dick Burow, Waheed Siddiqi, Steve Swanson, Ora Flaningam, David Morgan, Frank Bartlett, and Cheryl Clark.

Literature Cited

1. Swanson, S. P.; Cull, R. A. *Precision Cleaning* September 1997, p 15.
2. Environmental Protection Agency *Federal Register*, 1994, No. *192*, pp 50693-50696.
3. Atkinson, R. *Environmental Science and Technology* 1991, vol. *25*, p 863.
4. Sommerlade, R.; Parlar, H.; Wrobel, D; Kochs, P. *ibid* 1993, vol. *27*, p 2435.
5. Markgraf, S. J.; Wells, J. R. *J. Chem. Kinet.* 1997, vol. *29*, p 445.
6. Hobson, J. F.; Atkinson, R.; Carter, W. P. L. In *The Handbook of Environmental Chemistry: Organosilicon Materials*, Chandra, G., Ed., Springer-Verlag: New York, NY, 1997.
7. Cull, R. A.; Moore, J. A.; Williams, D. E. *Proceedings Int. SAMPE Tech. Conf.;* SAMPE: Covina, CA, 1995; Vol. 27, p 70.
8. Flaningam, O. L.; Williams, D. E. *US Patent 5,478,493*, 1995.
9. Morgan, D. L.; Williams, D. E. *US Patent 5834416*, 1998.
10. Flaningam, O. L.; Williams, D. E. *US Patent 5,454,970*, 1995.
11. Williams, D. E.; Flaningam, O. L. *US Patent 5,516,450*, 1996.
12. Flaningam, O. L.; Moore, J. A.; Williams, D. E. *US Patent 5,456,856*, 1995.
13. Flaningam, O. L.; Morgan, D. L.; Williams, D. E. *US Patent 5824632*, 1998.
14. Williams, D. E. *Proceedings of CleanTech '98;* Witter Publ. Corp.: Flemington NJ, 1998; p 374.

Chapter 21

Green Chemistry Through the Use of Supercritical Fluids and Free Radicals

J. M. Tanko, B. Fletcher, M. Sadeghipour, and N. K. Suleman

Department of Chemistry, Virginia Polytechnic Institute and State University, Blacksburg, VA 24061-0212

Supercritical carbon dioxide (SC-CO_2) is proving to be a suitable "environmentally benign" solvent for free radical reactions, providing a unique alternative to many conventional solvents for these reactions which are either carcinogenic or damaging to the environment. Part 1 of this paper examines the implications associated with the use of SC-CO_2 with regard to issues of solvent effects on chemical reactivity. In Part 2, a new "environmentally benign" chemical process is described which effects the conversion RH + C=C-C-Br → R-C-C=C + HBr via a free radical chain reaction.

Part 1: Free Radical Chemistry in Supercritical Carbon Dioxide

Solvents which can be used for free radical reactions must be inert towards free radicals (*i.e.*, they should not possess reactive functionalities such as abstractable hydrogens or reactive multiple bonds). Unfortunately, many of the typical solvents which fulfill this criterion are either carcinogenic (*e.g.*, benzene) or damaging to the environment (*e.g.*, CFCs, CCl_4).

There are a number of potential advantages associated with the use of supercritical carbon dioxide (SC-CO_2) as a solvent for free radical chemistry:

- CO_2 is essentially non-toxic and "environmentally benign"[1].
- Solvent properties of SC-CO_2 (*e.g.*, dielectric constant, solubility parameter, viscosity, density) can be dramatically altered in a manner not possible with conventional solvents--via manipulation of temperature and pressure [2,3].
- The properties of SC-CO_2 are intermediate between that of a liquid and gas.

The specific objective of this project was to examine the extent of the "cage-effect" in supercritical fluid solvents such as CO_2. Cage effects arise in situations where two reactive species are formed in close proximity within a solvent cage, and the species react with each other at a rate competitive with diffusion apart. Generally, different products are formed via these two competing pathways.

Cage effects in supercritical fluids have attracted a great deal of attention lately because there is some indication that near the critical point, the cage effect may be enhanced (*4 – 7*). The rationale for this enhancement is derived from the fact that near the critical point, the local density of solvent around a solute may be greater than the bulk density (a phenomenon referred to solvent/solute clustering) (*8 – 19*). This increased density would presumably lead to enhanced local viscosity, and as a result, an enhanced cage effect.

In order to test for the extent (and potential enhancement) of the cage effect in SC-CO_2, the "chlorine atom cage effect" was employed as a probe (*20*). The chlorine atom cage effect (Scheme 1) was discovered by Skell and Baxter in 1983, (*21*) and confirmed and extended by other investigators (*22 – 24*). To understand the origin of the cage effect, consider the free radical chlorination of an alkane (RH_2). The designation RH_2 rather than the traditional "RH" is used as a reminder that an alkane possesses more than one hydrogen, and thus, multiple substitution is possible.

Scheme 1

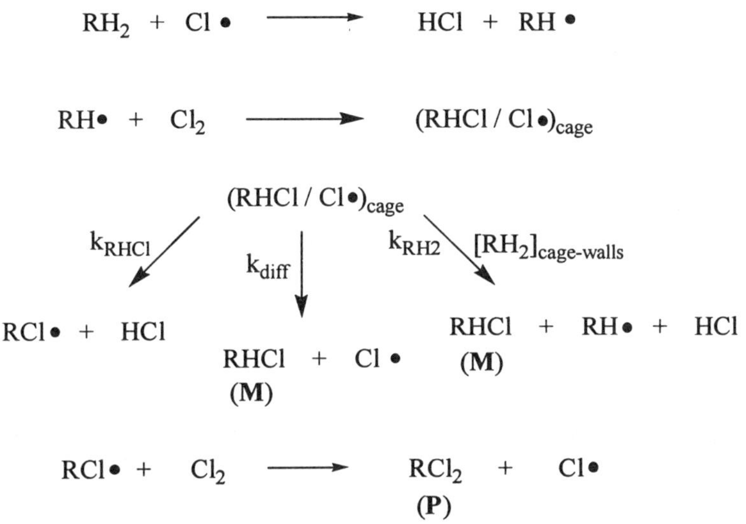

$$RH_2 + Cl\bullet \longrightarrow HCl + RH\bullet$$

$$RH\bullet + Cl_2 \longrightarrow (RHCl / Cl\bullet)_{cage}$$

$$(RHCl / Cl\bullet)_{cage}$$

k_{RHCl} $\quad k_{diff}$ $\quad k_{RH2}$ $\quad [RH_2]_{cage-walls}$

$$RCl\bullet + HCl \qquad\qquad RHCl + RH\bullet + HCl$$
$$\qquad\qquad RHCl + Cl\bullet \qquad (\mathbf{M})$$
$$\qquad\qquad (\mathbf{M})$$

$$RCl\bullet + Cl_2 \longrightarrow RCl_2 + Cl\bullet$$
$$(\mathbf{P})$$

The free radical chlorination of an alkane is a free radical chain process consisting of a hydrogen abstraction step ($RH_2 + Cl\bullet \rightarrow RH\bullet + HCl$) and a chlorine atom transfer step ($RH\bullet + Cl_2 \rightarrow RHCl + Cl\bullet$). It is important to realize that chlorine atom is a highly reactive species, and reacts with alkanes at nearly diffusion-controlled rates (*25*). In the second propagation step of the free radical chlorination of an alkane, the encounter of $RH\bullet$ with Cl_2 results (initially) in formation alkyl chloride / chlorine atom as a caged pair (Scheme 1). This is referred to as a *geminate* caged-pair because the two species are produced simultaneously within a solvent cage from a common precursor(s). Because of the extremely high reactivity of chlorine atom, $(RHCl/Cl\bullet)_{cage}$ partitions between three pathways: 1) Diffusion apart (k_{diff}) yielding monchloride (RHCl, designated as **M**), 2) An *in-cage* hydrogen abstraction from the alkyl chloride (k_{RHCl}), yielding $RCl\bullet$ which inevitably leads to a polychloride (RCl_2, designated as **P**), and 3) Hydrogen abstraction from RH_2 comprising the cage walls (k_{RH2}, a process which is only important at high alkane concentrations, and which yields RH•, HCl, and monochloride). Thus, only the cage reaction (k_{RHCl}) of the geminate caged-pair leads to polychloride.

Tanner (24) has shown that in *conventional* solvents, the ratio of mono- to polychlorides (**M/P**) varies as a function of the viscosity of the solvent. (The **M/P** ratio decreases with increasing viscosity, or put another way, the cage effect is more pronounced at higher solvent viscosities). This phenomenon arises because at high viscosities, the rate of diffusional separation of RHCl and Cl• out of $(RHCl/Cl•)_{cage}$ is diminished, resulting in a greater probability that reaction will occur in-cage.

The free radical chlorination of three alkanes (cyclohexane, 2,3-dimethylbutane, and neopentane) was examined both in $SC-CO_2$, and, for comparison purposes, in conventional organic solvents. All of these experiments were carried out at 40 °C, and at alkane concentrations ≤ 0.03 M (20). At such low alkane concentrations, little monochloride arises from reaction of Cl• with alkane comprising the cage walls, k_{RH2} (Scheme 1).

Figure 1 demonstrates how the **M/P** ratio for the chlorination of cyclohexane in SC CO_2 varies as a function of pressure. It is especially significant that the variation in **M/P** with pressure mirrors the change in viscosity as a function of pressure (dotted line), clearly showing that at higher viscosity, more polychloride is produced (*i.e.*, the cage effect becomes more significant).

Is there an *enhancement* of the cage effect near the critical point? Cage effects are typically treated in terms of the Noyes model (26), which predicts that ratio of products arising from in-cage reaction vs. cage escape should vary linearly with inverse viscosity ($1/\eta$). In Figure 2, **M/P** for the chlorination of cyclohexane in conventional solvents and $SC-CO_2$ is plotted as a function of $1/\eta$. Within experimental error, all the data points define a single line. In other words, the magnitude of the cage effect in $SC-CO_2$ at all pressures examined (and especially at pressures near the critical pressure) is consistent with what is observed in conventional solvents and there is no evidence of an enhanced cage effect near the critical point in $SC-CO_2$ (20).

The results for the chlorination of 2,3-dimethylbutane (23DMB) in conventional solvents and $SC-CO_2$ (Figure 3) leads to the identical conclusions (20). With 23DMB, there is *also* the issue of selectivity (*i.e.*, abstraction of the 3° vs. 1° hydrogen). In $SC-CO_2$, the relative reactivities of the 3° vs. 1° hydrogens of 23DMB towards Cl• (r(3°/1°)) were found to vary as a function of viscosity (Figure 4) and were intermediate between the gas phase (r(3°/1°) = 4.00) and solution phase (r(3°/1°) = 3.0) values (20).

The explanation for this phenomenon is related to the fact that in solution, these hydrogen abstraction reactions are nearly diffusion-controlled. Thus, the intrinsic selectivity (observed in the gas phase) is obscured in solution because of the onset of diffusion control: In solution because of increased viscosity, the rate of hydrogen abstraction is governed to a greater extent by the rate of encounter of 23DMB and

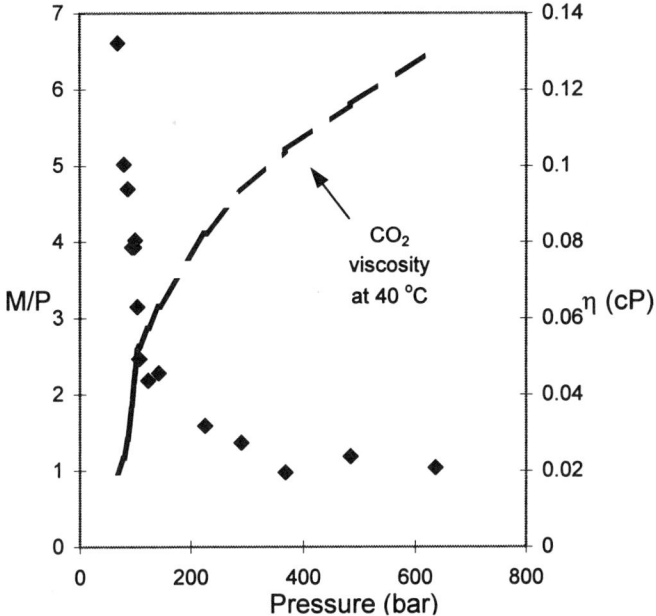

Figure 1. Ratio of mono- to polychlorides (M/P) produced in the free radical chlorination of cyclohexane in SC-CO$_2$ (40 °C).

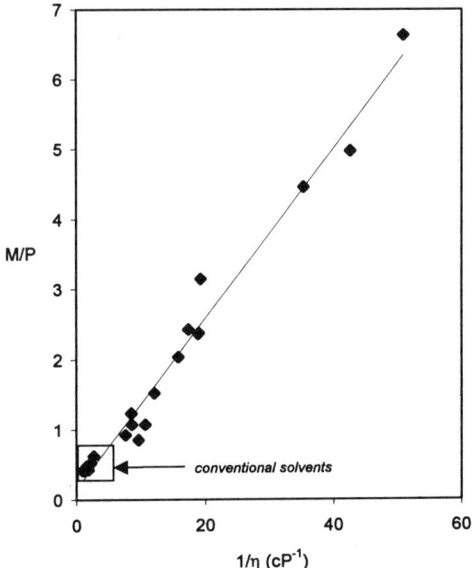

Figure 2. Ratio of mono- to polychlorides produced in the free radical chlorination of cyclohexane at 40 °C in conventional solvents and SC-CO$_2$ vs. inverse viscosity

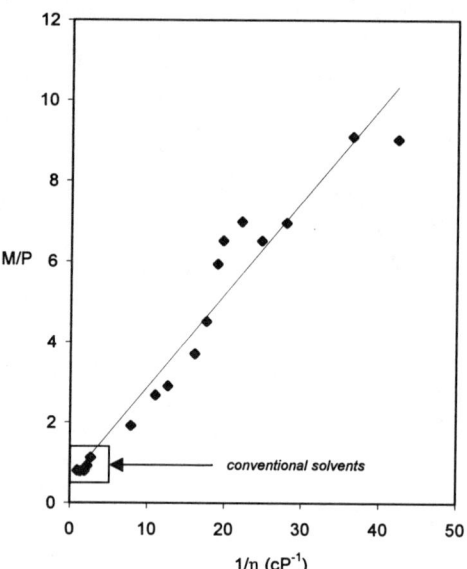

Figure 3. Ratio of mono- to polychlorides produced in the free radical chlorination of 2,3-dimethylbutane (23DMB) at 40 °C in conventional solvents and SC-CO$_2$ vs. inverse viscosity

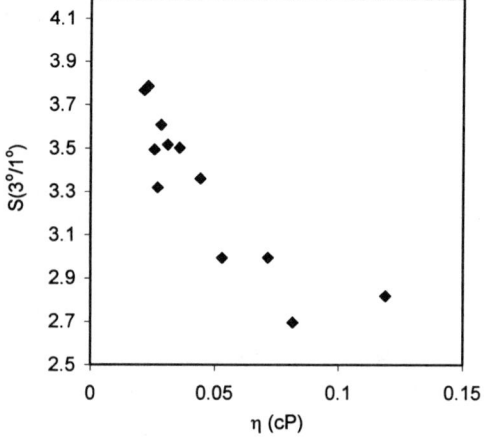

Figure 4. Variation of selectivity with viscosity observed in the free radical chlorination of 2,3-dimethylbutane in SC-CO$_2$ at 40 °C

Cl•, rather than the intrinsic reactivity of Cl•. The increase in r(3°/1°) with a decrease pressure in SC-CO$_2$ reflects the change in the medium from liquid-like to something approaching the gas phase.

The observed variation of M/P in the chlorination of neopentane in conventional solvents and SC-CO$_2$ as a function of 1/η also suggest that the magnitude of the cage effect in supercritical fluids is consistent with what is observed in conventional solvents (Figure 5) (20). For chlorination of neopentane, there are two isomeric dichlorides produced: 1,3-dichloro-2,2-dimethylpropane and 1,1-dichloro-2,2-dimethylpropane. At high viscosity (where the cage effect is greatest), these isomers are produced in nearly a 5 : 1 ratio (1,3/1,1). The ratio decreases smoothly with decreasing viscosity approaching a value of 2.5 : 1, similar to that observed in the gas phase chlorination of neopentyl chloride (Figure 6) (20).

The enhanced yield of the 1,3-dichloride at high viscosity is related to the fact that for the neopenytyl chloride/Cl• caged pair, hydrogen abstraction occurs on a time scale competitive with molecular rotation. At the moment of their "birth", neopentyl chloride and chlorine atom are in close proximity, and the methyl hydrogens are more accessible for abstraction within the cage (Scheme 2). As viscosity decreases, the cage effect becomes less important and the small amounts of dichlorides produced arise from diffusive encounters of Cl• and neopentyl chloride rather than the cage effect. Consistent with this notion is the fact that at low viscosity, the 1,3/1,1 ratio is similar to that found in the gas phase chlorination of neopentyl chloride (20).

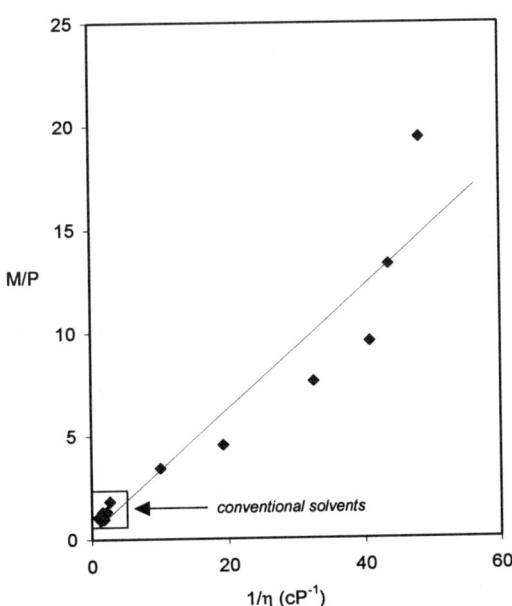

Figure 5. Ratio of mono- to polychlorides produced in the free radical chlorination of neopentane at 40 °C in conventional solvents and SC-CO$_2$ vs. inverse viscosity

Scheme 2

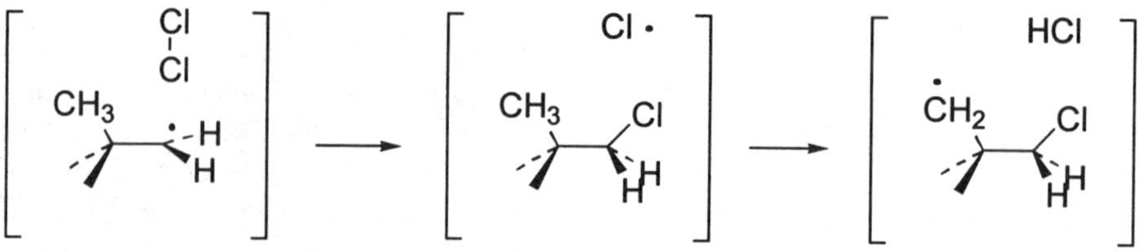

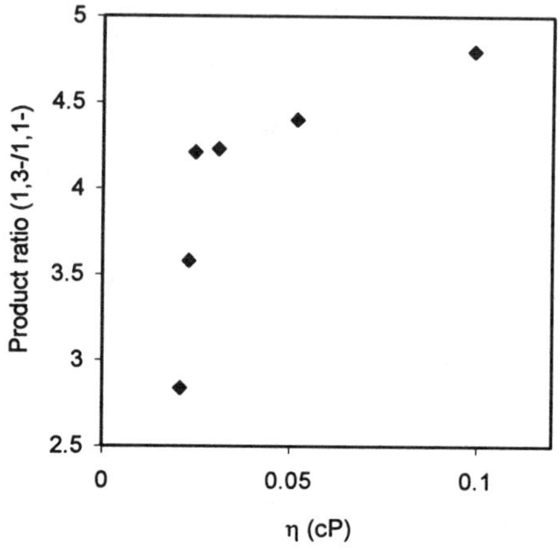

$$\left[\begin{array}{c} CH_3 \\ \overset{|}{C}-\overset{\cdot}{H} \\ H \end{array} \overset{Cl}{\underset{Cl}{|}} \right] \longrightarrow \left[\begin{array}{c} CH_3 \\ \diagup \diagdown \\ H \end{array} \overset{Cl \cdot}{\underset{H}{\diagup}} Cl \right] \longrightarrow \left[\begin{array}{c} \overset{\cdot}{C}H_2 \\ \diagup \diagdown \\ H \end{array} \overset{HCl}{\underset{H}{\diagup}} Cl \right]$$

Figure 6. Ratio of 1,3- to 1,1-dichloro-2,2-dimethylpropane produced in the free radical chlorination of neopentane in SC-CO_2 at 40 °C as a function of visosity

To summarize the results described in Part 1:
- The magnitude of the cage effect in SC-CO_2 is as anticipated based upon extrapolations from conventional liquid solvents. No evidence was found for an *enhanced* cage effect, which might be attributable to solvent/solute clustering.
- Selectivities for free radical chlorinations in SC-CO_2 are viscosity-dependent, and intermediate between vapor and liquid phase values.

With regard to the use of SC-CO_2 as a solvent for radical reactions:
- High product yields are not compromised by the use of this non-toxic, less environmentally-threatening medium.
- The "tunable" solvent properties of SC-CO_2 provide a means of "dialing up" reactivity/selectivity
- SC-CO_2 is an excellent (and perhaps superior) solvent for free radical reactions.

Part Two: Development of a New Synthetic Method for Hydrocarbon Functionalization

Another approach to the problem of toxic waste reduction in chemical synthesis and manufacture involves the development of new reactions which minimize or eliminate side products. Toward this end, a new chemical process has been developed which results in concomitant hydrocarbon functionalization and C-C bond formation.

The suggested mechanism of this reaction is the free radical chain process outlined in Scheme 3 (27). The key intermediate in this process is bromine atom, which abstracts (with high selectivity) a hydrogen atom from a hydrocarbon (typically an alkylaromatic such as toluene or cumene). The resulting benzylic radical subsequently adds to the double bond of an allylbromide (CH_2=C(Z)CH_2Br, where Z = Ph, CO_2R, or CN). The resulting radical adduct completes the chain by undergoing β-cleavage to eliminate bromine atom. The reaction requires a free radical initiator such as benzoyl peroxide or di-*t*-butyl peroxide; initial reactions were conducted in benzene as a solvent.

Scheme 3

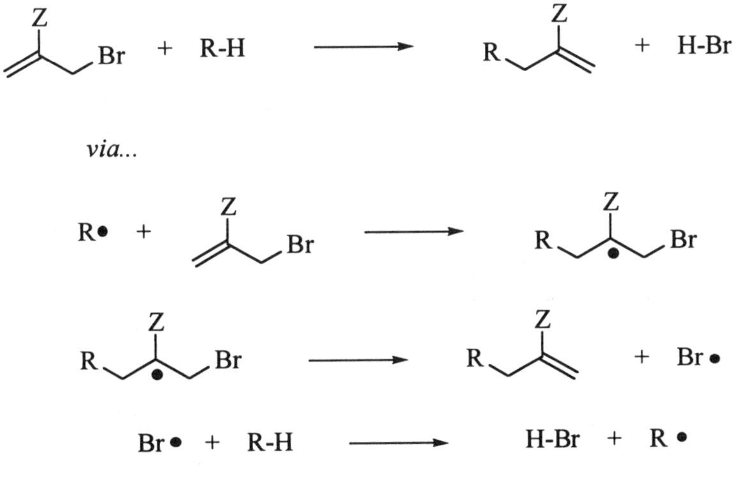

What is unique about this reaction is that it achieves hydrocarbon functionalization and C-C bond formation, while avoiding the use of strong bases, organometallic reagents, etc. Because Br• is the chain carrier, hydrogen abstractions proceed with high selectivity (*i.e.*, only the weakest C-H bond in the hydrocarbon is susceptible to attack), and based upon our experiences with radical reactions, this reaction should be fully amenable to SC-CO_2 as a reaction solvent.

Overall, yields in this reactions are excellent, as outlined in Table I (27).

Direct evidence for the intermediacy of Br• in this reaction was obtained by competition experiments. When a mixture of toluene and cumene is allowed to react with α-bromomethylstyrene (α-BMS), in the presence of a HBr scavenger, the observed selectivity (*i.e.*, 3° vs. 1°) is nearly identical to what is observed in the direct free radical bromination of toluene and cumene by Br_2 (27). The need for an HBr scavenger such as 1,2-epoxybutane is related to the fact that hydrogen abstractions by Br• are reversible; addition of a scavenger effectively blocks the reverse reaction by consuming the HBr.

Table I. Reactions of Alkylaromatics with Substituted Allyl Bromides

R	Z	Yield (%)
H	H	33
H	Ph	82
CH_3	Ph	100
H	CO_2Et	47
CH_3	CO_2Et	48
H	CN	66
CH_3	CN	80

The relative rates of addition of $PhCH_2$• to various allyl bromides were also determined via competition experiments (*i.e*, a mixture of two allyl bromides were allowed to react with toluene). The relative rates of addition (for $CH_2=C(Z)CH_2Br$, Z = CN (180) > CO_2Et (110) > Ph (65) > H (1.0) at 80 °C) (27) nicely parallel the relative rates of addition of $PhCH_2$• to similarly substituted alkenes (*e.g.*, $CH_2=C(Z)CH_3$) reported by Fischer) (28).

The initial chain lengths (rate of product formation relative to rate of initiator production) for these reactions are high and are summarized in Table II (27).

Table II. Chain Lengths for Reaction of Alkylaromatics with Allyl Bromides

Z	R	Chain length
H	H	10
Ph	H	400
CO_2Et	H	800
CN	H	700
Ph	CH_3	60
CO_2Et	CH_3	60
CN	CH_3	400

The fact that the effect of the substituent Z on the chain length tends to parallel the reactivity of the allyl bromide towards $PhCH_2$• (Z = CN ≥ CO_2R > Ph > H), and also the fact that the chain lengths tend to be greater for $PhCH_3$ than $PhCH(CH_3)_2$

suggest that the rate limiting step for this reaction is the addition the benzyl radical to the C=C.

Part 2 summary: A new, "environmentally benign" reaction has been developed for the allylation of a hydrocarbon via a free radical process involving bromine atom. Reaction yields and chain lengths for this process are greatest when the allyl bromide possesses a radical-stabilizing substituent (*e.g.*, Ph, CO_2R, or CN). This process is especially attractive because (unlike other procedures which effect this conversion) the transformation is accomplished in a single step and does not require strongly acidic or basic reaction conditions, or the use of metals. *Most notably, this procedure provides a viable means of achieving hydrocarbon functionalization and C-C bond formation in a radical reaction which does not use tri-n-butyl tin hydride.* Finally, it is envisioned that this reaction will prove amenable to $SC-CO_2$ solvent.

Acknowledgement. Financial support from the National Science Foundation (CHE-9524986) is acknowledged and appreciated.

Literature Cited

1. a) Collins, T. J., *Green Chemistry. Macmillan Encyclopedia of Chemistry.* Macmillan, Inc.: New York, 1997. b) Anastas, P. T.; Williamson, T. C., *Green Chemistry: Frontiers in Benign Chemical Syntheses and Processes.* Oxford University Press: Oxford, 1998. c) Anastas, P. T.; Williamson, T. C., *Green Chemistry: An Overview,* in *Green Chemistry: Designing Chemistry for the Environment,* American Chemical Society Symposium Series, No. 626, Anastas, P. T.; Williamson, T. C., Eds., American Chemical Society: Washington, D. C., 1996, pp. 1 – 17. d) Anastas, P. T.; Farris, C. A., *Benign by Design Chemistry* in *Benign by Design: Alternative Synthetic Design for Pollution Prevention,* American Chemical Society Symposium Series, No. 577, Anastas, P. T.; Farris, C. A., Eds., American Chemical Society: Washington, D. C., 1994, pp. 2 – 22. e) Hancock, K. G.; Cavanaugh, M. A., *Environmentally Benign Chemical Synthesis and Processing for the Economy and the Environment* in *Benign by Design Chemistry* in *Benign by Design: Alternative Synthetic Design for Pollution Prevention,* American Chemical Society Symposium Series, No. 577, Anastas, P. T.; Farris, C. A., Eds., American Chemical Society: Washington, D. C., 1994, pp. 23 – 30. f) Anastas, P. T.; Warner, J. C., Green Chemistry: Theory and Practice, Oxford University Press: Oxford, 1998.

2. McHugh, M.; Krukonis, V. *Supercritical Fluid Extraction, Principles and Practice*, Butterworths: Boston, 1986, pp. 1 – 11.

3. Johnston, K. P. in *Supercritical Fluid Science and Technology*, Johnston, K. P.; Penninger, J. M. L., Eds., American Chemical Society: Washington, D.C., 1989, pp. 1 – 12.

4. O'Shea, K. E.; Combes, J. R.; Fox, M. A.; Johnston, K. P. *Photochem. Photobiol.* **1991**, *54*, 571 – 576.

5. Roberts, C. B.; Chateauneuf, J. E.; Brennecke, J. F. *J. Am. Chem. Soc.* **1992**, *114*, 8455 – 8463. Roberts, C. B.; Zhang, J.; Chateauneuf, J. E.; Brennecke, J. F. *J. Am. Chem. Soc.* **1993**, *115*, 9576 – 9582. Roberts, C. B.; Zhang, J.; Brennecke, J. F.; Chateauneuf, J. E. *J. Phys. Chem.* **1993**, *97*, 5618. Roberts, C. B.; Zhang, J.; Chateauneuf, J. E.; Brennecke, J. F. *J. Am. Chem. Soc.* **1995**, *117*, 6553 – 6560.

6. Andrew, D.; Des Islet, B. T.; Margaritis, A.; Weedon, A. C. *J. Am. Chem. Soc.* **1995**, *117*, 6132 – 6133.

7. Bunker, C. E.; Rollins, H. W.; Gord, J. R.; Sun, Y.-P. *J. Org. Chem.* **1997**, *62*, 7324 – 7329.

8. Kim, S.; Johnston, K. P. *AIChE Journal* **1987**, *33*, 1603 - 1611. Kim, S.; Johnston, K. P. *Ind. Eng. Chem. Res.* **1987**, *26*, 1206 -1213. Johnston, K. P.; McFann, G. J.; Peck, D. G.; Lemert, R. M. *Fluid Phase Equilibria* **1989**, *52*, 337 – 346.

9. Kajimoto, O.; Futakami, M.; Kobayashi, T.; Yamasaki, K. *J. Phys. Chem.* **1988**, *92*, 1347.

10. Debenedetti, P. G.; Petsche, I. B.; Mohamed, R. S. *Fluid Phase Equilibria* **1989**, *52*, 347 – 356.

11. Brennecke, J. F.; Tomasko, D. L.; Peshkin, J.; Eckert, C. A. *Ind. Eng. Chem. Res.* **1990**, *29*, 1682 – 1690.

12. Betts, T. A.; Zagrobelny, J.; Bright, F. V. *J. Am. Chem. Soc.* **1992**, *114*, 8163 – 8171. Zagrobelny, J.; Betts, T. A.; Bright, F. V. *J. Am. Chem. Soc.* **1992**, *114*, 5249 – 5257. Heitz, M. P.; Bright, F. V. *J. Phys. Chem.* **1996**, *100*, 6889 – 6897.

13. Randolph, T. W.; Carlier, C. *J. Phys. Chem.* **1992**, *96*, 5146 – 5151. Ganapathy, S.; Carlier, C.; Randolph, T. W.; O'Brien, J. A. *Ind. Eng. Chem. Res.* **1996**, *35*, 19 – 27.

14. Sun, Y.-P.; Bunker, C. E.; Hamilton, N. B. *Chem. Phys. Lett.* **1993**, *210*, 111 – 117. Sun, Y.-P.; Bunker, C. E. *Ber. Bunsenges. Phys. Chem.* **1995**, *99*, 976 – 984.

15. Sun, Y.-P.; Fox, M. A.; Johnston, K. P. *J. Am. Chem. Soc.* **1992**, *114*, 1187 – 1194. Sun, Y.-P.; Bennett, G.; Johnston, K. P.; Fox, M. A. *J. Phys. Chem.* **1992**,

96, 10001 – 10007. Sun, Y.-P; Fox, M. A. *J. Am. Chem. Soc.* **1993**, *115*, 747 – 750. Rhodes, T. A.; Fox, M. A. *J. Phys. Chem.* **1996**, *100*, 17931 – 17939.

16. Anderton, R. M.; Kauffman, J. F. *J. Phys. Chem.* **1995**, *99*, 13759 – 13762.

17. Urdahl, R. S.; Rector, K. D.; Myers, D. J.; Davis, P. H.; Fayer, M. D. *J. Chem. Phys.* **1996**, *105*, 8973 – 8976.

18. Tucker, S. C.; Maddox, M. W. *J. Phys. Chem. B* **1998**, *102*, 2437 – 2453.

19. Takahashi, K.; Abe, K.; Sawamura, S.; Jonah, C. D. *Chem. Phys. Lett.* **1998**, *282*, 361 – 368.

20. Tanko, J. M.; Suleman, N. K.; Fletcher, B. *J. Am. Chem. Soc.* **1996**, *118*, 11958 - 11959. Fletcher, B.; Suleman, N. K.; Tanko, J. M. *J. Am. Chem. Soc.* **1998**, *120*, 11839 - 11844.

21. Skell, P. S.; Baxter, H. N., III *J. Am. Chem. Soc.* **1985**, *107*, 2823 – 2824.

22. Raner, K. D.; Lusztyk, J.; Ingold, K. U. *J. Am. Chem. Soc.* **1988**, *110*, 3519 – 3524.

23. Tanko, J. M.; Anderson, F. E., III *J. Am. Chem. Soc.* **1988**, *110*, 3525 – 3530.

24. Tanner, D. D.; Oumar-Mahamat, H.; Meintzer, C. P.; Tsai, E. C.; Lu, T. T.; Yang, D. *J. Am. Chem. Soc.* **1991**, *113*, 5397 – 5402.

25. Bunce, N. J.; Ingold, K. U.; Landers, J. P.; Lusztyk, J.; Scaiano, J. C. *J. Am. Chem. Soc.* **1985**, *107*, 5464.

26. For a review, see Koenig, T.; Fischer, H. in *Free Radicals, Vol. 1*, Kochi, J. K., Ed., Wiley: New York, 1973, pp. 157 - 189.

27. Tanko, J. M.; Sadeghipour, M. *Angew. Chem. Int. Ed.* **1999**, *38*, 159 - 161.
    1. Walbiner, M; Wu, J.-Q.; Fischer, H. *Helv. Chim. Acta* **1995**, *7*

Chapter 22

Supercritical Fluids as Solvent Replacements in Chemical Synthesis

Jefferson W. Tester[1], Rick L. Danheiser[2], Randy D. Weinstein[1,3], Adam Renslo[2], Joshua D. Taylor[1], and Jeffrey I. Steinfeld[2]

[1]Energy Laboratory and Department of Chemical Engineering, and
[2]Department of Chemistry, Massachusetts Institute of Technology, Cambridge, Massachusetts 02139,
[3]Department of Chemical Engineering, Villanova University, Villanova, PA 19085

This paper discusses the topic of using supercritical fluids as a replacement solvent for chemical synthesis. To begin, a review of what supercritical fluids are and how they work is presented, then a few selective examples of synthetic pathways and reactions are provided to illustrate the factors that control chemical kinetics and selectivity. Much of the content of this paper reflects an active inter-disciplinary collaboration between the Chemical Engineering and the Chemistry Departments at the Massachusetts Institute of Technology. Interactions between chemists and chemical engineers are crucial to the development of this field – particularly in understanding carbon-carbon bond forming reactions in supercritical fluids and in trying to optimize them to enable scaled-up designs for economically competitive processes. In closing, research and development needs and future opportunities are discussed.

Introduction and Motivation

The need for replacement solvents as reaction media for the synthesis of many specialty and commodity chemicals has been clear for some time. Anastas,

Williamson, and Warner (*1*), for example, provide considerable documentation of this need in their recent monographs on green chemistry. Many traditional organic liquid solvents such as benzene, toluene, methylene chloride, THF and DMF are being phased out as a result of US FDA and US EPA regulations to limit emissions of and exposures to potentially toxic and environmentally damaging compounds. Carbon dioxide and water in both liquid and supercritical states are candidates to replace the typical suite of liquid-phase organic solvents in use today because of their attractive physical and toxicological properties.

A supercritical fluid is simply defined as being above both its critical temperature and critical pressure. For a pure component, defining the supercritical region is straightforward, as shown in Figure 1, but for a mixture the situation is more complex, as a locus of critical points typically exists over a range of compositions. In other cases, when fluids having very different critical properties are mixed, multiphase systems may be favored in certain composition regions. For example, two environmentally attractive solvents, H_2O and CO_2, have very different critical temperatures and pressures; CO_2 being close to room temperature at 31.1 °C and a pressure of 74 bar, and H_2O being at a much higher temperature, 374 °C, and a higher pressure at 221 bar. Mixtures of H_2O and CO_2 exhibit distinctly non-ideal behavior with regions having immiscible fluid phases present at CO_2-rich compositions.

In chemical synthesis, mixtures of at least one solvent containing varying amounts of reactants, intermediates, and products are dealt with. In these situations, care has to be taken about whether or not one is operating in a homogeneous supercritical fluid regime. In addition to solubility considerations, many other important properties roughly correlate with fluid density changes which vary significantly in the near-critical region for relatively small changes in T and P.

Table I. lists a few possible advantages for conducting chemical synthesis in a supercritical environment. For example, supercritical fluids might provide a means to manipulate reaction environments by altering density and temperature to influence both reaction rate and selectivity. Futhermore, in a homogeneous supercritical phase, one can, in principle, eliminate interfacial transport limitations. Effectively, temperature and pressure are used to alter density in a way to influence both solvation dynamics and equilibrium solubility. With supercritical fluid solvents, there is the possibility of integrating both reaction and separation processes, which could lead to economic advantages over conventional synthetic processes carried out in liquid solvents.

If tunable reaction rates, adjustable selectivities, and the like could be obtained by adjusting *PVT* conditions, then one may be able to control conversion and yields of specific products. In addition, because of this flexibility in supercritical fluid properties, there may be an opportunity to increase catalyst life and efficiency by varying operating conditions. Even if only marginal gains in reaction performance are realized in supercritical fluids, the net result still would be positive in that some of the more toxic and environmentally unacceptable liquid solvents used widely today in organic synthesis could be replaced.

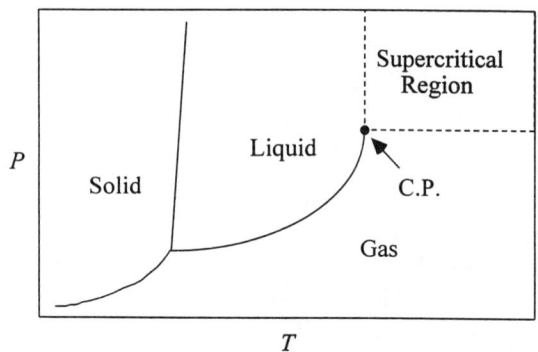

Fluid	T_c (°C)	P_c (bar)	ρ_c (g/cc)
Carbon Dioxide	31.1	73.4	0.46
Water	374	221	0.32

Figure 1. Phase conditions of a pure component where C.P. is the critical point at the end of the liquid-gas coexistence curve.

Table I. Possible advantages of supercritical fluids for waste remediation/treatment, and as an alternative solvent for waste minimization.

1. Enhanced solubilities of reactants or products
2. Elimination of interphase transport limitations
3. Integration of reaction and separation processes
4. Improved catalytic life and efficiency
5. Replacement of halogenated and other environmentally unacceptable solvents
6. Tunable reaction rates, selectivities, and conversions by varying temperature and pressure/density

How do supercritical fluids influence chemical reactions? To begin with, they alter properties (viscosity, diffusivity, dielectric constant, etc.), equilibrium states (solubilities, phase boundaries, etc.) and kinetics (energetic barriers, reaction pathways, etc.). Near critical points, density and pressure conditions have a profound effect at the molecular level, leading to large differences in local density versus bulk average density. The very fact that reactant and product molecules are solvated to different degrees depending on proximity to critical conditions causes a local compositional change suggesting that the solvent could play a critical role in altering mechanistic pathways.

Popular supercritical solvent choices include pure carbon dioxide (CO_2) and mixtures of CO_2 with various co-solvents. These CO_2–based solvents have reasonably low critical temperatures, which makes them ideal for the synthesis of complex molecules that might be more thermally labile at higher temperatures such as in supercritical water. Furthermore, the low costs of both CO_2 and H_2O, their ease in handling, environmental benignness and the ability to adjust solubility by changing density with pressure to induce phase separation are all attractive characteristics.

As illustrated in Figure 2, over a relatively narrow range of pressure, near the critical point, CO_2 displays a significant change in the density, viscosity and diffusivity. The viscosity is almost gas-like in the supercritical region while the self-diffusion coefficient is somewhere midway between that of a liquid and a gas. Just how these properties can be exploited for organic synthesis has defined a key global research objective.

Research groups in the U.S., Europe, and Japan are investigating a range of prospective chemical transformation reactions in supercritical fluids as a means of synthesizing useful compounds. Research typically is focused on obtaining measurements of reaction yields, rates and selectivities under well-defined conditions. Studies of physical chemistry effects at both macroscopic and molecular levels are also common. Figure 3 illustrates the connections among synthetic organic chemistry, physical chemistry and chemical engineering disciplines that characterize this field.

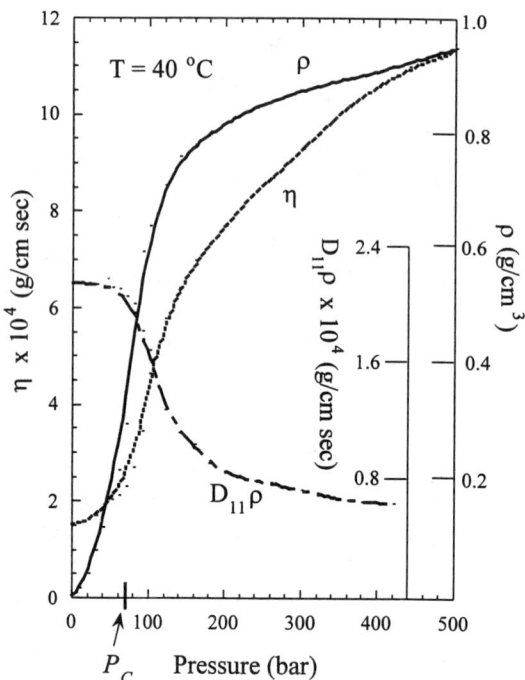

Figure 2. Isothermal change in the density (ρ), viscosity (η), and self-diffusivity (D_{11}) of Carbon Dioxide (adapted from Subramaniam and McHugh (38)).

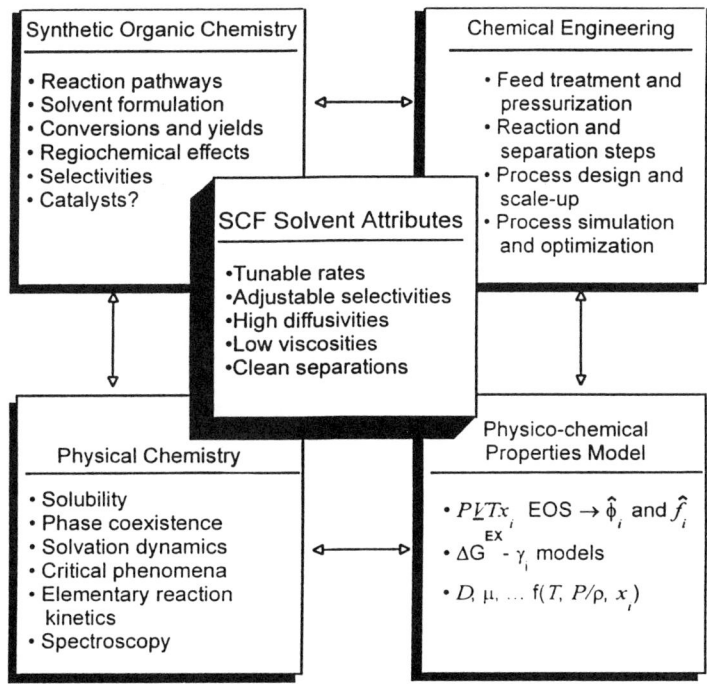

Figure 3. Connecting organic and physical chemistry to chemical engineering in utilizing supercritical fluids (SCF) for chemical synthesis

Carbon-carbon bond formation: Diels-Alder reaction

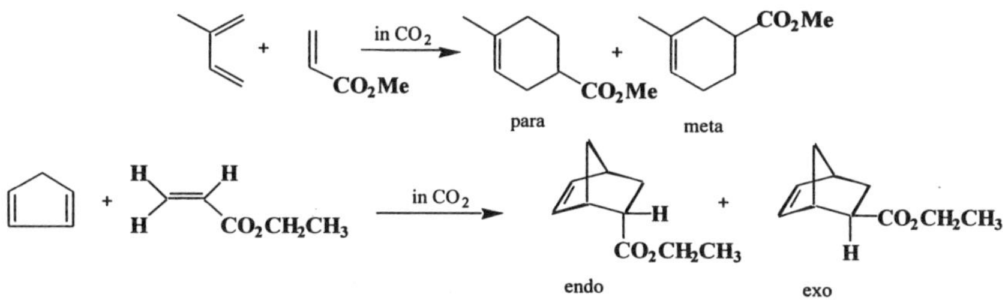

Carboxylation reaction: the Kolbe-Schmitt (Marasse modification) reaction

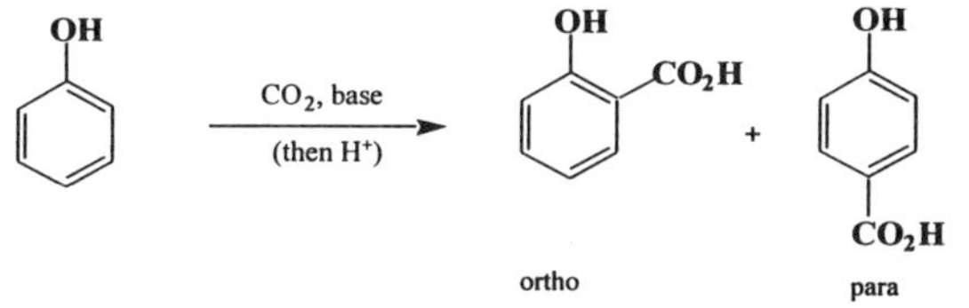

Figure 4(a). Representative synthetic transformations in scCO$_2$ (Under study by Tester, Danheiser, and co-workers (7, 26)).

Phase Transfer Catalysis
(examined by Eckert and co-workers (9,10))

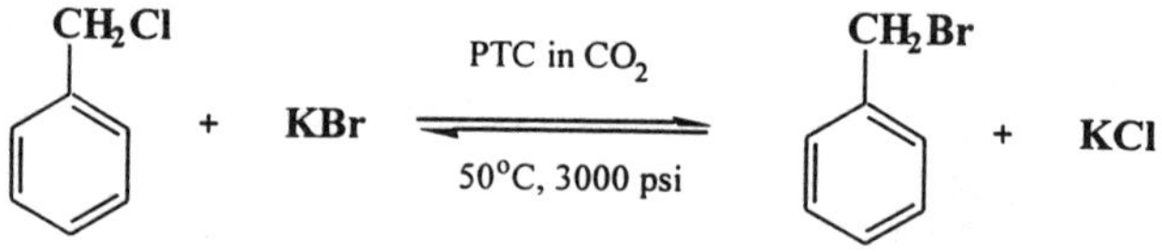

PhCH$_2$Cl + KBr $\underset{50°C, 3000\ psi}{\overset{PTC\ in\ CO_2}{\rightleftharpoons}}$ PhCH$_2$Br + KCl

PTC = tetra-n-heptylammonium bromide or 18-crown-6

Polymerization
(examined by DeSimone and co-workers (18))

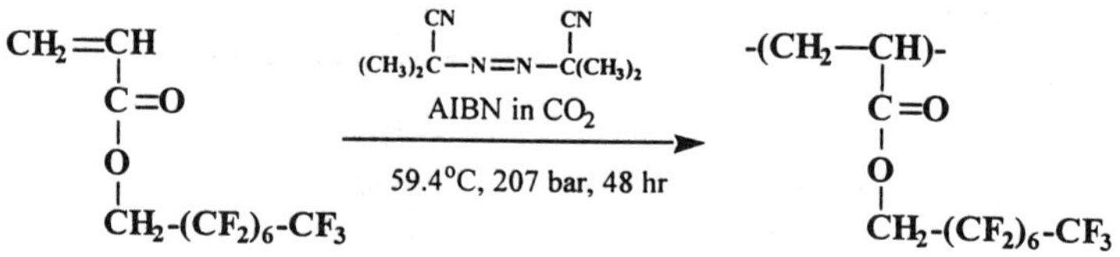

CH$_2$=CH–C(=O)–O–CH$_2$-(CF$_2$)$_6$-CF$_3$ $\xrightarrow[59.4°C,\ 207\ bar,\ 48\ hr]{(CH_3)_2C(CN)-N=N-C(CN)(CH_3)_2 \\ AIBN\ in\ CO_2}$ –(CH$_2$–CH)– with side chain C(=O)–O–CH$_2$-(CF$_2$)$_6$-CF$_3$

Friedel-Crafts reaction
(examined by Tumas, Burk, and co-workers (27,28))

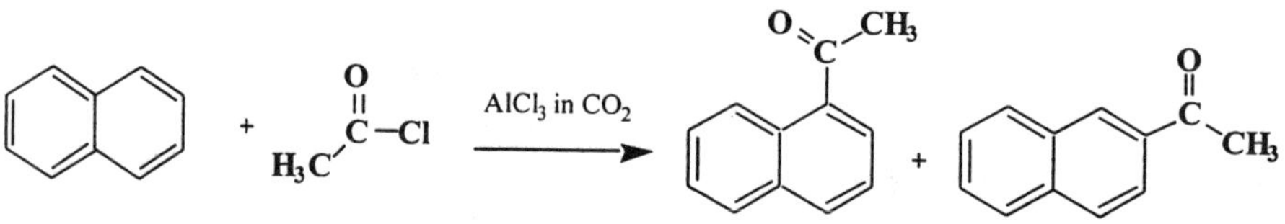

Naphthalene + H$_3$C–C(=O)–Cl $\xrightarrow{AlCl_3\ in\ CO_2}$ 1-acetylnaphthalene + 2-acetylnaphthalene

Figure 4(b). Representative synthetic transformations in CO$_2$

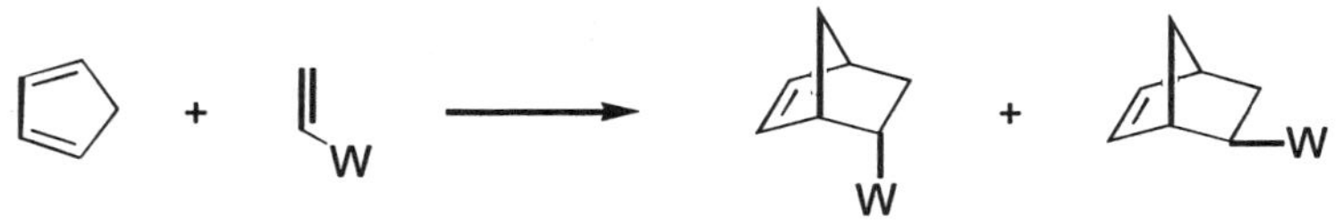

	Conditions	Promoter	Yield (%)	Endo:Exo
W=COMe				
1	CO_2, 50 °C, 4 h	none	29	82:18
2	CO_2, 50 °C, 4 h	0.5g SiO_2	82	92:8
W=CN				
3	CO_2, 50 °C, 4 h	none	5	57:43
4	CO_2, 50 °C, 4 h	0.5g SiO_2	14	59:41
W=CO_2Me				
5	CO_2, 50 °C, 4 h	none	5	72:28
6	CO_2, 50 °C, 4 h	0.5g SiO_2	21	85:15

Figure 4(c). Silica promoted Diels-Alder reactions in scCO2 (Under study by Tester, Danheiser, and co-workers (20)).

Table II. Regioselectivity of Diels-Alder reactions in Carbon Dioxide and conventional solvents (from Renslo et al. (7))

Entry	Conditions	Yield (%) [a]	Ratio [b]

[Reaction scheme: isoprene + methyl acrylate (CO₂Me) → product **1** (para, CO₂Me at 4-position) + product **2** (meta, CO₂Me at 3-position)]

Entry	Conditions	Yield (%) [a]	Ratio [b]
1	PhCH$_3$, 145 °, 15 h	78	71:29 (71:29)
2	PhCH$_3$, 50 °, 3 d	(7)	69:31 (72:28)
3[c]	CO$_2$, 49.5 bar, 50 °, 4 d	(11)	69:31 (73:27)
4[c]	CO$_2$, 74.5 bar, 50 °, 4 d	(5)	67:33 (73:27)
5	CO$_2$, 95.2 bar, 50 °, 7 d	(4)	71:29 (73:27)
6	CO$_2$, 117 bar, 50 °, 3 d	(3)	70:30 (72:28)

[Reaction scheme: t-Bu-substituted diene + methyl acrylate → product **3** (para) + product **4** (meta)]

Entry	Conditions	Yield (%) [a]	Ratio [b]
7	neat, 185 °, 16 h	78	63:37 (63:37)
8	PhCH$_3$, 50 °, 3 d	(19)	69:31 (68:32)
9	CO$_2$, 87 bar, 50 °, 3 d	(5)	71:29 (68:32)
10	CO$_2$, 117 bar, 50 °, 3 d	(4)	69:31 (68:32)
11	CO$_2$, 117 bar, 150 °, 24 h	54	65:35 (64:36)

Table II. Continued

		5	6
12	PhCH$_3$, 110 °, 45 h	48	87:13
13	CO$_2$, 90 bar, 50 °, 3 d	trace[d]	
14	CO$_2$, 117 bar, 50 °, 3 d	(<1)	92:8[e]
15	CO$_2$, 117 bar, 150 °, 24 h	(31)	85:15

		7	8
16	PhH, BHT, 60 °, 5 h	77	81:19 (81:19)
17	CO$_2$, 86 bar, 50 °, 3 d	(31)	84:16 (84:16)
18	CO$_2$, 117 bar, 50 °, 3 d	(14)	84:16 (84:16)

[a] Isolated yield (estimated by ^{1}H NMR).
[b] Ratio of isomers determined by ^{1}H NMR and GC analysis.
[c] Two-phase reaction mixture observed.
[d] Ratio could not be determined.
[e] Approximate ratio due to low conversion.

In this paper we select a few examples to illustrate the types of reactions being studied (see Figure 4a-c). Others are discussed in greater depth in the papers appearing in this volume, as well as in recent reviews [see Savage et al. (2) and numerous papers in Noyori (3)]. For example, cycloadditions, including Diels-Alder reactions, represent a popular synthetic pathway investigated in scCO_2. A number of research groups, including ours at MIT, have been interested in these types of reactions, partly because they are representative of carbon-carbon bond-forming reactions and partly because of their general importance in practical manufacturing operations.

Two specific Diels-Alder reactions that have been examined at MIT are shown in Figure 4(a): (1) isoprene and methyl acrylate to produce para and meta isomers, and (2) cyclopentadiene and ethyl acrylate to produce an endo-exo isomer product mixture. Others have also examined regioselectivity effects for a number of Diels-Alder reactions [for example, see Kim and Johnston (4) and Ikushima et al. (5,6)]. In the latter study of the cycloaddition of isoprene and methyl acrylate, Ikushima and co-workers reported a significant reversal of regioselectivity in supercritical carbon dioxide (scCO_2) near its critical pressure at 50 °C that they attributed to *steric* effects. Intrigued by this, MIT researchers re-examined the isoprene-methyl acrylate reaction with an improved reactor system that provided visual access, enabling us to identify the phases that were present [Renslo et al. (7), also see Table II]. This study yielded completely different results than reported by Ikushima et al.; namely lower conversions and no reversal of the normal Diels-Alder regioselectivity [para:ortho = (67 to 72%):(33 to 28%)] typically obtained in traditional liquid solvents such as toluene (see entries 1 and 2 in Table II). Furthermore, the MIT supercritical fluids research group's experiments indicated that Ikushima and co-workers may have been operating in a two-phase region rather than in a homogeneous supercritical phase.

In a separate study, the MIT group investigated the Diels-Alder reaction of methyl acrylate with 2-*t*-butyl-1,3-butadiene, assuming that the ratio of isomers produced from this diene in scCO_2 would be more sensitive to reaction conditions if steric interactions were important (7). Little variation in regioselectivity was observed relative to normal Diels-Alder synthesis.

Similar non-dramatic results were obtained when the MIT group varied the electronic character of the dienophile substituents [see Table 2, entries 12-18]. The reaction of 2-trimethylsiloxybutadiene with methyl acrylate had previously been reported to produce isomers **5** and **6** in a ratio of 98:2 by Jung et al. (8). Under identical conditions (refluxing toluene, 45 h, entry 12), **5** and **6** were formed in an 87:13 ratio as determined by ^{1}H NMR analysis. In scCO_2 this reaction proved very sluggish, yielding only traces of **5** and **6** after 3 days at 50 °C (entries 13, 14). At 150 °C, higher conversions were obtained (entry 15) with a selectivity similar to that observed in toluene (entry12). The last system examined involved the reaction of nitroethylene with isoprene. Surprisingly, under standard conditions (benzene, BHT, 60 °C, entry 16) the expected Diels-Alder adducts **7** and **8** were obtained in a ratio of 81:19 in contrast to the ratio of 95:5 reported previously under identical conditions. When the reaction was conducted in scCO_2, in the absence of a radical inhibitor such as BHT, very similar regioselectivity (84:16) was observed (entries 17, 18).

The MIT group also studied Kolbe-Schmitt synthesis as a model of direct carboxylation reaction in $scCO_2$. Here the focus was on understanding the kinetics of generating the set of intermediates – one of which, of course, leads to the production of aspirin. The ortho and para selectivity between isomers of hydroxylated benzoic acid are of particular interest.

Another important area of research in supercritical media is performed in Professor Eckert's laboratory at Georgia Tech, where phase transfer catalyzed reactions are being evaluated (*9, 10*). For example, consider the bromination displacement of a chlorinated aromatic in supercritical CO_2, shown in Figure 4(b).

Johnston and co-workers (see Hrnjez, *et al.*, (*11*)) report a small effect of pressure on the stereochemical pathway of the photochemical dimerzation of isophorone in $scCO_2$. Weedon and co-workers (*12*) have observed an interesting increase in the ratio of rearrangement to cage-escape products in the photo-Fries rearrangement of naphthyl acetate in near-critical CO_2. Noyori and co-workers have shown that the homogeneous catalytic hydrogenation of CO_2 to formic acid and its derivatives proceeds significantly faster in $scCO_2$ (in the presence of a co-solvent) than in traditional liquid solvents (*13, 14, 15, 16*). Although carbon dioxide has not yet been shown to provide dramatic advantages over conventional solvents with respect to improved reaction rates and selectivities, the potential environmental and process separation advantages of $scCO_2$ continue to fuel interest in this area.

A significant limitation in using $scCO_2$ as a replacement solvent is the low solubilities of many common reagents and reactants in carbon dioxide. The solvating ability of high density CO_2 (supercritical or liquid) is often compared to that of non-polar organic liquid solvents such as hexane. Given this inherent limitation, several innovative approaches have been proposed to improve the solubility of reactants in CO_2. These include Johnston's group's (*17*) investigation of water in carbon dioxide microemulsions and DeSimone *et al.'s* (*18*) work on self-assembled micelles of block copolymer amphiphiles.

Catalytic pathways in supercritical fluids are also being evaluated. For example, Burk *et al.* (*19*) examined rhodium-catalyzed asymmetric hydrogenations of enamide in supercritical carbon dioxide. Furthermore, since many standard synthetic reactions studied in $scCO_2$ occur too slowly to be practical for commercial application, catalyzed routes are of interest. While resorting to various Lewis acid promoters might provide a remedy to this problem, the use of such additives diminishes the environmental benefits associated with the use of $scCO_2$. For this reason, our group recently investigated the use of environmentally benign, reusable solids such as silica and alumina as promoters for organic reactions in $scCO_2$ (*20*). Among the many attractive features of these materials are their low cost, minimal environmental impact, easy separation from CO_2, and potential for reuse. For example, silica (SiO_2) is an environmentally benign promoter for the Diels-Alder reaction in $scCO_2$.

Despite considerable studies of silica promoters in traditional liquid solvents [Basiuk, (*21*), Kropp *et al* (*22*) and Kodomari *et al*, (*23*)], the use of silica or alumina to promote reactions in $scCO_2$ had not been reported as of 1997. Alternatively, Poliakoff and co-workers reported Friedel-Crafts (*24*) and hydrogenation (*25*) reactions in $scCO_2$ using sulfonylated polysiloxane and polysiloxane-supported palladium promoters, respectively. These reactions involve the use of gaseous reagents (propene and hydrogen), thereby taking advantage of the high solubility of these gases in $scCO_2$. Compared to the polysiloxane polymers used in their investigations, silica has some advantage in having lower costs.

In order to characterize the reaction system, the partitioning of species between solid silica and fluid phases was measured and modeled as a function of operating conditions. Also studied was the use of protic and Lewis acid dopants to further enhance reaction rates.

In gaseous and supercritical CO_2, the yield and selectivity of several Diels-Alder reactions are enhanced by fumed silica with a high specific area (≥ 400 m^2/g). Acid doping of silica promoters has been shown to greatly enhance their activating effect. The initial studies with these promoters lay the groundwork for their application to a wide range of other acid and Lewis acid promoted reactions in $scCO_2$. Although the selectivity of these silica-enhanced reactions are not affected by changes in pressure or fluid density; product yields generally decrease as pressure increases. This is probably caused by the changes in concentration of the reactants on the surface of the silica particles. At each reaction temperature, adsorption isotherms were obtained at selected pressures. As expected, when pressure is increased, the ratio of the amount of reactants adsorbed on the silica surface to amount of reactants in the fluid phase decreases, thus causing the yield to decrease. A Langmuir adsorption model fits the data well, primarily as a result of the decrease in the Langmuir equilibrium partitioning constant that occurs with increasing pressure [see Weinstein *et al.* (*20*) for details].

Professor DeSimone's group at North Carolina has focused on carrying out polymerization reactions in supercritical fluids (*18*). One representative example is shown in Figure 4(b), where a fluorinated monomer which is soluble in CO_2 is polymerized via free radical reactions. Control of molecular weight distribution has been achieved and the rates are comparable to other synthesis methods. As a direct result, there is considerable commercial interest in using supercritical CO_2 as solvent replacement for fluorinated polymer synthesis.

Dr. Bill Tumas's group at Los Alamos National Laboratory has been working for some time on a wide range of possible synthetic reactions in supercritical CO_2 and co-solvent mixtures thereof [Tumas *et al.* (*27, 28*)]. Figure 4(b) shows one reaction under investigation that would be particularly important if it could be developed and exploited; namely the idea of using a Lewis acid in dense and supercritical CO_2 to facilitate acrylations or alkylations.

Another area that has been of interest to our group at MIT as well as Eckert's group (*9, 10*) involves the hydrolysis and oxidation of compounds in near and supercritical water. In the early 1980s, Helling and Tester started to study the simple water-gas shift reaction with CO reacting with water to form hydrogen and CO_2 [Helling and Tester (*29, 30*)]. At reaction conditions near the critical point and slightly above, this reaction is roughly two to three orders of magnitude faster than one would predict by extrapolating low pressure homogeneous gas phase reaction kinetics. Clearly, water is influencing the kinetic pathway of the water-gas shift channel in a fundamentally different way at supercritical densities.

In addition, a number of other opportunities for waste treatment and destruction of very toxic compounds involve oxidation pathways in supercritical fluids, where a free radical mechanism is invoked to achieve complete destruction of a wide range of organic compounds. Mineralization to CO_2, H_2O, and inorganic acids and/or salts is complete at residence times of only 60 to 120 s at 550 °C and 230 bar in supercritical water for a substantial range of toxic organic compounds including dioxin precursors, PCBs, mixtures of toxic chlorinated solvents, or mustard or VX chemical warfare agents. Supercritical water oxidation (SCWO) is being pursued commercially for the detoxification of these types of compounds.

As an example of the complexity of SCWO, consider the (CH_2Cl_2) hydrolysis and oxidative pathways for methylene chloride [see Figure 5, and for further details Marrone et al. (*31*) and Tester et al. (*32*)]. In near-critical and subcritical water, the primary channel is the direct hydrolysis of CH_2Cl_2 to form hydrochloric acid (HCl) and formaldehyde (CH_2O). As the reaction mixture is heated, passing through the critical region, and entering the supercritical region at temperatures of approximately 400 °C and pressures of 230 to 250 bar, a free-radical pathway begins to appear. In this parallel pathway, carbon monoxide is produced and becomes an important refractory intermediate which is further oxidized to CO_2 in two main global reaction channels: (1) the water-gas shift reaction and (2) direct oxidation to CO_2. At temperatures above 450 °C at these pressures, the free-radical channel dominates the conversion of CH_2Cl_2.

Physical Chemistry Effects in Supercritical Fluids

As mentioned earlier, physical chemistry plays a significant role in providing understanding of phenomena important to developing supercritical fluid technology for chemical synthesis applications (see Figure 3). A previous study of Diels-Alder reactions with cyclopentadiene and ethyl acrylate in our group illustrates this important role. As mentioned above, of paramount interest was the intrinsic rate of this bimolecular reaction as well as the endo to exo selectivity in $scCO_2$ (*26*). For this particular reaction, modest control of rate was possible by adjusting pressure to alter density as shown in Figure 6. A 30% increase in the bimolecular rate constant, k^*, resulted from increasing the pressure from 80 to 210 bar at 38 °C.

k^* is a second-order rate constant that can be represented by a simplified transition state theory approach. It contains parameters that involve the free energy change in the reaction ΔG_{rx}^{o} in a pre-selected standard reference condition between the transition state and the reactants, a term that captures non-ideal behavior using an activity coefficient for each component, and an explicit term in density. Using standard notation, k^* is given on a concentration basis by:

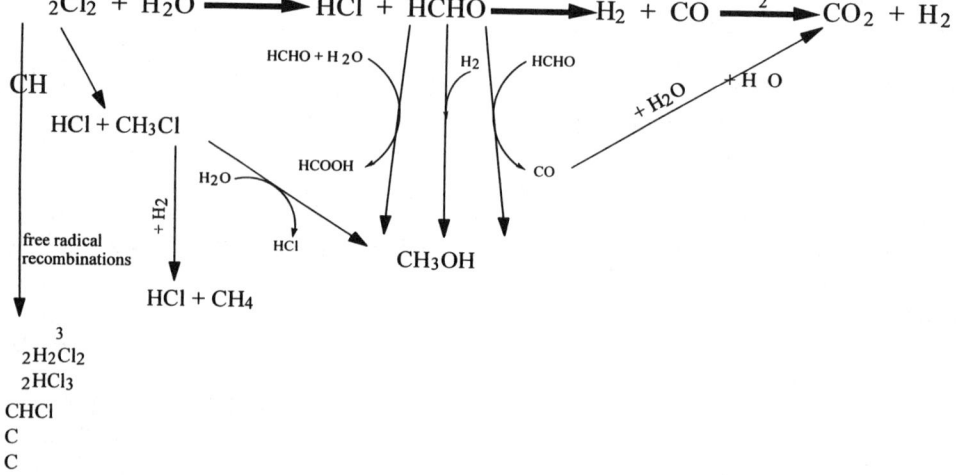

Figure 5. Proposed reaction network for CH_2Cl_2 breakdown under hydrolysis conditions (see Marrone, et al. (37)).

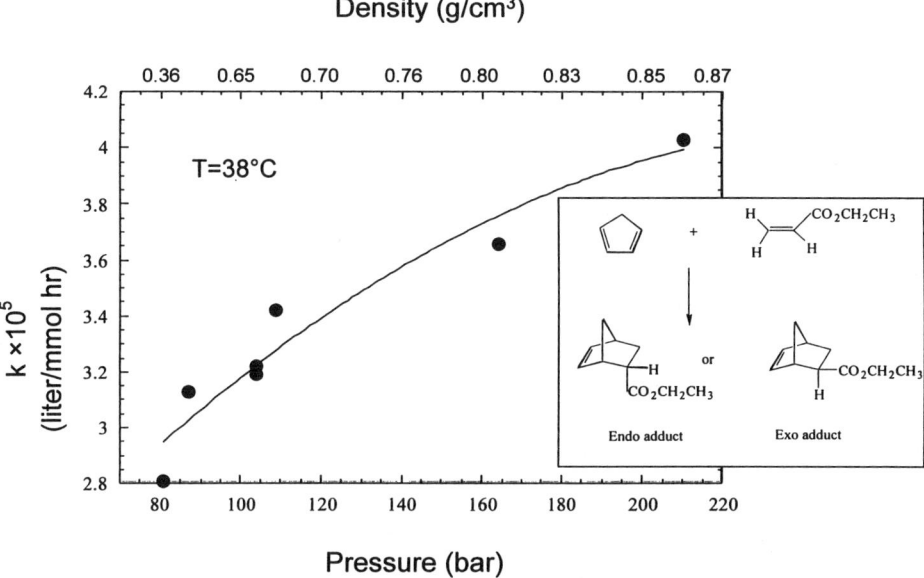

Figure 6. Diels-Alder bimolecular rate constant based on total conversion of cyclopentadiene and ethyl acrylate to products at 38 °C in $scCO_2$ (26).

$$k^* = \kappa \frac{k_B T}{h} \frac{K_a^{\neq}}{K_\gamma^{\neq}} \frac{1}{\rho} \qquad (1)$$

where

κ = transmission coefficient
k_B = Boltzmann's constant
h = Planck's constant
T = temperature
ρ = fluid density

$K_a^{\neq}$ = activity-based thermodynamic equilibrium constant =

$$\Pi \left(\frac{\hat{f}_i}{\hat{f}_i^+} \right)^{\upsilon_i} = \exp\left[\frac{-\Delta G_{rx}^o}{RT} \right]$$

for the reaction

$A + B = (A\text{-}B)^{\neq}_{\text{transition state}}$ where $i = A, B, \text{ or } \neq$

where

$K_\gamma^{\neq}$ = activity-coefficient-based equation $\equiv \Pi(\gamma_i)^{\upsilon_i}$

$\hat{f}_i$ = mixture fugacity of component i at T, P or ρ, x_i

$\hat{f}_i^+$ = reference state fugacity of component at T, P^+, x_i^+

Using Eq. (1) and a mole fraction basis, the ratio k/k_o is given as:

$$\ln\left(\frac{k_x(T,\rho)}{k_x(T,\rho_o)} \right) = \ln\left(\frac{k_x}{k_{x,o}} \right) = \ln\left(\frac{\kappa}{\kappa_o} \right) - \left(\frac{\Delta \overline{G}_{\neq}^+ - \Delta \overline{G}_{\neq,o}}{RT} \right) - \ln\frac{K_\gamma^{\neq}}{K_{\gamma,o}^{\neq}} \qquad (2)$$

$$\ln\left(\frac{k_x}{k_{x,o}} \right) = \ln\left(\frac{k}{k_o} \right) + \ln\left(\frac{\rho}{\rho_o} \right) \qquad (3)$$

In Eqs. (2) and (3) and in Figure 7, we have normalized k using its value k_o at a fixed density of $\rho_o = 0.5$ g/cm^3 and T. For several reactions analyzed, most of the density

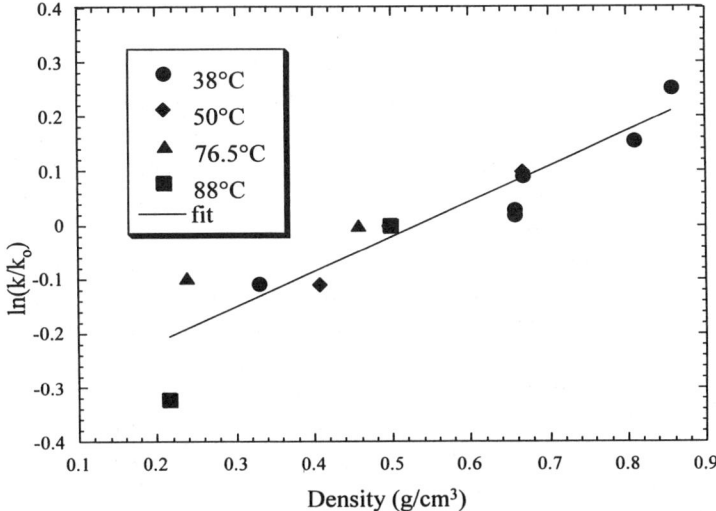

Figure 7. Ratio of bimolecular rate constants for the Diels-Alder reaction of cyclopentadiene and ethyl acrylate in CO_2 versus solvent density. k_o is the normalization factor as the rate constant at $\rho = 0.5$ g/cm^3 and T (26).

dependence of $k_X/k_{X,o}$ is accounted for in the (ρ/ρ_o) term and not a strong function of density-dependent effects on κ and $K_\gamma^{\neq}$.

One might expect that if we plot this normalized rate constant as a function of density, a reasonable empirical correlation for each temperature would be obtained. Figure 7 represents k/k_o as a function of ρ for a particular reaction between cyclopentadiene and ethyl acrylate over a range of temperatures from 38 to 88 °C.

Research and Development Needs and Opportunities

As illustrated by the examples discussed, molecular-level understanding and ultimately the proper design and control of practical processes in supercritical fluids requires strong linkages between synthetic organic and physical chemistry. These include the selection of reaction pathways; the choice of particular solvents and mixtures of solvents; the need for co-solvents and identifying operating conditions to obtain desired conversions and yields; selectivity (regiochemical or stereochemical); and whether or not selective catalysts are required. In addition, correlating global and elementary reaction kinetics and spectroscopic behavior is of importance in predicting dominant reaction channels and rates. Proper design of separation steps requires quantitative understanding of solubility, including the appearance and disappearance of phase boundaries and conditions for coexistence, solvation dynamics, and critical phenomena themselves – particularly if one is close to a critical point.

Ultimately, chemical engineering methods and technology are needed to transform this molecular and empirical chemistry into a real process. This chemistry must be understood in terms of many scale-up, design and simulation issues that will influence the economics of the process. For example, process equipment and operating procedures in supercritical fluids require special attention to safety issues because of the high pressures involved. There is an economic penalty in achieving the compression required to reach operating pressures and to recover products at ambient conditions. Sometimes extensive pretreatment of feeds is required or a specialized co-solvent may have to be added.

Of course, the ability to carry out accurate process design and simulation calculations depends on the availability of high quality data and the robustness of property models that are accessible. Quantitative predictions of the temperature, density or pressure and compositional dependence of phase equilibria and kinetics in multicomponent mixtures are needed. On a molecular level, it would be useful to be able to connect the structure of transition states with solvation effectiveness (structure and energetics) properly accounting for effects on chemical kinetics and selectivities.

Enhanced diagnostics to probe into the molecular-level events that are occurring in supercritical fluids is an area of enormous opportunity that needs to be expanded. There are several approaches and methods that could help with these ambitious goals, and there is interesting work going on in a number of research groups, but much more remains to be done. A few examples where additional research could help include: (1) using Raman spectroscopy to improve the understanding of reaction pathways and solvation dynamics in supercritical fluids in an *in-situ* mode; (2) molecular modeling employing *ab initio* quantum chemistry and molecular dynamics methods to provide better understanding of solvation processes as well as to probe the influence of solvent interaction on reaction rates; and (3) using NMR imaging and spectroscopy as a way to probe solution structure and measure properties.

Careful analysis of phase coexistence limits and solubility boundaries in supercritical fluid mixtures with and without co-solvents is very important. To accurately predict or extrapolate behavior to commercial-scale systems, more data has to be collected on candidate systems. Although there is active research in this area, much of the work done so far is very preliminary. More detailed studies with duplication in other laboratories are needed before commercialization will be realized.

Finally, for practical chemical processing, good constitutive models for physical and chemical properties and the ability to predict or correlate reaction kinetics are prerequisites for competent process modeling and flowsheet simulation in order to explore operability and economic feasibility.

Conclusions

The central theme of this review has been to provide an overview of how supercritical fluids might serve as green solvent replacements in a wide range of chemical synthesis applications. While the evaluation of their suitability is incomplete, several examples from ongoing research were discussed to illustrate the potential of using the unique properties of supercritical fluids involving CO_2 and H_2O to achieve acceptable conversions and selectivities for practical processes.

Acknowledgements

We are very grateful to many colleagues at MIT and elsewhere with interests in chemistry of supercritical fluids. Particular thanks also goes to Kenneth Smith, Alan Hatton, Jack Howard, Phillip Marrone, and Bill Peters at MIT, Keith Johnston at Texas, Joseph DeSimone at North Carolina, Robert Shaw at the U.S. Army Research Office, Paul Anastas, Lauren Bartlett, Barbara Karn, and Tracy Williamson at the U.S. EPA/NSF, and William Tumas and Steve Buelow at Los Alamos National Laboratory. Partial support is provided by the U.S. EPA/NSF under their Green Chemistry Program (grant #: R826738-01-0), the U.S. Army Research Office under the University Research Initiative Program (grant #s: DAAL03-92-G-0177-URI, DAAG04-94-G-0145, DAAH04-96-1-0174), the Martin Foundation, Clean Harbors Corporation, the Emissions Reduction Research Center, and the Alliance for Global Sustainability is appreciated. We thank Gillian Kiley for her help in preparing the manuscript for publication.

Literature Cited

1. (a) Anastas, P. T.; Williamson, T. C. *ACS Symp. Ser.* **1996**, *626*, Chapter 1ff, 1,; (b) Anastas, P.T; Williamson, T.C. *Green Chemistry: Frontiers in Benign Chemical Syntheses and Processes.* Oxford University Press: Oxford, 1998 (c) Anastas, P.T.; Warner, J.C. *Green Chemistry: Theory and Practice.* Oxford University Press: Oxford, 1998.
2. Savage, P. E.; Gopalan, S.; Mizan, T. I.; Martino, C. J.; Brock, E. E. *AIChE J.* **1995**, *47*, 1723.
3. Noyori, R. "Supercritical Fluids", *Chem. Reviews* **1999**, *99*, 353 (10 papers included).
4. Kim, S.; Johnston, K. P. *Chem. Engr. Commun.* **1988**, *63*, 49.
5. Ikushima, Y.; Ito, S.; Asano, T.; Yokoyama, T.; Saito, N.; Hatakeda, K.; Goto, T. *J. Chem. Eng. Jpn.* **1990**, *23*, 96.
6. Ikushima, Y.; Saito, N.; Arai, M. *J. Phys. Chem.* **1992**, *96*, 2293.
7. Renslo, A. R.; Weinstein, R. D.; Tester, J. W.; Danheiser, R. L. *J. Org. Chem.* **1997**, *62*, 4530.
8. Jung, M. E.; McCombs, C. A.; Takeda, Y.; Pan, Y-G. *J. Am. Chem. Soc.* **1981**, *103*, 6677.
9. Eckert, C. A. *Ann. Rev. Phys. Chem.* **1972**, *23*, 239.
10. Eckert, C. A.; Hsieh, C. K.; McCabe, J. R. *AIChE J.*, **1974**, *20*, 20.
11. Hrnjez, B. J.; Mehta, A. J.; Fox, M. A.; Johnston, K. P. *J. Amer. Chem. Soc.* **1989**, *111*, 2662.
12. Andrew, D.; Des Islet, B. T.; Margaritis, A.; Weedon, A. C. *J. Am. Chem. Soc.* **1995**, *117*, 6132.
13. Jessop, P. G.; Ikariya, T.; Noyori, R. *Nature*, **1994**, *368*, 231.
14. Jessop, P. G.; Ikariya, T.; Noyori, R. *Science*, **1995**, *269*, 1065.
15. Jessop, P. G.; Hsiao, Y.; Ikariya, T.; Noyori, R. *J. Am. Chem. Soc.* **1996**, *118*, 344.
16. Jessop, P. G. *Top. Catalysus*, **1998**, *5*, 95.
17. Johnston, K. P.; Harrison, K. L.; Clarke, M. J.; Howdle, S. M.; Heitz, M. P.; Bright, F. V.; Carlier, C.; Randolph, T. W. *Science*, **1996**, *271*, 624.
18. DeSimone, J. M.; Guan, Z; Elsbernd, C. S. *Science*, **1992**, *257*, 945.
19. Burk, M. J.; Feng, S.; Gross, M. F.; Tumas, W. *J. Amer. Chem. Soc.* **1995**, *117*, 8277.
20. Weinstein, R. D.; Renslo, A. R.; Danheiser, R. L.; Tester, J. W. *J. Phys. Chem.B* **103** (15) 2878-2887 **1999**.
21. Basiuk, V. A. *Russ. Chem. Rev.* **1995**, *64*, 1003.
22. Kropp, P. J.; Breton, G. W.; Craig, S. L.; Crawford, S. D.; Durland, W. F.; Jones, J. E.; Raleigh, J. S. *J. Org. Chem.* **1995**, *60*, 4246.
23. Kodomari, M.; Nawa, S.; Miyoshi, T. *J. Chem. Soc., Chem. Commun.* **1995**, 1895.
24. Hitzler, M.G.; Poliakoff, M; J. "Chem. Soc., Chem. Commun. **1998**, 259.

25. Hitzler, M. G.; Poliakoff, M. *J. Chem. Soc., Chem. Commun.* **1997**, 1667.
26. Weinstein, R. D.; Renslo, A. R.; Danheiser, R. L.; Harris, J. G.; Tester, J. W. *J. Phys. Chem.* **1996**, *100*, 12337.
27. Morgenstern, D. A.; LeLacheur, R. M.; Morita, D. K.; Borowsky, s. L.; Feng, S.; Brown, G. H.; Luan, L.; Gross, M. F.; Burk, M. J.; Tumas, W. *ACS Symp. Ser.* **1996**, *626*, 132.
29. Burk, M.J.; Feng, S.; Gross, M. G.; Tumas, W. *J. Am. Chem. Soc.* **1995**, *117*, 8277.
30. Helling, R. K.; Tester, J. W. *Energy and Fuels* **1987**, *1*, 417-423.
31. Helling, R. K.; Tester, J. W. *Environ. Sci. Technol.* **1988**, *22* (11), 1319-1324.
32. Marrone, P. A.; Gschwend, P. M.; Swallow, K. C.; Peters, W. A.; Tester, J. W. *J. Phys. Chem. A.* **1998**, *102*, 7013-7128.
33. Tester, J. W.; Marrone, P. A.; DiPippo, M. M.; Sako, K.; Reagan, M. T.; Arias, T.; Peters, W. A. *J. Supercritical Fluids* **1998**, *13*, 225-240.
34. Black, H. *Environ. Sci. Technol.* **1996**, *30*, 124A.
35. Clifford, T.; Bartle, K. *Chem. & Ind. 1996*, 449.
36. Hitzler, M. G.; Smail, F. R.; Ross, S. K.; Poliakoff, M. *J. Chem. Soc., Chem. Commun.* **1998**, *259*.
37. Hitzler, M. G.; Poliakoff, M. *J. Chem. Soc., Chem. Commun.* **1997**, 1667.
38. McHugh, M. A.; Krukonis, V. J. *Supercritical Fluid Extraction*, 2nd ed., Butterworth-Heinmann: Boston, 1994.
39. Subramaniam, B.; McHugh, M. A. *Ind. Eng. Chem. Process Des. Dev.* **1986**, *25*, 1.

Chapter 23

Expeditious Solvent-Free Organic Syntheses Using Microwave Irradiation

Rajender S. Varma

National Risk Management Research Laboratory, U.S. Environmental Protection Agency, 26 West Martin Luther King Drive, Cincinnati, OH 45268

Microwave-expedited solvent-free synthetic processes involve the exposure of neat reactants to microwave (MW) irradiation in the presence of supported reagents or catalysts on mineral oxides. Recent developments are described and the salient features of these high yield protocols, namely the enhanced reaction rates, greater selectivity and the experimental ease of manipulation are highlighted. Research from our laboratory on the use of supported reagents or recyclable mineral oxides such as $Fe(NO_3)_3$-clay (clayfen), $Cu(NO_3)_2$-clay (claycop), NH_4NO_3-clay (clayan), NH_2OH-clay, $PhI(OAc)_2$-alumina, $NaIO_4$-silica, CrO_3-alumina, MnO_2-silica, and $NaBH_4$-clay etc., and their role in MW-promoted deprotection, condensation, cyclization, rearrangement, oxidation and reduction reactions including the rapid one-pot assembly of useful heterocyclic compounds from *in situ* generated intermediates, are described.

Introduction

Heterogeneous organic reactions facilitated by reagents immobilized on porous solid supports have advantages over conventional solution phase reactions because of the good dispersion of active reagent sites, associated selectivity and easier work-up procedures. Although first described in 1924 (1) the technique did not receive attention until the late 1970s with the appearance of two reviews (2,3), as well as other books and journal articles (4-11). The recyclability of some of these mineral supports and their application under solvent-free condition render these processes truly eco-friendly clean protocols (11). The usage of solvents, which are used in larger amounts in organic synthesis, is reduced or eliminated thus preventing pollution 'at-source'.

The recent use of microwave (MW) irradiation techniques for the acceleration of organic reactions had profound impact on these heterogeneous reactions. Since the appearance of the first article on the application of microwaves for chemical reactions in polar solvents (12), the approach has blossomed into a useful technique for a variety of applications in organic synthesis and functional group transformation (*13-59*). The focus has now shifted to less cumbersome and more practical solvent-free methods (*21-60*) wherein the neat reactants, often in the presence of mineral oxides or supported catalysts, undergo facile reactions to provide high yields of pure products thus eliminating or minimizing the use of organic solvents.

Microwave reactions involve selective absorption of MW energy by polar molecules, non-polar molecules being inert to MW dielectric loss. The initial experiments with microwaves used high dielectric solvents such as dimethyl sulfoxide (DMSO) and dimethylformamide (DMF). The rate enhancements in these reactions are now believed to be due to rapid superheating of the polar solvents. In solution-phase reactions, the development of high pressures, and the need for specialized Teflon vessels and sealed containers are some of the limitations for using MW processes. Recently, a practical dimension to the microwave heating protocols has been added. Reactions can be performed under solvent-free conditions by adsorbing the organic compounds on to the surface of inorganic oxides, such as alumina, silica and clay, or 'doped' supports. While the organic compound absorbs microwaves, the solid support does not absorb or restrict their transmission. These solvent-free reactions occur at relatively low bulk temperature although higher localized temperatures may be reached during microwave irradiation. These solvent-free MW-assisted reactions allow one to work with open vessels, thus avoiding the risk of high pressure development and increasing the potential of such reactions for commercial scale-up.

Microwave Accelerated Solvent-Free Organic Reactions

The practical utility of microwave-assisted solvent-free protocols has been realized in several synthetic operations such as protection/deprotection, condensation, oxidation, reduction, rearrangement reactions, and in the rapid synthesis of various heterocyclic systems on inorganic solid supports. Herein, we describe our results using this eco-friendly microwave approach for the synthesis of a wide variety of industrially important compounds and intermediates including, enones, imines, enamines, nitroalkenes, oxidized sulfur species and heterocycles. These compounds, obtained by conventional procedures would contribute to the burden of chemical pollution. The reactions described are performed in open glass containers (test tubes, beakers and round-bottomed flasks) using neat reactants under solvent-free conditions in an unmodified household MW oven operating at 2450 MHz. Comparison of MW-accelerated reactions to conventional methods have been made by conducting the same reaction in an oil bath at the same bulk temperature. The problems associated with waste disposal of solvents (used several fold in chemical reactions) and excess chemicals are avoided or minimized. Some of the supported reagents namely clay-supported iron(III) nitrate (clayfen), and copper(II) nitrate (claycop) are prepared according to literature procedure (*61*). The general procedure involves simple mixing of neat reactants with the catalyst/promoter or their adsorption on mineral or 'doped' supports.

Protection/Deprotection Reactions

The protection-deprotection reaction is an integral part of organic synthesis namely in the preparation of monomer building blocks, fine chemicals and precursors for pharmaceuticals. These reactions often involve the use of acidic, basic, or otherwise hazardous and corrosive reagents, and toxic metal salts. A variety of MW-activated reactions involving protection/deprotection of functional groups under solvent-free conditions are highlighted below.

Deacylation Reactions

The orthogonal deprotection of alcohols has been achieved using neutral alumina under microwave irradiation (Scheme 1). Chemoselectivity between alcoholic and phenolic groups in the same molecule can be achieved simply by varying the reaction time; as alcoholic acetates are deacetylated slower than the corresponding phenolic analoges (24).

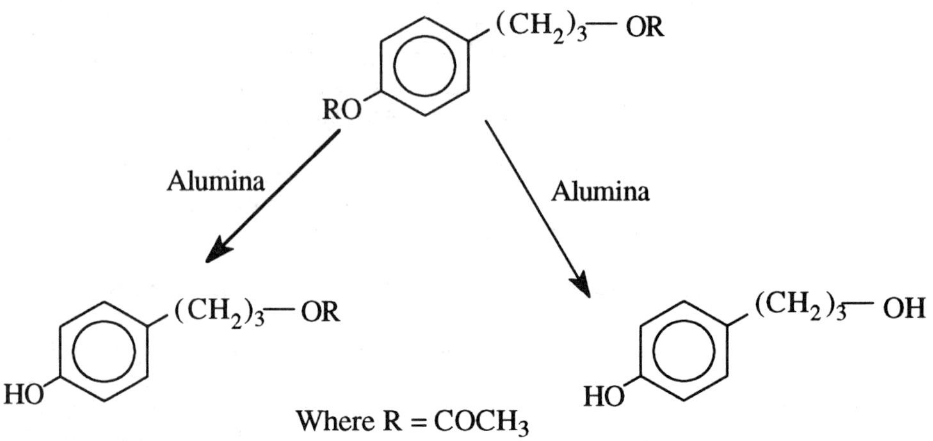

Scheme 1

Cleavage of Aldehyde Diacetates

The diacetate derivatives of aromatic aldehydes are very rapidly cleaved on neutral alumina surfaces (25) upon brief exposure to MW irradiation (Scheme 2). The selectivity in these deprotection reactions is attained simply by adjusting the time of irradiation. In molecules possessing acetoxy functionality (R=OCOCH3), the aldehyde diacetate group is selectively removed in 30 sec; whereas an extended period of 2 min is required to cleave both the diacetate and ester groups. The protocol is applicable to compounds such as cinnamaldehyde diacetate and yields obtained are better than conventional methods.

Debenzylation of Carboxylic Esters

The solvent-free efficient debenzylation of esters (Table I) can be extended to protected amines by changing the surface characteristics of the solid support under

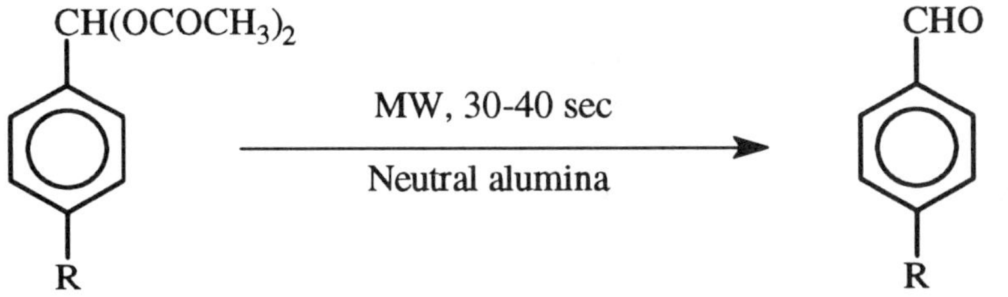

Scheme 2

Ar-CH(OCOCH$_3$)$_2$ → Ar-CHO

MW, 30-40 sec, Neutral alumina

R = H, Me, CN, NO$_2$, OCOCH$_3$ (88-98%)

solvent-free conditions (27); the cleavage of *N*-protected moieties require the use of basic alumina and extended irradiation time at ~130-140 °C.

This approach may find application in peptide bond formation that would eliminate the use of irritating and corrosive chemicals such as trifluoroacetic acid and piperidine.

Table I. Debenzylation of Carboxylic Esters on Alumina Surface

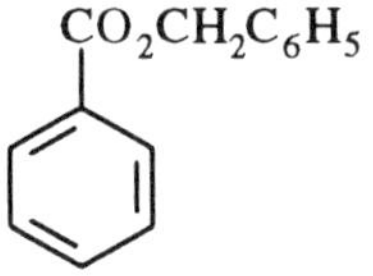

C$_6$H$_5$-CO$_2$CH$_2$C$_6$H$_5$	4-MeO-C$_6$H$_4$-CO$_2$CH$_2$C$_6$H$_5$	4-HO-C$_6$H$_4$-CH$_2$CH$_2$CO$_2$CH$_2$C$_6$H$_5$
7 min/Acidic, 92% (20 min)	7 min/Acidic, 89% (20 min)	10 min/Acidic, (92%)

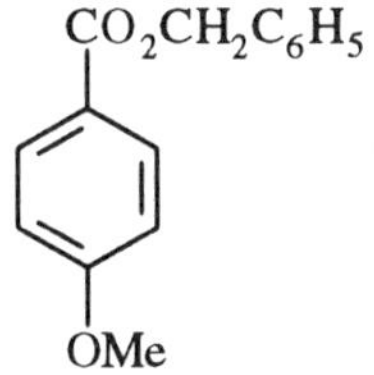

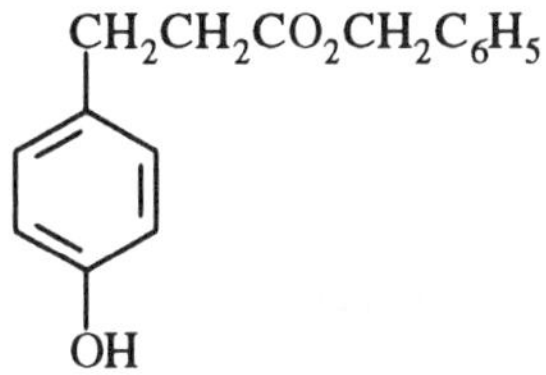

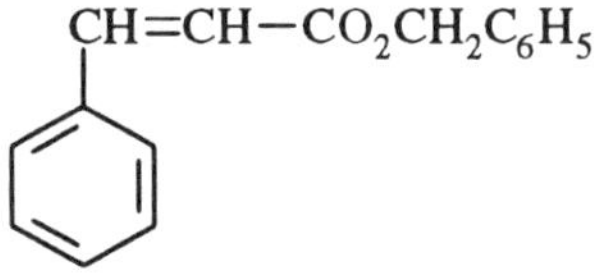

C$_6$H$_5$-CH=CH-CO$_2$CH$_2$C$_6$H$_5$	NH$_2$CH$_2$CO$_2$CH$_2$C$_6$H$_5$	NH$_2$CH(CH$_2$OH)CO$_2$CH$_2$C$_6$H$_5$
10 min/Acidic, 90% (20 min)	4 min/Neutral, (95%) (10 min)	3 min/Neutral, (92%) (10 min)

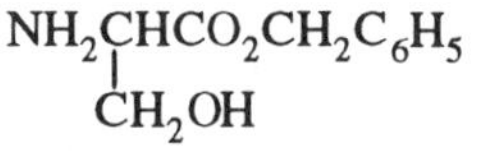

NOTE: Time in parentheses refer to deprotection in oil bath at the same temperature

Desilylation Reactions

t-Butyldimethyl silyl (TBDMS) ether derivatives of a variety of alcohols have traditionally been deprotected using a source of 'naked' fluorides. In contrast, TBDMS ethers are rapidly cleaved on alumina surfaces (*26*). Under MW irradiation (Table II). This approach circumvents the use of corrosive fluoride ions (*26*).

Table II. Microwave-Assisted Desilylation on Alumina Surface

(11 min, 75%)	(18 min, 68%)
(11 min, 93%)	(11 min, 78%)
(10 min, 91%)	(10 min, 93%)

Dethioacetalization Reaction

The cleavage of thioacetals and thioketals which are stable in acid and base is quite a challenging problem and invariably requires the use of toxic heavy metals such as Hg^{+2}, Ag^{+2}, Ti^{+4}, Cd^{+2}, Tl^{+3}, or reagents such as benzeneseleninic anhydride (23). We have achieved the solid state dethioacetalization reaction in high yield (9) using clay doped with iron(III) nitrate, clayfen (Scheme 3)

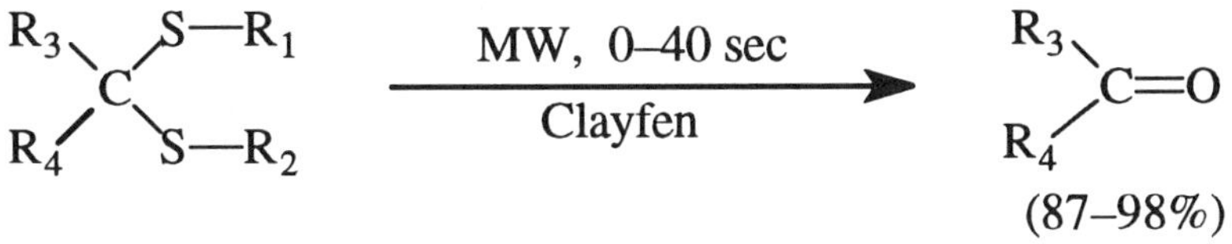

$$\underset{R_4}{\overset{R_3}{>}}C\underset{S-R_2}{\overset{S-R_1}{<}} \xrightarrow[\text{Clayfen}]{\text{MW, 0-40 sec}} \underset{R_4}{\overset{R_3}{>}}C=O$$

(87–98%)

R_3 = Ph, p-anisyl, p-$NO_2C_6H_4$; R_4 = H ; R_1–R_2 = –$(CH_2)_2$–

R_3 = R_4 = Et ; R_1–R_2 = –$(CH_2)_2$– ; R_3 = R_4 = Ph ; R_1 = R_2 = Et

R_3 = Ph ; R_4 = Me ; R_1–R_2 = –$(CH_2)_2$–

R_3–R_4 = 2-Methylcyclohexyl ; R_1–R_2 = –$(CH_2)_2$–

R_3–R_4 = Isoflavanolyl ; R_1–R_2 = –$(CH_2)_3$–

Scheme 3

Deoximation Reactions

Oximes play an important role as protecting groups due to their hydrolytic stability. Their frequent use has provided incentives for the development of newer deoximation reagents such as pyridinium chlorochromate, Raney nickel, triethylammonium chlorochromate, pyridinium chlorochromate-H_2O_2, dinitrogen tetroxide, trimethylsilyl chlorochromate, H_2O_2 over titanium silicalite-1, Dowex-50, dimethyl dioxirane, zirconium sulfophenyl phosphonate, N-haloamides, and bismuth chloride (22).

The solvent-free cleavage of oximes has been successfully demonstrated (22) using relatively benign ammonium persulfate on silica (Scheme 4). Neat oximes are admixed with solid supported reagent and the contents are irradiated at full power in a MW oven to regenerate both aldehydes and ketones. The role of the solid surface is critical since the same reagent supported on a clay surface delivers predominantly the Beckmann rearrangement products, the amides (58).

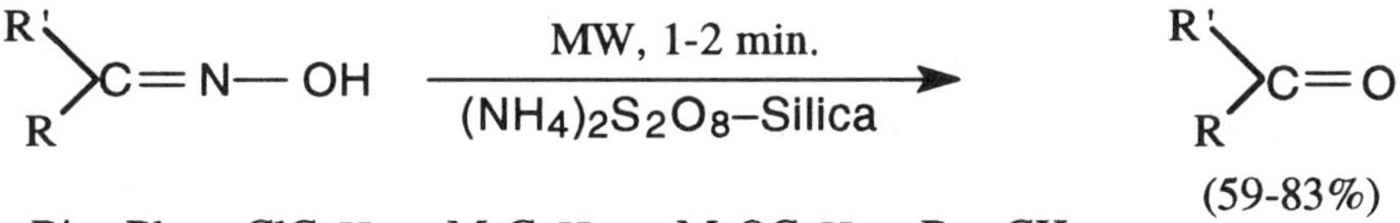

$$\underset{R}{\overset{R'}{>}}C=N-OH \xrightarrow[(NH_4)_2S_2O_8-\text{Silica}]{\text{MW, 1-2 min.}} \underset{R}{\overset{R'}{>}}C=O$$

(59-83%)

R' = Ph, p-ClC_6H_4, p-MeC_6H_4, p-$MeOC_6H_4$; R = CH_3

R' = Ph, p-$NO_2C_6H_4$, m,p-$(MeO)_2C_6H_3$, 2-thienyl, 1-naphthyl ; R = H

R = R' = ⌬

Scheme 4

A facile deoximation protocol with sodium periodate impregnated on premoistened silica (Scheme 5) has been introduced that is applicable exclusively to ketoximes (28).

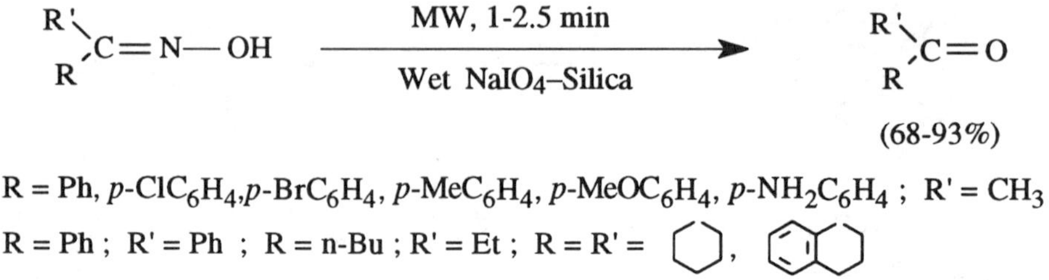

$$\underset{R}{\overset{R'}{>}}C=N-OH \xrightarrow[\text{Wet NaIO}_4\text{-Silica}]{\text{MW, 1-2.5 min}} \underset{R}{\overset{R'}{>}}C=O$$

(68-93%)

R = Ph, p-ClC$_6$H$_4$, p-BrC$_6$H$_4$, p-MeC$_6$H$_4$, p-MeOC$_6$H$_4$, p-NH$_2$C$_6$H$_4$; R' = CH$_3$

R = Ph ; R' = Ph ; R = n-Bu ; R' = Et ; R = R' = ⌬, ⌬⌬

Scheme 5

Cleavage of Semicarbazones and Phenylhydrazones

Aldehydes and ketones can also be regenerated expeditiously from the corresponding semicarbazones and phenylhydrazones using ammonium persulfate impregnated montmorillonite K 10 clay (Scheme 6) under either microwave or ultrasound irradiation conditions (29). However, in these solvent-free procedures the microwave-expedited process requires only minutes compared to ultrasound-promoted reactions that take 1-3 h for completion of the deprotection reaction.

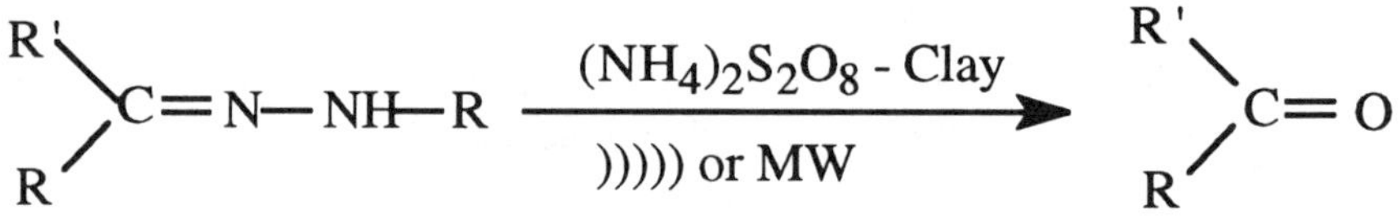

$$\underset{R}{\overset{R'}{>}}C=N-NH-R \xrightarrow[\text{))))) or MW}]{\text{(NH}_4\text{)}_2\text{S}_2\text{O}_8 \text{ - Clay}} \underset{R}{\overset{R'}{>}}C=O$$

Scheme 6

Condensation Reactions

Synthesis of Imines, Enamines and Nitroalkenes

The azeotropic removal of water from the reaction intermediate is the driving force in the preparation of imines, enamines and nitroalkenes in processes that are normally catalyzed by p-toluenesulphonic acid, titanium(IV) chloride and montmorillonite K 10 clay. In conventional protocols, a Dean Stark apparatus is used which requires a large excess of aromatic hydrocarbons such as benzene or toluene for azeotropic removal.

MW-promoted acceleration of such dehydration reactions using montmorillonite K 10 clay (40) or Envirocat reagent (41), EPZG® (Scheme 7, 8) has been demonstrated in the preparation of imines and enamines. Microwaves at 2450 MHz frequency are ideally suited to remove water in imine or enamine forming reactions. For low boiling starting materials, lower power intensity levels have been used (*40,41*).

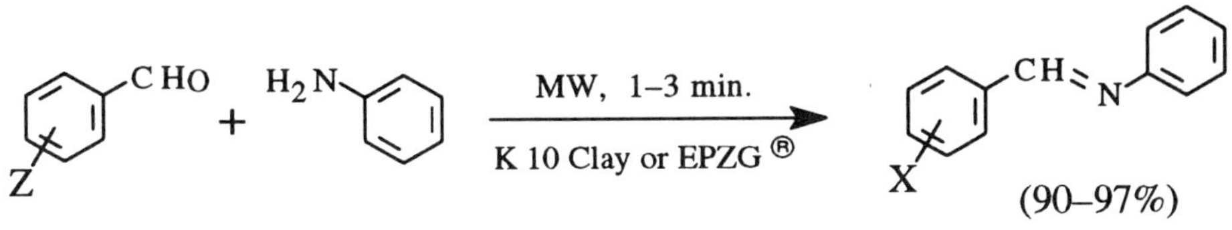

Scheme 7

Z = H, *o*-OH, *p*-OH, *p*-Me, *p*-OMe, *p*-NMe$_2$

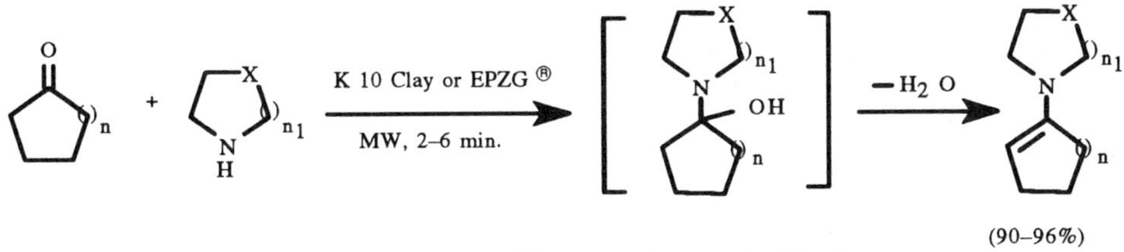

$n = 1$; $n_1 = 2$; $X = CH_2$; $n = 1$; $n_1 = 2$; $X = O$
$n = 2$; $n_1 = 1$; $X = CH_2$; $n = 2$; $n_1 = 2$; $X = CH_2$
$n = 2$; $n_1 = 2$; $X = O$

Scheme 8

The condensation of carbonyl compounds with nitroalkanes to afford nitroalkenes (Henry reaction) can also be expedited via this MW approach. A catalytic amount of ammonium acetate is sufficient for these reactions (*42*) thus avoiding the use of polluting nitrohydrocarbons in large excess which are normally employed (Scheme 9).

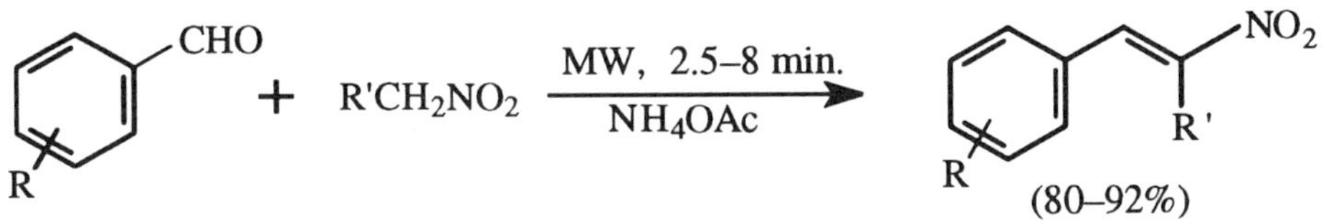

R = H, *p*-OH, *m,p*-(OMe)$_2$, *m*-OMe-*p*-OH, 1-naphthyl, 2-naphthyl ; R' = H
R = H, *p*-OH, *p*-OMe, *m,p*-(OMe)$_2$, *m*-OMe-*p*-OH ; R' = Me

Scheme 9

The oxidation, reduction and cycloaddition reactions originating from α,β-unsaturated nitroalkenes provide easy access to an array of functionalities that include nitroalkanes, *N*-substituted hydroxylamines, amines, ketones, oximes, and α-substituted oximes and ketones *(64-66)*. Consequently, there are numerous possibilities for using these *in situ* generated nitroalkenes for the preparation of valuable building blocks and synthetic precursors *(66)*.

Oxidation Reactions

Oxidation of Alcohols and Sulfides

The common oxidizing reagents employed for organic functional groups include peroxides, peracids, potassium permanganate ($KMnO_4$), manganese dioxide (MnO_2), chromium trioxide (CrO_3), potassium dichromate ($K_2Cr_2O_7$) and potassium chromate (K_2CrO_4) *(62)*.

Although used extensively in organic synthesis, the utility of metal-based reagents in the oxidative transformation is compromised due to their inherent toxicity, the need for cumbersome work-up procedures, potential danger (explosion or ignition) in handling of their complexes and difficulties in terms of product isolation and waste disposal. The introduction of metallic reagents on solid supports has solved some of these problems and provides an attractive alternative in organic synthesis because of the selectivity and relative ease of manipulative.

Selective and Solvent-Free Oxidation with Clayfen

A facile method for the oxidation of alcohols to carbonyl compounds has been developed (Scheme 10) using montmorillonite K 10 clay-supported iron(III) nitrate (clayfen) in amounts that are half that of used by Laszlo *et al.* under heterogeneous conditions *(61)*. This solid state reaction is accelerated enormously by exposure to MW irradiation *(31)* and presumably proceeds *via* the intermediacy of nitrosonium ions. Remarkably, no carboxylic acids are formed in the oxidation of primary alcohols. The experimental procedure simply involves mixing neat alcohols with clayfen and brief irradiation of the reaction mixture in a MW oven for 15-60 seconds *(31)* in the absence of solvent. This extremely rapid, inexpensive, and selective protocol is simple to perform and avoids the use of excess solvents and toxic oxidants.

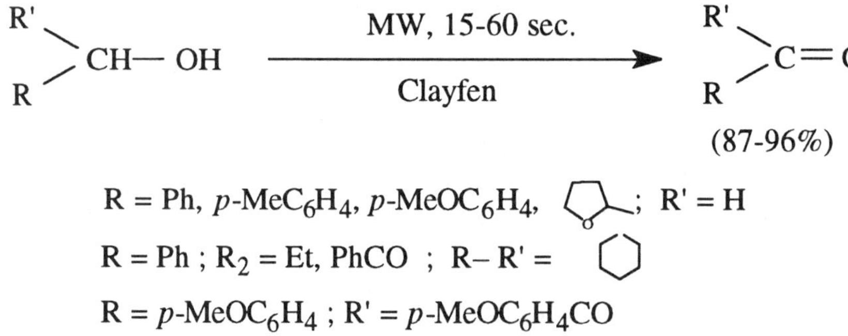

$$\underset{R}{\overset{R'}{>}}CH-OH \xrightarrow[\text{Clayfen}]{\text{MW, 15-60 sec.}} \underset{R}{\overset{R'}{>}}C=O$$

(87-96%)

R = Ph, *p*-MeC$_6$H$_4$, *p*-MeOC$_6$H$_4$, ⟨furyl⟩ ; R' = H

R = Ph ; R$_2$ = Et, PhCO ; R–R' = ⟨cyclohexyl⟩

R = *p*-MeOC$_6$H$_4$; R' = *p*-MeOC$_6$H$_4$CO

Scheme 10

Typical experimental procedure for oxidation of alcohols

The oxidation of benzoin is representative of the general procedure used for the oxidation of alcohols. Clayfen (0.125 g) was thoroughly mixed with benzoin (0.106 g, 0.5 mmol) in the solid state using a vortex mixer and the material was placed in an alumina bath inside a household microwave oven equipped with a turntable and irradiated at full power (900 Watts). The reaction was monitored by TLC using, hexane:AcOEt (10:1, v/v). Upon completion, the product was extracted into methylene chloride. The bulk temperature of the alumina bath, which serves as a heat sink and convenient site to hold the reaction vessel inside the microwave oven, reached ~65 °C after 30 seconds of irradiation as measured by inserting a thermometer into the alumina bath.

Claycop-Hydrogen Peroxide

Metal ions play a significant role in oxidative reactions as well as in biological dioxygen metabolism. Hydroperoxy copper(II) compound, generated from copper(II) acetate and hydrogen peroxide, is obtainable from copper(II) nitrate and hydrogen peroxide (Eqn. 1). The ensuing nitric acid in this reaction requires neutralization by a base to maintain a pH ~5.

$$2Cu(NO_3)_2 + 2H_2O + H_2O_2 \longrightarrow 2CuO_2H + 4HNO_3 \qquad Eqn.\ 1$$

Copper(II) nitrate impregnated on K 10 clay (claycop)-hydrogen peroxide system is an effective reagent for the oxidation of a variety of substrates (*33*) and provides excellent yields of products (Scheme 11) wherein the maintenance of pH of the reaction mixture is not required.

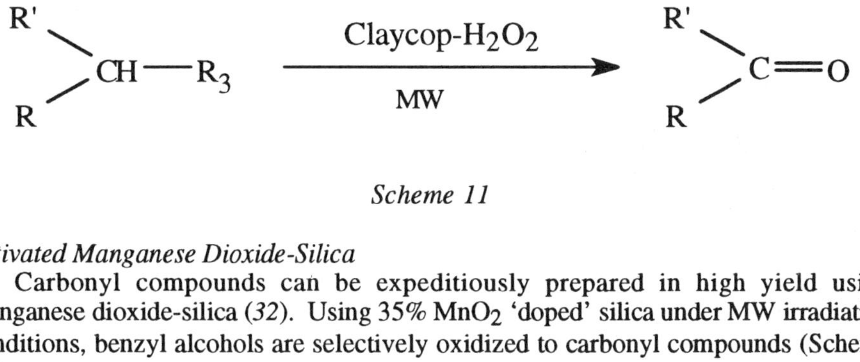

Scheme 11

Activated Manganese Dioxide-Silica

Carbonyl compounds can be expeditiously prepared in high yield using manganese dioxide-silica (*32*). Using 35% MnO_2 'doped' silica under MW irradiation conditions, benzyl alcohols are selectively oxidized to carbonyl compounds (Scheme 12).

Chromium Trioxide Impregnated Wet Alumina

The use of chromium(VI) reagents for oxidative transformations is compromised due to Cr(VI) toxicity. Preparation of its various complexes can involve cumbersome

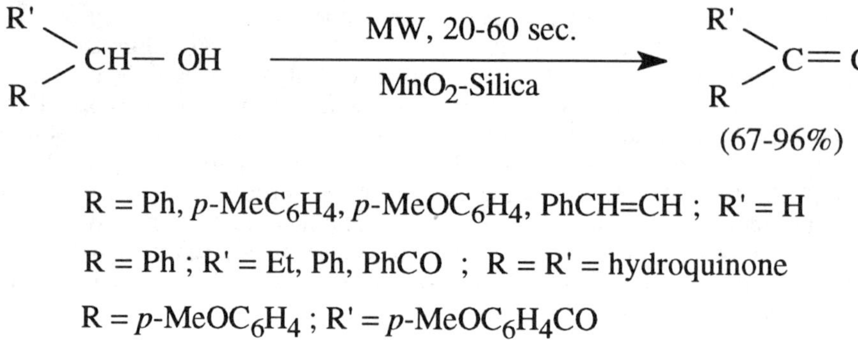

$$\begin{array}{c} R' \\ \\ R \end{array}\!\!\!\!\!\!>\!\!CH\!-\!OH \quad \xrightarrow[\text{MnO}_2\text{-Silica}]{\text{MW, 20-60 sec.}} \quad \begin{array}{c} R' \\ \\ R \end{array}\!\!\!\!\!\!>\!\!C\!=\!O$$

(67-96%)

R = Ph, p-MeC$_6$H$_4$, p-MeOC$_6$H$_4$, PhCH=CH ; R' = H

R = Ph ; R' = Et, Ph, PhCO ; R = R' = hydroquinone

R = p-MeOC$_6$H$_4$; R' = p-MeOC$_6$H$_4$CO

Scheme 12

work-ups and associated waste disposal problems. Alternatively, chromium trioxide (CrO$_3$) impregnated pre-moistened alumina is an efficient oxidizing system which converts benzyl alcohols to carbonyl compounds by simply admixing substrates with the reagent (*34*) at room temperature (Scheme 13). No overoxidation to carboxylic acids and no tar formation is observed that is typical of many CrO$_3$ oxidations conducted in pyridine or acetic acid as solvents. The stabilization of the inorganic anhydride species on alumina surface and their thermal activation may account for this facile oxidation.

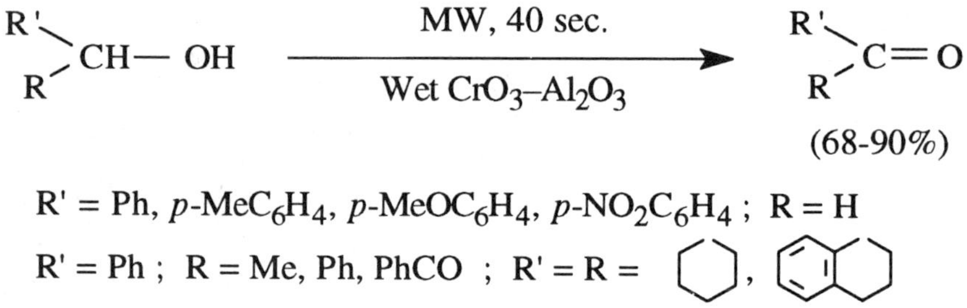

$$\begin{array}{c} R' \\ \\ R \end{array}\!\!\!\!\!\!>\!\!CH\!-\!OH \quad \xrightarrow[\text{Wet CrO}_3\text{-Al}_2\text{O}_3]{\text{MW, 40 sec.}} \quad \begin{array}{c} R' \\ \\ R \end{array}\!\!\!\!\!\!>\!\!C\!=\!O$$

(68-90%)

R' = Ph, p-MeC$_6$H$_4$, p-MeOC$_6$H$_4$, p-NO$_2$C$_6$H$_4$; R = H

R' = Ph ; R = Me, Ph, PhCO ; R' = R = cyclohexyl, tetralin

Scheme 13

Nonmetallic Oxidants – Iodobenzene Diacetate (IBD) 'Doped' Alumina
Several organohypervalent iodine reagents such as iodoxybenzene, *o*-iodoxybenzoic acid (IBX), bis(trifluoroacetoxy)-iodobenzene (BTI), and Dess-Martin periodinane have been used for the oxidation of alcohols and phenols but the use of iodobenzene diacetate (IBD), in spite of its low cost, has not been fully realized. Most of these reactions are normally conducted in solvents such as high boiling DMSO and toxic acetonitrile resulting in environmental pollution.
We have discovered that a rapid oxidation of alcohols to carbonyl compounds occurs in almost quantitative yields with alumina-supported IBD using MW

irradiation under solvent-free conditions (*35*). The advantage of using alumina as a support results in marked improvement in yields obtained with alumina-IBD system as compared to neat IBD (Scheme 14). These oxidative protocols avoid the use of metallic oxidants and consequently help reduce the burden of pollution.

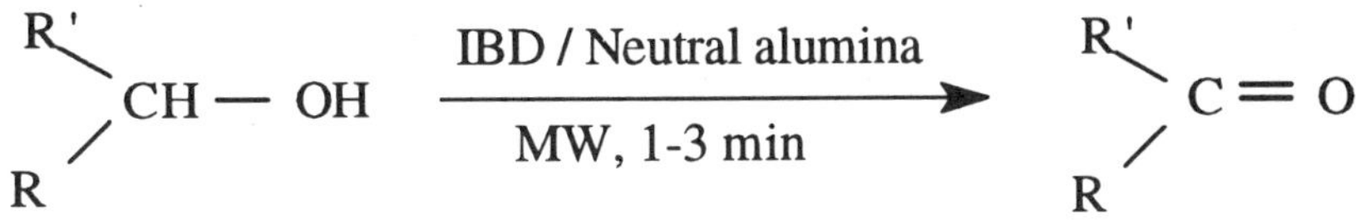

$$\underset{R}{\overset{R'}{\diagdown}}CH-OH \xrightarrow[\text{MW, 1-3 min}]{\text{IBD / Neutral alumina}} \underset{R}{\overset{R'}{\diagdown}}C=O$$

R = Substituted aryl, R' = H, Et, CH$_2$Ph

Scheme 14

Oxidation of Sulfides

Sodium Periodate 'doped' Silica-Oxidation of sulfides to Sulfoxides and Sulfones

Strong oxidants such as hydrogen peroxide, nitric acid, chromic acid, periodate, and peracids are usually employed for the oxidation of sulfides to sulfoxides under strenuous conditions (*38*). The selective oxidation to obtain either sulfoxides or sulfones is achievable using MW irradiation in conjunction with silica 'doped' with 10% sodium periodate at appropriate power levels and reaction times (*38*). Consequently, much less of the active oxidizing agent is employed resulting in improved safety and handling (Scheme 15).

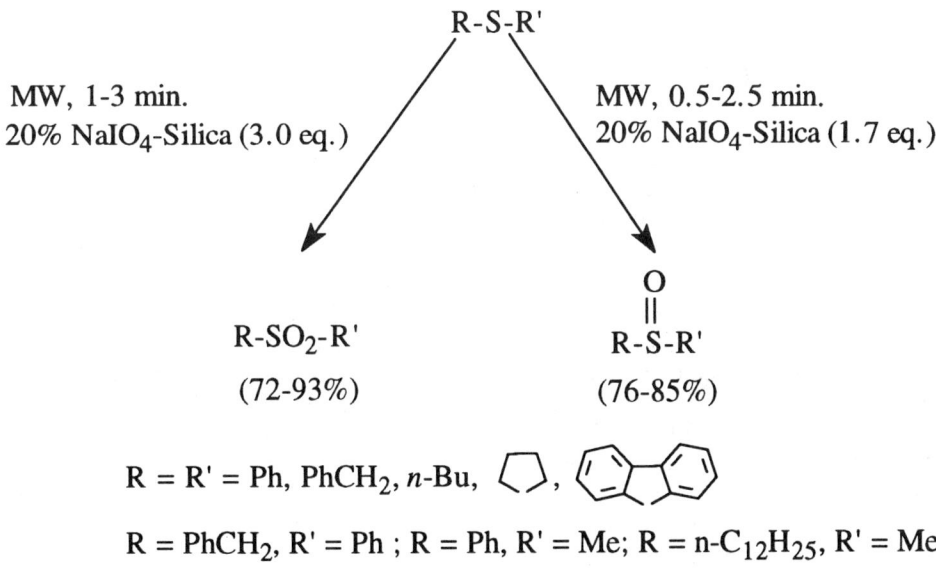

Scheme 15

Various refractory thiophenes that are often not reductively removable by conventional refining processes cab be oxidized under these conditions. For example, benzothiophenes are oxidized to the corresponding sulfoxides and sulfones using microwave or ultrasonic irradiation, respectively, in the presence of $NaIO_4$-silica (*38*).

Iodobenzene Diacetate–Alumina
As described earlier, the solid reagent system, IBD-alumina, is a useful oxidizing agent and its use can be extended to expeditious, high yield and selective oxidation of alkyl, aryl and cyclic sulfides to the corresponding sulfoxides (*39*) upon exposure to microwaves (Scheme 16).

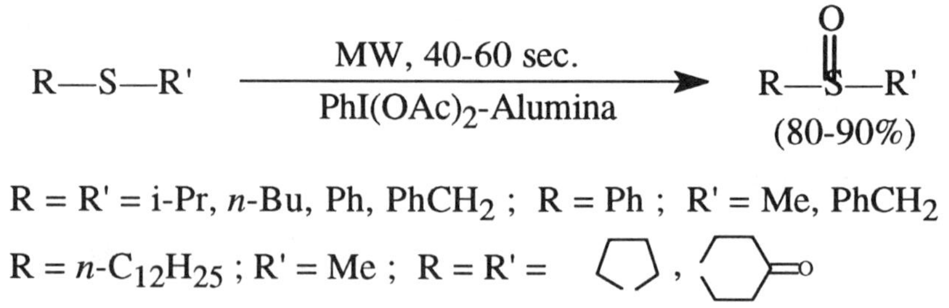

$$R\text{—}S\text{—}R' \xrightarrow[\text{PhI(OAc)}_2\text{-Alumina}]{\text{MW, 40-60 sec.}} R\text{—}\overset{\overset{O}{\|}}{S}\text{—}R' \quad (80\text{-}90\%)$$

R = R' = i-Pr, *n*-Bu, Ph, $PhCH_2$; R = Ph ; R' = Me, $PhCH_2$
R = *n*-$C_{12}H_{25}$; R' = Me ; R = R' = cyclopentyl, cyclohexanone

Scheme 16

Reduction Reactions

Borohydride Reduction of Carbonyl Compounds to Alcohols
Among various reducing agents, relatively inexpensive sodium borohydride ($NaBH_4$) has been used extensively in view of its compatibility with protic solvents and its safer nature. The solid state reduction of ketones has been achieved by mixing them with $NaBH_4$ and storing the mixture in a dry box for five days. The major disadvantage of the heterogeneous reaction with $NaBH_4$ is that the use of solvent slows down the reaction rate while in the solid state, the reaction time required is too long (5 days) for it to be of any practical utility (*46*).

We have developed a clean reduction protocol for aldehydes and ketones that uses alumina supported $NaBH_4$ under microwaves (*46*). No side product formation is observed in any of the reactions investigated and the reaction does not occur in the absence of alumina. The process in its entirety involves a simple mixing of carbonyl compound with (10%) $NaBH_4$-alumina under MW irradiation for 0.5-2 min (Scheme 17). The useful chemoselective feature of the reaction is apparent from the reduction of *trans*-cinnamaldehyde (cinnamaldehyde/$NaBH_4$-alumina, 1:1 mol equivalent) wherein the olefinic moiety remains intact and only the aldehyde functionality is reduced rapidly at room temperature.

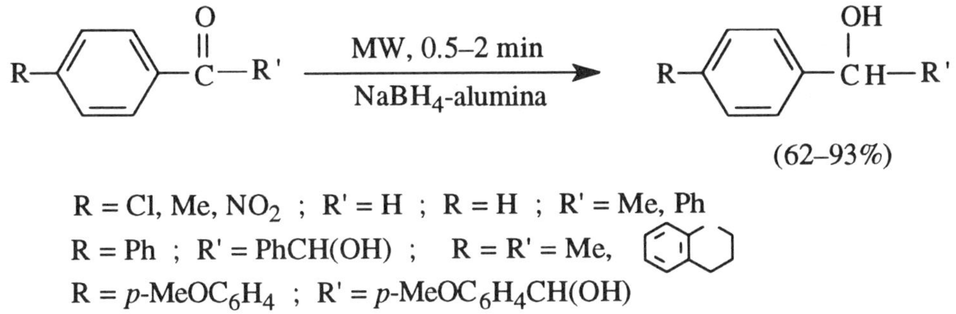

$$\text{R}-\text{C}_6\text{H}_4-\text{C(=O)}-\text{R}' \xrightarrow[\text{NaBH}_4\text{-alumina}]{\text{MW, 0.5–2 min}} \text{R}-\text{C}_6\text{H}_4-\text{CH(OH)}-\text{R}'$$

(62–93%)

R = Cl, Me, NO_2 ; R' = H ; R = H ; R' = Me, Ph
R = Ph ; R' = PhCH(OH) ; R = R' = Me, (tetralin)
R = p-MeOC$_6$H$_4$; R' = p-MeOC$_6$H$_4$CH(OH)

Scheme 17

The reaction rate improves in the presence of moisture. Alumina absorbs enough moisture during the recovery of the product that it can be recycled again by mixing with fresh borohydride and reused for subsequent reductions without any loss in activity.

Typical Experimental Procedure for Reduction of Carbonyl Compounds

The reduction of acetophenone illustrates the typical procedure employed for the reduction of carbonyl compounds (Scheme 17). Neat acetophenone (0.36 g, 3.0 mmol) was thoroughly mixed with freshly prepared NaBH$_4$-alumina[a] (1.13 g, 3.0 mmol of NaBH$_4$) in a test tube and placed in an alumina bath inside the household microwave oven operating at 2450 MHz at full power (900 Watts) and irradiated for 30 sec. The bulk temperature of the alumina bath (heat sink) inside the microwave oven reached ~70 °C after 30 sec of irradiation. Upon completion of the reaction, monitored by TLC (hexane:EtOAc, 8:2, v/v), the product was extracted into methylene chloride (2 x 15 mL). The removal of solvent under reduced pressure essentially provided pure *sec*-phenethyl alcohol in 87% yield.

Reductive Alkylation of Amines

Reductive amination of carbonyl compounds has been well documented using sodium cyanoborohydride and sodium triacetoxyborohydride in processes that produce waste streams. The environmentally benign methods developed in our laboratory have now been extended to a solvent-free reductive amination procedure for carbonyl compounds using wet montmorillonite K 10 clay supported sodium borohydride (47) that is facilitated by exposure to microwave irradiation (Scheme 18).

---

[a](10%) NaBH$_4$-Alumina, was prepared by thoroughly mixing NaBH$_4$ (5.0 g) with neutral alumina (45.0 g) in solid state using a pestle and mortar; admixing three components, carbonyl substrate, NaBH$_4$ and alumina together was equally efficient. The use of premoistened alumina further accelerated the reaction.

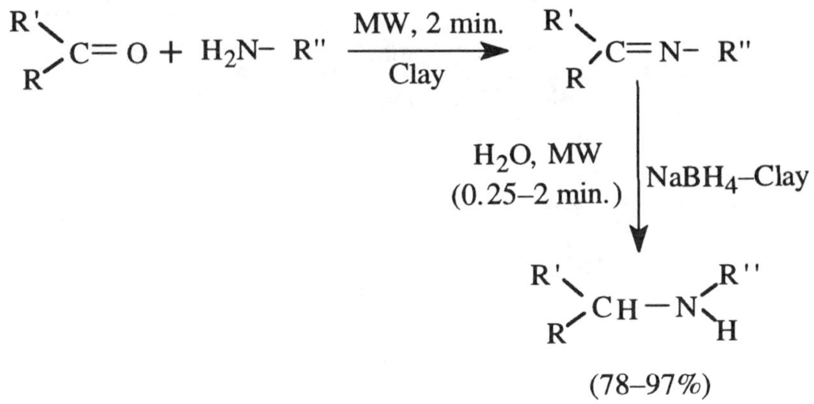

$$\begin{array}{c} R' \\ R \end{array}\!\!C\!\!=\!\!O + H_2N\!-\!R'' \xrightarrow[\text{Clay}]{\text{MW, 2 min.}} \begin{array}{c} R' \\ R \end{array}\!\!C\!\!=\!\!N\!-\!R''$$

$$\downarrow \begin{array}{c} H_2O, MW \\ (0.25\text{--}2 \text{ min.}) \end{array} NaBH_4\text{--Clay}$$

$$\begin{array}{c} R' \\ R \end{array}\!\!CH\!-\!N\!\!\begin{array}{c} R'' \\ H \end{array}$$

(78–97%)

R = i-Pr, Ph, o-HOC$_6$H$_4$, p-MeOC$_6$H$_4$, p-NO$_2$C$_6$H$_4$; R' = H ; R" = Ph
R & R' = –(CH$_2$)$_5$– ; R" = Ph ; R & R' = –(CH$_2$)$_6$– ; R" = n-Pr
R = p-ClC$_6$H$_4$; R' = H ; R" = o-HOC$_6$H$_4$; R = R' = Et ; R" = Ph
R = n-C$_5$H$_{11}$; R' = Me ; R" = morpholine, piperidine ; R = i-Pr ; R' = H ;
R" = n-C$_{10}$H$_{21}$

Scheme 18

While a variety of inorganic solid supports such as alumina, clay, silica and montmorillonite K 10 clay surfaces have been explored, but the *in situ* generated Schiff's bases have been successfully reduced using clay to deliver rapidly secondary and tertiary amines (47). Clay not only behaves as a Lewis acid but provides water from its interlayers that enhances the reducing ability of NaBH$_4$.

Caution: The air used for cooling the magnetron ventilates the microwave cavity thus preventing any ensuing species from reaching explosive concentrations. Although we did not encounter any accident during these reduction reactions, we recommend extreme caution for studies on larger scale.

Synthesis of Heterocylic Compounds

Synthesis of Flavones

Flavonoids are a class of widely distributed naturally occurring phenolic compounds, the most abundant being the flavones. Members of this class display a wide variety of pharmacological activities and have been useful in the treatment of various diseases. Flavones have been prepared by a variety of methods such as Allan-Robinson synthesis and synthesis from hydroxychalcones *via* an intramolecular Wittig strategy. The most popular approach, however, involves the Baker-

Venkataraman rearrangement, wherein *o*-hydroxyacetophenone is benzoylated to form the benzoyl ester followed by the treatment with base (pyridine/KOH) to effect an acyl group migration, forming a 1,3-diketone (*44*). The diketone formed is then cyclized under strongly acidic conditions using sulfuric acid and acetic acid to deliver the flavone. Therefore, opportunity exists for the development of an expedient approach using benign and readily available starting materials that avoids the use of pyridine or sulfuric acid as reagents.

We have achieved a solvent-free synthesis of flavones which simply involves the microwave irradiation of *o*-hydroxydibenzoylmethanes adsorbed on montmorillonite K 10 clay for 1-1.5 min (*44*). A rapid and exclusive formation of cyclized flavones occurs in good yields (Scheme 19).

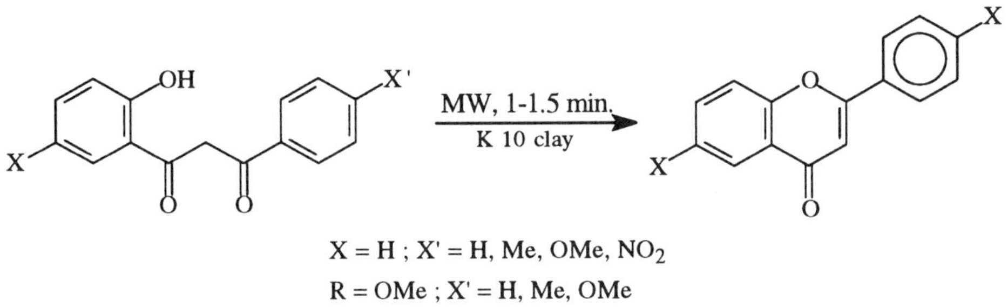
OH
X'
X
MW, 1-1.5 min.
K 10 clay
O
X
X
O
O
O
X = H ; X' = H, Me, OMe, NO2
R = OMe ; X' = H, Me, OMe

Scheme 19

Synthesis of 2-Aryl-1,2,3,4-Tetrahydro-4-Hydroquinolones

In an analogous solventless cyclization reaction using montmorillonite K 10 clay under microwave irradiation conditions, readily available 2'-aminochalcones provide easy access to 2-aryl-1,2,3,4-tetrahydro-4-quinolones (*45*) which are valuable precursors for the medicinally important quinolones (Scheme 20).

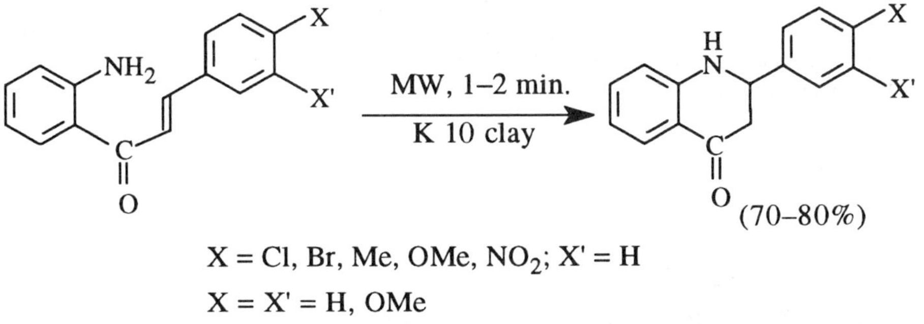
NH2
X
X'
C
O
MW, 1–2 min.
K 10 clay
H
N
X
X'
C
O
(70–80%)
X = Cl, Br, Me, OMe, NO2; X' = H
X = X' = H, OMe

Scheme 20

Synthesis of Substituted Isoflav-3-enes

Isoflav-3-enes encompassing the chromene nucleus, are well known estrogens and several derivatives of these oxygen heterocycles have attracted the attention of medicinal chemists. We have discovered a general method (43) for the rapid assembly of isoflav-3-enes substituted with basic moieties at the 2 position (Scheme 21). The results are especially promising in view of this convergent one-pot approach wherein the generation of the enamine derivatives occur *in situ* and subsequent reactions with *o*-hydroxyaldehydes are induced in the same pot (Scheme 21).

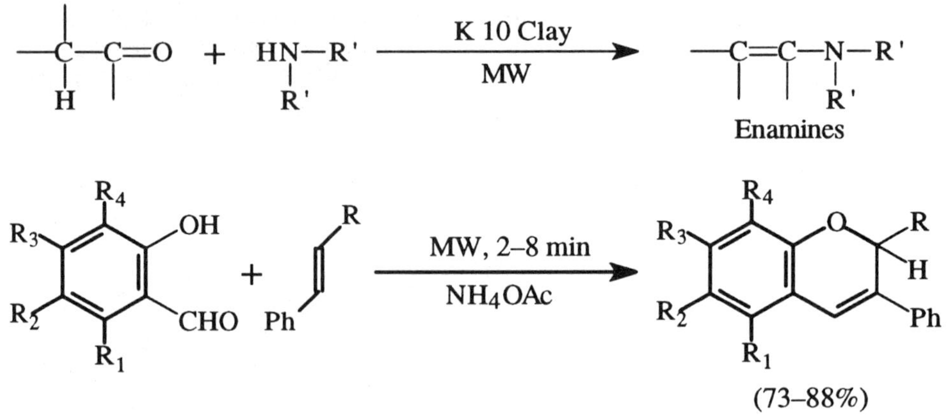

R' = morpholinyl or piperidinyl or pyrrolidinyl
$R_1, R_3, R_4 = H$; $R_2 = H, Cl, NO_2$

Scheme 21

Synthesis of Substituted Thiazoles

Thiazole and its derivatives are obtained by the reaction of α-tosyloxyketones with thioamides in the presence of K 10 clay using microwave irradiation (Scheme 22). α-Tosyloxyketones are generated *in situ* from arylmethyl ketones and [hydroxy(tosyloxy)iodo]benzene (HTIB), thus rendering the entire process solvent-free in both steps (67).

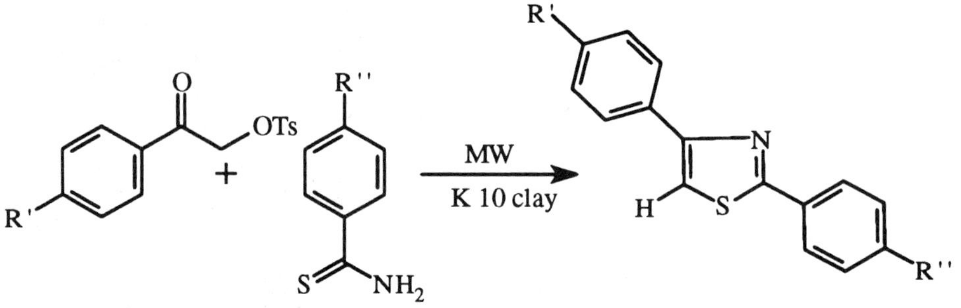

Scheme 22

The case of corresponding bridgehead heterocycles is one in which microwave effect is especially pronounced. The reaction of α-tosyloxyketones with ethylenethioureas remains incomplete in an oil bath (*67*). However, in a microwave oven, it is completed in a 3 min (Scheme 23).

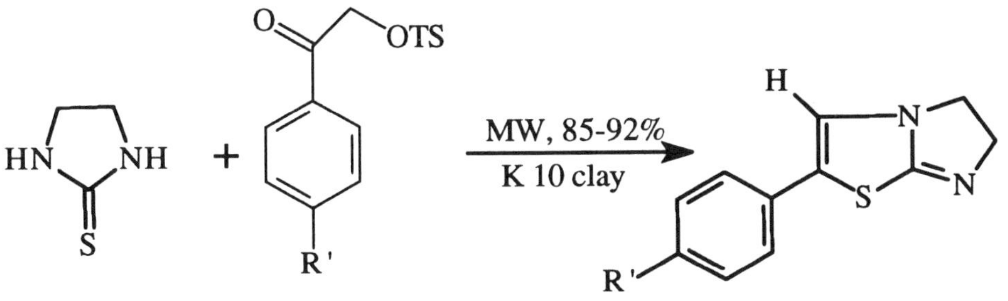

Scheme 23

Synthesis of 2-Aroylbenzofurans

Pharmacologically significant naturally occurring 2-aroylbenzofurans are easily obtainable in the solid state from α-tosyloxyketones and salicylaldehydes in the presence of a base such as potassium fluoride doped alumina (*67*) using microwave irradiation (Scheme 24).

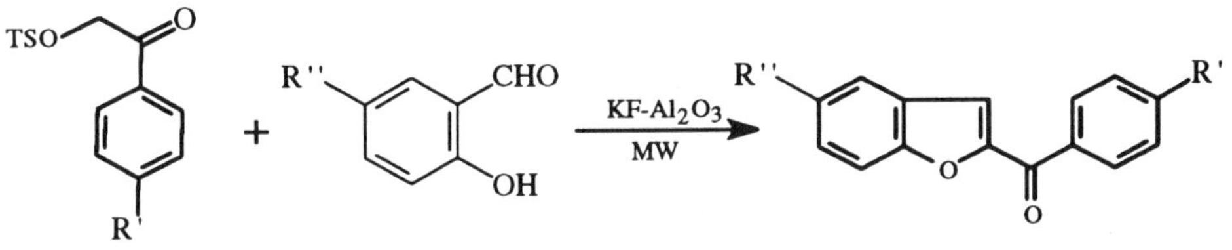

Scheme 24

Conclusion

This article summarizes our recent activity in the development of eco-friendly solvent-free reactions that are activated by exposure to microwave irradiation. The solventless approach opens up numerous possibilities for conducting selective organic functional group transformations more efficiently and expeditiously using a variety of supported reagents on mineral oxides. Our work, performed using an unmodified household microwave oven (multimode applicator), demonstrates the immediate practical applications in laboratory scale experiments. The engineering and scale-up aspects for the chemical process development have already been discussed (*68*). The major industrial applications of MW-enhanced clean chemistry include the preparation of hydrogen cyanide, a chlorination plant, drying of pharmaceutical powders and pasteurization of food products.

There are distinct advantages to these solvent-free protocols since they provide reduction or elimination of solvents thereby preventing pollution in organic synthesis 'at source'. Although not delineated completely, the reaction rate enhancements achieved in these methods may be attributable to non-thermal effects. The chemo-, regio- or stereoselective synthesis of high-value chemical entities may see the translation of these 'curious' laboratory experiments to large-scale operation pending the design of bigger microwave reactors and the participation of multi-disciplinary teams of chemical and electrical engineers to harness the true potential of this clean technology.

Acknowledgments

I am grateful for financial support to the Texas Research Institute for Environmental Studies (TRIES) and indebted to contributions from several research associates whose names appear in the references and especially to Dr. Kannan P. Naicker for his help in the preparation of this manuscript.

Literature Cited

1. Using chemical reagents on porous carriers, *Akt.-Ges. Fur Chemiewerte. Brit. Pat.*, 1924, 231,901. (*Chem. Abstr.* 1925, **19**, 3571).
2. Posner, G. H. *Angew Chem. Int. Ed. Engl.*, **1978**, *17*, 487-496.
3. McKillop, A.; Young, K. W. *Synthesis*, **1979**, 401-422 and 481-500.
4. Cornelis, A.; Laszlo, P. *Synthesis*, **1985**, 909-918.
5. Laszlo, P. *Preparative Chemistry Using Supported Reagents*, Academic Press, Inc. San Diego, CA, 1987.
6. Smith, K. *Solid Supports and Catalyst in Organic Synthesis*, Ellis Horwood, Chichester, 1992.
7. Balogh, M.; Laszlo, P. *Organic Chemistry Using Clays*, Springer-Verlag, Berlin, 1993.
8. Clark, J. H. *Catalysis of Organic Reactions by Supported Inorganic Reagents*, VCH publisher, Inc., NY, 1994.
9. Clark, J. H.; Macquarrie, D. J. *Chem. Commun.*, **1998**, 853-859.
10. Kabalka, G. W.; Pagni, R. M. *Tetrahedron*, **1997**, *53*, 7999-8065.
11. *Benign by Design. Alternative Synthetic Design for Pollution Prevention;* Anastas, P. T.; Farris, C. A., Eds.; American Chemical Society: Washington, DC, 1994.
12. Gedye, R.; Smith, F.; Westaway, K.; Ali, H.; Baldisera, L.; Laberge, L.; Rousell, J. *Tetrahedron Lett.* **1986**, *27*, 279-282.
13. Abramovich, R. A. *Organic Preperation Proceedings International*, **1991**, *23*, 683-711.
14. Majetich, G.; Hicks, R. *J. Microwave Power Electromagnatic Energy*, **1995**, *30*, 27-45.
15. Caddick, S. *Tetrahedron*, **1995**, *51*, 10403-10432.

16. Strauss, C. R.; Trainor, R. W. *Australian J. Chemistry*, **1995**, *48*, 1665-1692.
17. Bose, A. K.; Banik, B. K.; Lavlinskaia, N.; Jayaraman, M.; Manhas, M. S. *Chemtech*, **1997**, *27*, 18-24.
18. Varma, R. S. *Green Chemistry*, **1999**, 43-55.
19. Varma, R. S. *Clean Products and Processes*, **1999**, *1*, 132-147.
20. Giguere, R. J.; Namen, A. M.; Lopez, B. O.; Arepally, A.; Ramos, D. E.; Majetich, G.; Defrauw, J. *Tetrahedron Lett.*, **1987**, *28*, 6553-6556.
21. Varma, R. S. In *Microwaves: theory and application in material processing IV;* Clark, D. E.; Sutton, W. H.; Lewis, D. A., Eds., American Ceramic Society, Ceramic Transactions, Westerville, OH, 1997, Vol. 80, pp. 357-365.
22. Varma, R. S.; Meshram, H. M. *Tetrahedron Lett.*, **1997**, *38*, 5427-5428.
23. Varma, R. S.; Saini, R. K. *Tetrahedron Lett.*, **1997**, *38*, 2623-2624.
24. Varma, R. S.; Varma, M.; Chatterjee, A. K. *J. Chem. Soc., Perkin Transactions 1*, **1993**, 999-1000.
25. Varma, R. S.; Chatterjee, A. K.; Varma, M. *Tetrahedron Lett.*, **1993**, *34*, 3207-3210.
26. Varma, R. S.; Lamture, J. B.; Varma, M. *Tetrahedron Lett.*, **1993**, *34*, 3029-3032.
27. Varma, R. S.; Chatterjee, A. K.; Varma, M. *Tetrahedron Lett.*, **1993**, *34*, 4603-4606.
28. Varma, R. S.; Dahiya, R.; Saini, R. K. *Tetrahedron Lett.*, **1997**, *38*, 8819-8820.
29. Varma, R. S.; Meshram, H. M. *Tetrahedron Lett.*, **1997**, *38*, 7973-7976.
30. Varma, R. S.; Kumar, D. *Synth. Commun.*, **1999**, *29*, 1333-1340.
31. Varma, R. S.; Dahiya, R. *Tetrahedron Lett.*, **1997**, *38*, 2043-2044.
32. Varma, R. S.; Saini, R. K.; Dahiya, R. *Tetrahedron Lett.*, **1997**, *38*, 7823-7824.
33. Varma, R. S.; Dahiya, R. *Tetrahedron Lett.*, **1998**, *39*, 1307-1308.
34. Varma, R. S.; Saini, R. K. *Tetrahedron Lett.*, **1998**, *39*, 1481-1482.
35. Varma, R. S.; Dahiya, R.; Saini, R. K. *Tetrahedron Lett.*, **1997**, *38*, 7029-7032.
36. Varma, R. S.; Kumar, D.; Dahiya, R. *J. Chem. Res (S)*, **1998**, 324-325.
37. Varma, R. S.; Dahiya, R.; Kumar, D. *Molecules Online*, **1998**, *2*, 82-85.
38. Varma, R. S.; Saini, R. K.; Meshram, H. M. *Tetrahedron Lett.*, **1997**, *38*, 6525-6528.
39. Varma, R. S.; Saini, R. K.; Dahiya, R. *J. Chem. Res (S)*, **1998**, 120-121.
40. Varma, R. S.; Dahiya, R.; Kumar, S. *Tetrahedron Lett.*, **1997**, *38*, 2039-2040.
41. Varma, R. S.; Dahiya, R. *Synlett*, **1997**, 1245-1246.
42. Varma, R. S.; Dahiya, R.; Kumar, S. *Tetrahedron Lett.*, **1997**, *38*, 5131-5134.
43. Varma, R. S.; Dahiya, R. *J. Org. Chem.*, **1998**, *63*, 8038-8041.
44. Varma, R. S.; Saini, R. K.; Kumar, D. *J. Chem. Res (S)*, **1998**, 348-349.
45. Varma, R. S.; Saini, R. K. *Synlett*, **1997**, 857-858.
46. Varma, R. S.; Saini, R. K. *Tetrahedron Lett.*, **1998**, *38*, 4337-4338.
47. Varma, R. S.; Dahiya, R. *Tetrahedron*, **1998**, *54*, 6293-6298.
48. Villemin, D.; Labiad, B. *Synth. Commun.*, **1990**, *20*, 3333-3337.
49. Villemin, D.; Alloum, A. B. *Synth. Commun.*, **1990**, *20*, 3325-3331.

50. Villemin, D.; Alloum, A. B. *Synth. Commun.*, **1991**, *21*, 63-68.
51. Lerestif, J. M.; Bazureau, J. P.; Hamelin, J. *Synlett*, **1995**, 647-649.
52. Csiba, M.; Cleophax, J.; Loupy, A.; Malthete, J.; Gero, S. D. *Tetrahedron Lett.*, **1993**, *34*, 1787-1790.
53. Villemin, D.; Sauvaget, F. *Synlett*, **1994**, 435-436.
54. Benhaliliba, H.; Derdour, A.; Bazureau, J. P.; Texier-Boullet, F.; Hameline, J. *Tetrahedron Lett.*, **1998**, *39*, 541-542.
55. Rahmouni, M.; Derdour, A.; Bazureau, J. P.; Hameline, J. *Tetrahedron Lett.*, **1994**, *35*, 4563- 4564 and *Synth. Commun.*, **1996**, *26*, 453-458.
56. Gutierrez, E.; Loupy, A.; Bram, G.; Ruiz-Hitzky, E. *Tetrahedron Lett.*, **1989**, *30*, 945-948.
57. Villemin, D.; Martin, B.; Khalid, M. *Synth. Commun.* , **1998**, *28*, 3195-3200.
58. Bosch, A. I.; Cruez de la, P.; Diez-Barra, E.; Loupy, A.; Langa, F. *Synlett* , **1995**, 1259-1260.
59. Villemin, D.; Martin, B. *J. Chem. Res. (S)* **1994**, 146-147.
60. Nelson, D. A.; Devin, C.; Hoffmann, S.; Lau, A. Division of Chemical Education, *Abstr. No.* 101, ACS National Meeting, San Francisco, **1997**, April 13-17.
61. Balogh, M.; Laszlo, P. *Organic Chemistry Using Clays*, Springer, Berlin **1993**.
62. Trost, B. M.; ed.,*Comprehensive Organic Synthesis (Oxidation)*, Pergamon, New York, 1991, Vol 7.
63. Oussaid, A.; Loupy, A. *J. Chem. Res. (S)*, **1997**, 342-343.
64. Varma, R. S.; Kabalka, G. W. *Heterocycles*, **1986**, *24*, 2645-2677.
65. Kabalka, G. W.; Varma, R. S. *Org. Prep. Proc. International*, **1987**, *19*, 283-328.
66. Kabalka, G. W.; Guindi, L. H. M.; Varma, R. S. *Tetrahedron*, **1990**,*46*, 7443-7457.
67. Varma, R. S.; Kumar, D.; Liesen, P. J. *J. Chem. Res., Perkin Trans 1*, **1999**, 4093-4096.
68. Mehdizadeh, M. *Engineering and scale-up considerations for microwave induced reactions*, in *Proceedings: Microwave-Induced Reactions Workshop*, Electric Power Research Institute, Palo Alto, CA, 1993, p A-7.

Chapter 24

Choosing Solvents That Promote Green Chemistry

William M. Nelson

Waste Management and Research Center, 1 East Hazelwood Drive, Champaign, IL 61820

At this time, information on the present state of the science of solvents appropriate for use in green chemistry comes from many disparate sources. Simply stated, the aim of this chapter is to describe and promote the use of green solvents throughout chemistry and to provide a methodological approach to the selection of solvents currently applicable in green chemistry. The chapter will examine solvents in world chemical perspective. Highlights will include defining characteristics of green solvents, their current usage and their desired characteristics (ecologically and economically). It discusses the use of solvents in specific commercial and non-commercial practices, as a starting point for further discussions on how these solvents can be found or made.

Introduction

Green chemistry must become an integral part of all areas of chemistry, theoretical through practical. The ultimate value of green chemistry lies in its applicability for the new millenium. It embraces the design, manufacture, and use of chemicals and chemical processes with the goal of preventing the production of pollution. While this may seem an unobtainable goal, the vast chemical enterprise shows that it is economically and technologically desirable and feasible. Further, as scientists who are members of the global community, chemists cannot be ignorant nor can they claim to be unaware of the effects of the science they practice. Ultimately, the opportunities and challenges present in green chemistry will help define chemistry and the chemical industry well into the 21st century.

The ideal synthesis (1) or any chemical endeavor must include an element of environmental consciousness. While the ideal of synthesis, or any chemical enterprise, will contain laudatory goals for the individual enterprise, it will be the environmental component which will determine the appropriateness of the enterprise in the global village. What is remarkable is that historically, the decision to use one particular chemical over another largely never considered the effects on the planet, the local environment, human health, or wildlife (2). As our chemical knowledge has expanded, science has elucidated more fundamental principles in the relationships between structure and function. This knowledge, expressed in green chemistry, is the unique and invaluable contribution chemistry will make in pollution prevention.

Three drivers (economic, technical, and regulatory) directly affect the impact of green chemistry in the 21st century. Solvents (their use or non-use) are so integral to chemical practices, their employment is also dictated by the above drivers. Solvents directly impact the practice of chemistry and the design of safer chemicals for use in research, industry, and commerce. Modern day chemists, as citizens of the world, are expected to develop and use chemical substances that are safe to human health and the environment, as well as being efficacious. Green chemistry principles provide the guidelines (in broad terms) for scientists engaged in the discipline of chemistry to move in this direction.

The three drivers (economic, technical and regulatory) provide the practical milieu in which green chemistry pursues the mandate for pollution prevention (3). Green chemistry will be successful if it balances these pressures. Furthermore, the pursuit of environmentally sound strategies are consonant with the principle goals of chemistry and offer the opportunity to achieve good science. This chapter will demonstrate the principle considerations and options in the area of choosing solvents that will enable us to integrate their deployment in the thinking, planning, development and implementation in all phases of the design, production and use of chemicals (4).

Solvents in chemistry

The worldwide market for solvents is estimated at over 30 billion pounds per year. Solvents are used ubiquitously in a wide range of industries and applications including: metal cleaning and degreasing, dry-cleaning operations, automotive and aviation fuel additives, paints, varnishes, lacquers, paint removers, plastics and rubber products, adhesives, textiles, printing inks, pharmaceuticals, and food processing. The largest volume is used in the manufacturing of coatings such as paints, lacquers, varnishes, and printing inks. Substantial amounts are also used in the manufacturing of synthetic fibers and other polymeric materials. Additionally, solvents are used for the purification of the chemical substances and for cleaning the equipment used in the chemical process.

Often the quality and suitability of reactions and chemical processes are heavily dependant upon the identity of the solvent utilized. Solvents intervene in chemical processes by producing species from solutes that are more accessible than if the solvent were not present. Solvents serve several roles in the world of chemistry, including reactions, separations, extractions, and cleaning, just to name a few. To appreciate the magnitude of solvent use, the following table provides representative examples.

Table I. Solvent Roles in Chemistry

Broad Chemical Applications	Examples of Roles
Green Solvents for Academic Chemistry	
Early chemistries: Elementary through High School	demonstrations
Organic	chemical reactions
Inorganic	chemical reactions
Biochemistry	chemical reactions
Medicinal	chemical reactions
Teaching Laboratories	chemical reactions
Green Solvents for Industrial Chemistry	
Solvents for Industrial Reaction Process	reaction media
Solvents for the Health Fields	drug carriers
Solvents for Analyses & Special Solvents	separations
Military solvents	automotive maintenance
Green Solvents for Practical and Ordinary Chemical Usage	
Household solvents	cleaning
Recreational solvents	fuels
Transportation solvents	maintenance
Hygiene solvents	cleaning
Food, cooking & nourishment solvents	food preparation

The chemical industry (including the broad areas of academic, industrial and practical chemical usage shown in Table I) produces efficacious solvents to satisfy the roles shown in Table I. The development of our knowledge of solutions reflects to some extent the development of chemistry itself (6). Since this is true and since green chemistry is attempting to make chemical practices more environmentally benign, solvents must be evaluated to assay whether they are necessary (solvent-less or less solvent), and if so, how can their use be made more environmentally benign. This environmental criterion has assumed a critical role in the determination of appropriate chemistry for the 21st century.

The challenge then becomes to critically evaluate the understanding of solvent functions in light of environmental concerns.(7),(8). The medium effects, as seen in Table II, are the traditional concerns found in solution chemistry. It is logical that solvent classification, whether obtained through extensive empirical results or calculated entirely using statistical methods, correlates strongly with the chemist's intuition. This becomes the point of departure for green chemists, as the new direction of the science demand new properties to be incorporated into the "chemist's intuition". These will include not only chemical properties, but safety properties, and environmental properties as well.

Table II. Traditional Solvation Concerns

Medium Effects
Pure solvents, solvent mixture, no solvent;
Classification of solvents;
Solvent polarity;
Solvent as reagent;
Intermolecular forces;
Chemical interactions

Green Solvents: Integral part of the solution

Solvent usage is seemingly inseparable from the practice of chemistry. The use of solvents in all areas of chemistry is pervasive. Its presence and environmental properties also make solvents an area of focus in green chemistry. A stated principle of green chemistry is that the use of auxiliary substances (e.g. solvents) should be made unnecessary wherever possible and innocuous when used (5). This ambitious goal warrants a realistic appraisal of current chemical practices with regard to solvents and a rigorous methodology to guide the evaluation and selection of green solvents.

Solvents in Green Chemistry

What are appropriate green solvents? First (according to the technological and economic drivers) they are solvation media, which must allow completion of the task. Secondly, as chemicals their human and environmental toxicity must be understood and minimized. Finally, they must disposed of in ways that do not contribute to pollution (9). These considerations, novel due to today's environmentally conscious atmosphere, set green solvents apart from traditional solvents.

The modern chemist is expected, indeed has a moral responsibility, to consider these criteria when deciding the occasions of use and identity of solvents. The goal of green chemistry in general is to reduce the inherent hazard of chemicals and chemical processes. As a guidepost, we can evaluate solvents, using parameters that are in harmony with the principles of green chemistry. The considerations for what constitutes a safe solvent might contain some of the general considerations given in Table III.

Table III. General Considerations for Green Solvents

Characteristics
Must have reduced human absorption and bioavailability
The toxicity of it should be understood
Knowledge of its environmental fate is understood
Minimized hazardous properties (e.g., explosivity)
Minimal chemical effects due to reactivity (e.g., oxidation)
Reduced synergism with known environmental toxins

The appropriateness of a solvent is determined by many factors, such as bioavailability, metabolism, functional groups, and the presence of structural features that attenuate or enhance the reactivity of the parent molecule. Despite the structure-activity data available for many classes of commercial chemical substances, chemists have not recognized the use of structure-activity relations (SARs) as a rational approach for choosing or designing new, less toxic commercial chemical substances, even though it can be an exceptionally powerful tool. This probably results from the use of SARs to evaluate known compounds, but not to selectively predict desirable possible solvents. With qualitative structure-activity relationships, the correlation of toxic effect and structure is made by comparing the structural features to pharmacological effects. Through SARs, it may be possible to predict a relationship between structure, toxicity, and chemical properties and identify the least toxic members of a class of compounds as possible commercial alternatives to the more toxic relatives. Chemists who can recognize toxicophoric substituents (the structural portion of a substance that causes toxicity), have knowledge of the potential impact of the chemicals they use. They are in a much better position to respond in harmony with the mandates of green chemistry, because they can envision the potential dangers particular solvents pose to the global ecology or to the individuals who will contact the chemicals directly.

Solvent Toxicology

The effects that chemicals have on humans and the environment are studied in the field of environmental toxicology. Environmental toxicology encompasses many divergent disciplines (see Table IV). These factors all influence the choice of appropriate solvents.

Evaluating the toxicity of solvents provides one metric for comparing potential candidates. This evaluation (usually the LD50 or PEL (permissable exposure level)) allows one to screen out solvents that are environmentally undesirable. In actual practice, the preliminary ranking should incorporate the criteria given in Table V. In general, despite the data available in all the areas listed in Table V, chemists have not

Table IV. Components of Solvent Toxicology

Element	Description
Chemistry	the characterization of toxins;
Pharmacology	the mode of entry and distribution of toxins in the body;
Biochemistry	the metabolism and interaction of toxins with cell components;
Physiology	the effect of toxins on body organs;
Biology	the effect of toxins on the environment;
Epidemiology	the effect on the population as a whole as a function of chronic exposure to small quantities of suspected agents;
Law	regulation of the use or release into the environment of toxic substances;
Economics	evaluation of the environmental cost vs. benefit of economic development and the determination of trade-offs among economy, health, and the environment..

Table V. Novel Criteria for Green Solvents: Toxicity concerns

- Recognize toxicophoric or toxicogenic substituents;
- Obtain information pertaining to a toxic mechanism;
- Conduct literature searches on the original substance to identify toxicity studies;
- Use of Structure-Activity (toxicity) Relationships (SAR);
- Degradation and solvolysis products

recognized their importance as a rational approach for choosing or designing new, less toxic commercial chemical substances (*10*). While initially incorporating the ideals (principles) of pollution prevention into all its operating paradigms creates a new step in the process, it will eventually reveal opportunities for chemists in the general global movement toward greener strategies. Moreover, the chemist plays a critical role in the evolutionary utilization of practical toxicology.

Origin of Green Solvents

Green solvents are solvating media which promote pollution prevention, because they are inherently less hazardous. The following sections will list ways that the use of solvents can be made "greener". More fundamentally we can methodically describe the important characteristics of green solvents. This will lead us directly into the selection or development of green solvents for today's chemistries. The summary of internal considerations for a good solvent are provided in Table VI. These should be standard criteria for an effective solvating medium.

Table VI. Summary of internal considerations

Desirable Characteristics	*Chemical Insights*
Facilitation of solution	• mechanistic understanding of solvated species
	• condition of solution
	a. temperature
	b. agitation
	c. pressure
	• time scale
Reactivity of solvated species	• acceptable reactivity
	• desired product
Avoidance of undesired side reactions	• mechanistic understanding of new media
	• fine-tuning reaction

These considerations seemingly represent the traditional concerns for solvent selection in any process or reaction. For green solvents, in addition, the criteria are used to improve the solvent's performance so that it has a minimal effect on the environment. We might rephrase the above outline in terms of the following questions:

i. *Does the facilitation of solution lead to less energy expenditure? (lower temperature, less agitation, less time, etc.)*
ii. *Does the solvated species react more efficaciously and selectively? (chemo, enantio, regioselectively)*
iii. *Does the solvent assist in the development of an atom-economical process or reaction (17)?*

To incorporate the Green Chemistry paradigm into a methodical framework, the mechanisms and goals of desired reactions (both existing and future) must be known. In a parallel manner, the solvents must be scrutinized in terms of the desirable characteristics according to criteria of toxicology. This forms the the information provided in Table VII. This *modus operandi* is novel and it has not been rigorously nor universally applied to the conscious design and execution of new reactions nor solvent selection. The development of new environmentally favorable routes for the chemical processes which span the breadth of chemistry is an area of considerable interest, whose importance will only expand as the pressures of the ecological movement increase. Let us provide a summary of these considerations.

The above Table VII presents the novel considerations, which must become a part of chemist's solvent evaluation in the selection process in the new *modus operandi*. These are not as familiar in the routine considerations of the practicing chemist, but they represent the importance of toxicology in chemistry. The information might be rephrased in terms of the following questions:

i. *How does the solvent distribute itself upon release into the environment?*
ii. *Will the solvent be absorbed by organisms and how will it affect them?*
iii. *How will the solvent be absorbed?*
iv. *Is the solvent toxic, and if so, can it be detoxified?*

Systematically Selecting Green Solvents

It is widely accepted that solvents are critical to the success/failure of a reaction, but the task now becomes molding green chemistry goals into a useful methodology for solvent selection, while simultaneously achieving the mandates of economic and technological drivers. The process of selecting a solvent that meets the criteria of the principles of green chemistry is not trivial. In this section we attempt to provide a workable guideline to success in this area (as guided by the previously highlighted drivers.) This is not an exhaustive guide, but it provides the framework for selection of green solvents.

Table VII. Summary of Solvent Toxicology Criteria

Desirable Properties	Chemical Considerations

Properties related to environmental distribution/dispersion	• volatility/density/melting point
	 • water solubility
	 • persistence/biodegradation
	   a. oxidation
	   b. hydrolysis
	   c. photolysis
	   d. microbial degradation
	 • conversion to biologically active substances.
	 • conversion to biologically inactive substances

Properties related to uptake by organisms	• volatility
	 • lipophilicity
	 • molecular size
	 • degradation
	   a. hydrolysis
	   b. effect of pH
	   c. susceptibility to digestive enzymes

Consideration of routes of absorption by man, animals or aquatic life	• skin/eyes • lungs • gastrointestinal tract • gills or other species-specific routes
Reduction/elimination of impurities	• generation of impurities of different chemical classes • presence of toxic homologs • presence of toxic geometric, conformational or stereoisomers

Preparation: Laws and Global Regulations Governing Solvents.

Solvent use in our society is seemingly indispensible (*11*). Over time, our knowledge has grown regarding how solvents affect the environment and human health. In order to reflect this awareness, society has increased the number of regulations that govern the use of these chemicals (*12*). The Occupational Safety and Health Administration (OSHA) has regulated solvents that pose dangers to workers or handlers, by setting strict regulations called permissible exposure limits (PEL), for chemical concentrations to which one may be exposed without detrimental health effects (*12*). PELs are available for most commercial (or commonly used) chemicals. These limits inform chemists of the relative hazards dealing with diverse chemicals. A low PEL can raise cautionary flags on use or consideration of specific solvents.

Table VIII. Ozone-depleting solvents restricted under the Montreal Protocol (12).

Class I	Class II
CFCs (chlorofluorocarbons)	HCFCs
Halon 1211 (bromochlorodifluoromethane)	(hydrochlorofluorocarbons)
Halon 1301 (bromotrifluoromethane)	
Halon 2402 (dibromotetrafluoroethane)	
CCl_4 (carbon tetrachloride)	
Methyl chloroform (1,1,1-trichloroethane)	
CH_3Br (methyl bromide)	
$C_jH_jBr_kF_l$ (hydrobromofluorocarbons)	

A second area of regulations restricting potential solvents is based on laws, such as the Montreal Protocol. The Montreal Protocol bans the manufacture and use of stratospheric ozone-depleting solvents (12). Solvents restricted under this law are listed as class I or class II compounds. Class I compounds have already undergone a major phase out, whereas Class II will be completely phased out of use by 2030. The compounds are listed in Table VIII. Prior knowledge regarding regulated solvents should guide use or selection of solvents.

Hazardous air pollutants such as hexane and methanol are regulated under the Clean Air Act (CAA) (13). In addition, volatile organics defined by the U. S. Environmental Protection Agency (EPA) are compounds that evaporate at the temperature of use and react with oxygen to form tropospheric ozone, are also restricted under the CAA. Additionally, Section 313 of the Emergency Planning and Community Right-to-Know Act requires users of solvents to record their releases and waste carefully on the Toxics Release Inventory. A summary of many of the regulations at the federal, state, and local levels is found in a review by Breen and Dellarco (14).

Following the Pollution Prevention Act of 1990, EPA compiled a list of 17 priority pollutants, including some solvents and heavy metal compounds, whose use was to be voluntarily reduced by 50% by 1995. This program, known as the U.S. EPA 33/50 Program, was based on the criteria that the listed chemicals are used in large volume, are detrimental either to the environment or human health, and have available methods to reduce their use (A partial list is given in Table IX) (15). The 33/50 Program in conjunction with the previously mentioned global and federal regulations has provided a solid starting point for scientists and engineers to reevaluate traditional methods and to discover ways to limit or eliminate the use of hazardous solvents.

Taken together, the regulations become the backbone of the regulatory driver. They also serve to rule out or restrict solvents or classes of solvents.

Evaluating Present Solvent Usage

The next step in selecting greener solvents is to critically evaluate the present solvent use. Many methods for solvent pollution prevention have been developed.

Table IX. Eleven Compounds on the U.S. EPA 33/50 list

Eleven Compounds	Solvent use
Benzene	Synthesis, feedstock, waxes, oils, resins
Chloroform	Degreasers, rubbers, resins, waxes
Carbon tetrachloride	Degreasers, rubbers, resins, waxes, feedstock
Methhyl ethyl ketone	Coatings, resins
Methyl isobutyl ketone	Similar to MEK
Methylele chloride	Polymers, synthesis, cleaning and degreasing
Perchloroethylene	Dry cleaning, degreasing
Toluene	Feedstock, paints, resins, extractants, gasoline additive
1,1,1-trichloroethane	Adhesives, coatings, inks, degreasers, wases, alkaloids
Trichloroethylene	Cleaning, degreasers, waxes, resins, rubbers, paints, extractants, feedstock
xylenes	General solvent, feedstock

Pollution prevention tactics employed range from limiting/eliminating solvent use to seeking engineering solutions to minimize solvent loss/release. Table X presents some representative approaches to evaluating present solvent usage.

Present Examples of green solvents

The chemical industry is broad, and it would not be possible to adequately cover all the areas here. However, some representative examples illustrate the earlier concepts and provide incentive for future work. The solvents presently used in chemical reactions and processes were chosen based upon on performance criteria, including optimal yield, chemical reactivity and selectivity. Each industry or chemical sector must determine how the regulatory, economic and technological drivers affect their solvent supply in the near and far term, and motivate them to use green solvents.

Table X. Representative Approaches to Evaluating Current Solvent Usage

Problem	Adjustment
Hazardous solvents	• new synthetic pathways; • new chemical processes; • more selective/efficient processes
Fugitive air emissions	• adjust process equipment
Waste generation	• process integration techniques; • optimize solvent recovery; • recycling
Banned or environmentally discouraged solvents	• solvent "drop-in" replacement

Development of Green Solvents

The goal in this area of green chemistry is to find or develop replacement solvents that are attractive for reasons of safety, toxicity, emissions, and other aspects covered increasingly by the many government regulations. As a minimum expectation, new chemicals and techniques would allow chemistries to occur in the absence of solvents, or with solvents that are environmentally benign (for example, carbon dioxide or water.) Since most synthetic reactions and chemical processes are solvent based, and will continue to be so for the foreseeable future, there is a need for alternative solvents that are more environmentally friendly. Given this fact, examples of green solvents will include process redesign and actual solvent replacements. Since green chemistry offers the goal of pollution prevention, the cited examples represent improvements in waste minimization.

As environmental regulations and concern increase, a variety of methods have been developed by the chemical industry to minimize chemical release and its effects (*18*). The processes using a large amount of solvents are often those processes consuming large amounts of energy for purification and recycling of the solvents. Many researchers and industries are seeking to improve pollution prevention by reducing solvent usage through process modification. An interesting example of an approach to reduction in the use of solvents is to modify reaction sequences so as to shortcut multistep syntheses. Nonselective reactions can be replaced by highly selective reaction systems. An interesting example of research in the area of substituting alternative reagents and thus creating a more friendly process see the work being done on dimethyl carbonate (*19*). The scientists in this work found that the reaction of arylacetonitriles and methyl arylacetates with dimethylcarbonate (DMAC) at 180-220 °C in the presence of a weak base (K_2CO_3), produces the monomethylderivatives (2-arylpropionitriles and methyl 2-aryl propionates, respectively) with a selectivity higher than 99%. The selection of the best synthesis and processes must be guided by reduced environmental effects (20).

One further approach to incorporating the principles of green chemistry into solvent selection with process redesign (beyond tables of which solvents are or are not acceptable) is to evaluate the chemical characteristics of the desired solvent (or the solvent it is replacing) within the life-cycle of the entire chemical process (21). In this analysis disposal economics and safety concerns become critical considerations. Once these factors are summed, including a similar evaluation of the solvent in question, the true environmental cost of the process is found. Based upon the results of the analysis, new information is gained toward selection of the appropriate solvent or chemical procress.

Although the idea of designing safer chemicals is not new, the *concept* of how this idea can be developed, introduced and integrated into the real world of commercially viable industrial chemicals is new. Historically, the knowledge and skill of the chemist was developed and focused primarily on the physical and chemical properties of chemical classes and chemical substituents and the molecular manipulation of these factors to attain the desired chemical properties of the end product. The concept of designing safer chemicals brings a new dimension to molecular design; namely, a greater consideration of the impact of a new or existing chemical on human health and the environment throughout the chemical's life-cycle of manufacture, use and disposal. This involves not only important considerations by chemists regarding the design of new chemicals, but, equally important, the reconsideration and redesign of a wide range of existing chemicals (9).

Drop-in Replacements

In the short term, the possibility exists in some situations to exchange a greener solvent for a more hazardous one (direct "swap"). Success in finding green solvents as drop-in replacements in existing processes is extremely variable. Inevitably, some adjustments need to be made, and some alternations in the chemistry are necessary. Listed below in Table XI are some examples of possible solvent alternatives. In this area a chemist's knowledge of solvents and chemistry are invaluable in utilizing alternative solvents.

Future directions in solvents

Carbon dioxide and water are emerging as potential green solvents. Carbon dioxide may be a greenhouse gas, but it has several qualities that make it more environmentally benign than other fluids. CO_2 is isolated from the environment and offers low toxicity danger to workers or the environment. Supercritical carbon dioxide

Table XI. Drop-in solvents for chemical synthesis and processes

Solvent	Replacement	Reference
Diethyl ether	methyl t-butyl ether	[12]
benzene	toluene	[2]
n-Hexane	2,5-Dimethylhexane	[2]
Tetrachloroethylene	hydrofluorocarbons	[12]
Ethylene glycol monomethyl ether	1-(Methoxy)-2-propanol	[2]
methylene chloride	benzotrifluoride	[12]
polar solvents	1-ethyl-3-methyl imidazolium chloride-aluminum(III) chlorides	[22]
chloroform	dimethoxy ethane	[12]

(SC CO2) is an inexpensive, environmentally benign alternative to conventional solvents. Based on solubility considerations, supercritical CO_2 can potentially replace many chemical reactions and processes.

Johnston and colleagues recently reported the findings of their microemulsion research. Using supercritical CO_2 they overcame its solubility limitations by using nontoxic ammonium carboxylate perfluoropolyether as surfactant (23). Work that is on-going in this area is indicative of the potential this media holds (24),(25). Eliminating the need for associated toxic substances has been a recognized approach to designing safer chemicals for years, particularly in coating applications. A classic example of this is the replacement of oil-based paint with water-based, environmentally friendlier products. Likewise, supercritical carbon dioxide has been used to replace volatile organic compounds (VOCs) in spray paints (26).

In nature water provides the milieu for numerous biochemical organic and inorganic reactions within living systems. Yet, most laboratory and industrial organic reactions are carried out in organic solvents. Current research is demonstrating that it is indeed possible to replace organic solvents with water. Examples are given in two current book sections. Ronald Breslow discusses his work with the Diels-Alder reaction in water and studies on artificial enzymes that act in water (27). Chao-Jun Li presents work on Indium-mediated bisallyation of carbonyl compounds and metal-mediated allenylation-propargylation reactions of carbonyl compounds done in water

(*28*). The references contained in these two chapters illustrate the possibilities and highlight the advantages to using water as a green solvent.

Conclusion

Admittedly the challenges facing chemists to comply with the economic, regulatory and technological drivers are daunting. Green chemistry provides opportunities to move the goals of pollution prevention closer to reality. The evaluation and choice of solvents are integral to reaching the goal. The evaluation, development and ultimate selection of green solvents (more environmentally benign media and chemical processes) necessitate the application of traditional and novel considerations to mundane chemical choices in this area.

Literature Cited

1. Wender, P. A.; Scott, H.; Wright, D. L. *Chemistry & Industry* **1997**, *19*, 765-768.
2. DeVito, S.C. In *Designing Safer Chemicals: Green Chemistry for Pollution Prevention*; DeVito, S. C.; Garrett, R. L. Eds.; American Chemical Society: Washington, DC, 1996; pp. 17-59.
3. *Pollution Prevention Act of 1990;* U. S. Environmental Protection Agency, *U.S.E.P.A.*: Washington, DC, 1990; pp. 13101-13109.
4. Garrett, R.L. In *Designing Safer Chemicals: Green Chemistry for Pollution Prevention*; DeVito, S. C.; Garrett, R. L. Eds., American Chemical Society: Washington, DC, 1996; pp. 2-15.
5. Anastas, P. T.; Warner, J. C. *Green Chemistry: Theory and Practice*; Oxford: Oxford University Press: Oxford, UK, 1998, p 135.
6. Walden, P., In *Sammlung chemischer und chemisch-technischer Vortrage.* Chem. Zentralbl. 1910; p. 1352.
7. Connors, K. A. *Chemical Kinetics: The Study of Reaction Rates in Solution*; VCH Publishers, Inc, New York, NY, 1990.
8. Reichardt, C. *Solvents and Solvent Effects in Organic Chemistry.* 2nd ed.; VCH Verlagsgesellschaft mbH: Weinheim, 1988.
9. DeVito, S. C. *Chemtech* **November 1996**, 34-47.
10. Nelson, W. M. In *Green Chemistry: Frontiers in Benign Chemical Syntheses and Processes*; Anastas, P. T.; Williamson, T. C., Eds.; Oxford University Press: Oxford, UK, 1998, pp. 200-224.
11. Parrish, C.F. In *Kirk-Othmer Concise Encyclopedia of Chemical Technology* Grayson, M., Ed.; John Wiley & Sons: New York, NY, 1985, pp. 1091-1092.
12. Sherman, J., *et al. Envir. Health Perspectives* **1998**, *106(Suppl 1)*:, 253-271.
13. *Clean Air Act of 1991;* U. S. Environmental Protection Agency, U.S.E.P.A.: Washington, DC, 1991.

14. Breen, J. J.; Dellarco, M. J. In *Pollution Prevention in Industrial Processes*; Breen, J. J.; Dellarco, M. J. Eds.; American Chemical Society: Washington, DC, 1992, pp. 2-12.
15. *33/50 Program EPA-741-K-92-001;* U. S. Environmental Protection Agency U.S.E.P.A.: Washington, DC, 1992.
16. Kirschner, E. M. *Chem Eng Prog*, **1994.** *72:*, 13-20.
17. Trost, B. M. *Science* **1991**. *254,* 1471-1477.
18. Draths, K. M.; Frost, J. W. In *Green Chemistry: Frontiers in Benign Chemical Syntheses and Processes*; Anastas, P. T.; Williamson, T. C. Eds.; Oxford University Press: Oxford, UK, 1998, pp. 150-165.
19. Tundo, P.; Selva, M.; Marques, C. A. In *Green Chemistry: Designing Chemistry for the Environment*; Anastas, P. T.; Williamson, T. C., Eds.; *ACS Symposium Series* 626; American Chemical Society: Washington, DC, 1996; pp. 81-91.
20. Anastas, P. T. In *Benign by Design: Alternative Synthetic Design for Pollution Prevention*; Anastas, P. T.; Farris, C. A., Eds.; American Chemical Society: Washington, DC, 1994, pp. 2-22.
21. *Design for Environment: Creating Eco-Efficient Products and Processes*; Fiksel, J., Ed.; McGraw-Hill: New York, NY, 1996.
22. Freemantle, M. *Chem. Eng. News*, **1998**. *76*, 32-37.
23. Gwynne, P. *R&D Magazine*, **May 1996**, 27-30.
24. Morgenstern, D. A., *et al.* In *Green Chemistry: Designing Chemistry for the Environment*; Anastas, P. T.; Williamson, T. C., Eds.; *ACS Symposium Series* 626; American Chemical Society: Washington, DC, 1996, pp. 132-151.
25. Brennecke, J. F.; Chateauneuf, J. E. *Chem. Rev.*, **1999**. *99*, 433-452.
26. Donohue, M.D., *et al.* In *Green Chemistry: Designing Chemistry for the Environment*; Anastas, P.T.; Williamson, T. C., Eds.; *ACS Symposium Series* 626; American Chemical Society: Washington, DC, 1996, pp. 152-167.
27. Breslow, R., In *Green Chemistry: Frontiers in Benign Chemical Syntheses and Processes;* Anastas, P. T.; Williamson, T. C. Eds.; Oxford University Press: Oxford, UK, 1998, pp. 225-233.
28. Li, C.-J. In *Green Chemistry: Frontiers in Benign Chemical Syntheses and Processes*; Anastas, P. T.; Williamson, T. C. Eds.; Oxford University Press: Oxford, UK, 1998, pp. 233-245.

INDEXES

Author Index

Anastas, Paul T., 1
Bergbreiter, David E., 182
Brainard, Adam P., 143
Cabezas, Heriberto, 230
Carlson, Glenn R., 8
Chen, Zhendao, 143
Cooper, Susan M., 114
Culp, R. D., 194
Danheiser, Rick L., 270
Devon, T. J., 194
DiCosimo, Robert, 114
Edwards, Laura C., 18
Eisenberg, Amy, 114
Fager, Susan K., 114
Fish, Richard H., 172
Fletcher, B., 258
Freeman, Harold S., 18
Gallagher, F. Glenn, 114
Gavagan, John E., 114
Grassian, V. H., 206
Gonzalez, Michael A., 217
Hann, Eugenia C., 114
Heine, Lauren G., 1
Hennessey, Susan M., 114
Ho, Nancy W. Y., 143
Jang, B. W-L., 194
Knipple, Douglas, C., 33
Larsen, S. C., 206
Li, Chao-Jun, 62, 74
Lima, André O.S., 143
Lin, Chhiu-Tsu, 43
Marsella-Herrick, P., 33
Memoli, Sofia, 87
Miller, Jr., Edward S., 126
Miller, S. J., 33
Muñoz, Carmen L., 43

Myli, K. B., 206
Nelson, William M., 313
Nikolopoulos, A. A., 194
Nishtala, Subba A., 230
Olsen, D. J., 194
Panov, A. G., 206
Paquette, Leo A., 100
Perkins, Neal E., 114
Peretti, Steven W., 126
Pizzirani-Kleiner, Aline A., 160
Rabion, Alain, 172
Renslo, Adam, 270
Roelofs, W. L., 33
Rosenfield, C.-L., 33
Sadeghipour, M., 258
Sahle-Demessie, Endalkachew, 217
Sedlak, Miroslav, 143
Selva, Maurizio, 87
Steinfeld, Jeffrey I., 270
Stieglitz, Barry, 114
Spivey, J. J., 194
Subramanian, R., 194
Suleman, N. K., 258
Tanko, J. M., 258
Taylor, Joshua D., 270
Tester, Jefferson, W., 270
Tundo, Pietro, 87
Varma, Rajender S., 292
Vincent, Jean-Marc, 172
Weintstein, Randy D., 270
Whitten, Mary C., 43
Williams, Dwight E., 244
Williamson, Tracy C., 1
Xiang, Y., 206
Yu, Tao, 43
Zhao, Renhong, 230

Subject Index

A

Academic chemistry, green solvents, 315t
Acetone condensation. *See* Condensation of acetone
Acetylene gas, coupling of aryl halides water as solvent, 69, 71
1,2-Additions
  fdiastereoselective, to α-oxygenated ketones, 108–109
  facial selectivity to nucleophilic additions, 108t
  stereochemical course involving bromo esters and functionalized aldehydes, 103t
Adipamide (ADAM). *See* 5-Cyanovaleramide (5CVAM) production
Adiponitrile (ADN). *See* 5-Cyanovaleramide (5CVAM) production
Air Index
  description, 234–236
  methyl ethyl ketone (MEK) and replacement mixture, 241
Alcohols
  borohydride reduction of carbonyl compounds, 304–305
  oxidation reagents, 300
  typical experimental procedure for oxidation, 301
Aldehyde diacetates, cleavage, microwave irradiation, 294, 295
Aldehydes
  chiral, diastereoselective allylation, 101–102
  epoxidation reactions by fluorous biphasic catalysis, 178–180
  functionalized, stereochemical course of 1,2-additions involving bromo esters, 103t
  selective alkylation of aromatic over aliphatic, 75, 76

Aldol condensation
  efforts to replace conventional homogeneous catalysts, 194
  industrial practice, 195
  See also Condensation of acetone
Algorithm, PARIS II, 237–239
Aliphatic ketones, reactions with dimethylcarbonate (DMC), 94, 95t
Alkali oxides, acetone condensation, 195
Alkaline earth oxides, acetone condensation, 195
Alkane and alkene oxidations
  perfluorocarbon soluble metalloporphyrins as catalysts, 173–174
  perfluorocarbon soluble non-porphyrin catalysts, 175–177
  See also Fluorous biphasic catalysis (FBC)
Alkanes
  chlorination, 259–260
  oxidation industrially and in organic synthesis, 218–219
  See also Free radical chemistry
Alkylarylsulfones, mono-C-methylation with dimethylcarbonate (DMC), 91–92
Alkylation reactions
  carbohydrate derivatives, 77, 78
  chemoselectivity in water, 75
  reductive of amines, 305–306
Allylation reactions
  benzaldehyde with allyl bromide and iodide, 77, 78
  competitive indium-promoted allylations in water, 101t
  cyclohexanone selectively in presence of cyclopentanone, 75, 76
  diastereoselective allylation of chiral aldehydes, 101–102
  enolizable 1,3-dicarbonyl compounds, 77, 78
  indium-catalyzed organometallic reaction in water, 79, 80

331

Allyl bromides
  chain lengths for reaction of alkylaromatics with, 266t
  coupling involving geometrically biased, 105–106
  effect of Z substituent on chain length, 266–267
  hydrocarbon functionalization and C–C bond formation, 265–267
  reactions of alkylaromatics with substituted, 266t
  See also Free radical chemistry; Hydrocarbon functionalization
Alternatives
  dimethylcarbonate (DMC) alternative to toxic intermediates, 87, 98
  See also Dimethylcarbonate (DMC)
Alternative solvents. *See* Environmentally benign solvent substitutes
Alumina, promoting reactions in supercritical carbon dioxide, 281–282
Alumina, chromium trioxide impregnated wet, oxidation, 301–302
Alumina, iodobenzene diacetate (IBD) doped
  oxidation, 302–303
  oxidation of sulfides, 304
Amblyomma, binding affinity of tebufenozide relative to 20-hydroxyecdysone (20E), 16t
Ames test, toxicity of dyes by mammalian mutagenicity test with Prival modification, 21–22
Amines, reductive alkylation, 305–306
Amines, primary aromatic, selective mono-N-methylation with dimethylcarbonate (DMC), 92–93
Anilines, reactions with dimethylcarbonate (DMC), 93t
Anthonomus, binding affinity of tebufenozide relative to 20-hydroxyecdysone (20E), 16t
Aquatic toxicity
  Fe-complexed dyes, 30, 31f
  plant protocol involving duckweed (*lemna minor*), 22–23
  structure relationships in dyes, 28–30
  testing metal salts employed in dye complexation, 29f
  See also Iron-complexed dyes
Aromatic aldehydes
  exclusive pinacol coupling, 75, 76
  selective alkylation over aliphatic, 75, 76

Aromatic amines, primary, selective mono-N-methylation with dimethylcarbonate (DMC), 92–93
Aroxyacetates, monomethylation with dimethylcarbonate (DMC), 90–91
Aroxyacetonitriles, monomethylation with dimethylcarbonate (DMC), 90–91
2-Aroylbenzofurans, synthesis, 309
Aryl halides, coupling with acetylene gas, water as solvent, 69, 71
Arylacetoesters, selective monomethylation, 89
Arylacetonitriles, selective monomethylation, 89
2-Aryl-1,2,3,4-tetrahydro-4-hydroquinolones, synthesis, 307
Aspergillus nidulans
  biomass production and pollution reduction, 165, 166t
  chemical oxygen demand (COD) reduction and biomass production, 167–169
  consuming vinasse, 161–162
  dry biomass production and COD reduction by *A. nidulans* strains, 168f
  effect of temperature on radial growth of *Aspergillus* species, 165
  effect on radial growth by addition of fertilizer (NPK) solution, 163, 164t
  effect on radial growth by addition of sugar cane molasses to vinasse, 163, 164t
  growth kinetics of *A. nidulans* strains, 166, 167f
  medium establishment, 163
  vinasse biodegradation by *Aspergillus* species, 164–165
  vinasse conversion by recombinant strains of *A. nidulans*, 166–169
  See also Sugar cane vinasse, bioconversion
1,4-Asymmetric induction, coupling reaction in water, 106–108
Avian species, relative acute oral toxicity of tebufenozide, 13t
Azamacrocycles, candidates as non-porphyrin ligands, 175–176
Azeotropes
  determinations, 256
  volatile methyl siloxane (VMS) fluids, 247, 249

B

Barbier–Grignard reaction, magnesium reactions in water, 77, 79

Barbier-type reaction, ring expansions in water, 79, 80
Base catalysts, solid, acetone condensation, 195
Benign by design, philosophy, 100
Benign catalysis, environmentally, green chemistry research, 4
Benzaldehyde
  allylation in aqueous medium, 77, 78
  high selectivity in toluene oxidation without overoxidation, 222, 223t
  mechanism for formation on BaX and BaY zeolites, 213–214
  See also Photooxidation of toluene
Benzenes, mono-substituted with increasing side chain length, bioconversion candidates, 135, 138
Benzylarylsulfone, mono-C-methylation with dimethylcarbonate (DMC), 91–92
Benzylation reactions, dibenzylcarbonate, 97–98
Benzylic ketones, reactions with dimethylcarbonate (DMC), 94, 95t
(+)-Bergenin, synthesis in water, 81, 84
Biocatalysis, green chemistry research, 3–4
Biocatalytic process. See 5-Cyanovaleramide (5CVAM) production
Biochemical oxygen demand (BOD)
  pollution index, 161
  pollution reduction of vinasse by Aspergillus species, 165, 166t
  See also Sugar cane vinasse, bioconversion
Biochemistry, solvent toxicology, 318t
Bioconversion of toluene. See p-Hydroxybenzoate (HBA), bioconversion of toluene
Biology, solvent toxicology, 318t
Biomass hydrolysates. See Genetically engineered Saccharomyces yeasts
Biosynthesis, green chemistry research, 3–4
Biosynthetic pathways, pheromones, 36–37
Biphasic catalysis. See Fluorous biphasic catalysis (FBC); Polymer-facilitated biphasic catalysis
Boll weevil, binding affinity of tebufenozide relative to 20-hydroxyecdysone (20E), 16t
Borohydride reduction, carbonyl compounds to alcohols, 304–305

Bromine atom
  allylation of hydrocarbon via free radical process involving, 267
  See also Hydrocarbon functionalization
Bromo esters, stereochemical course of 1,2-additions involving functionalized aldehydes, 103t
Brønsted acidity
  colorimetric detection of sites, 211, 213
  correlation of loss of product selectivity and Brønsted acidity in photooxidations in zeolites, 214–215
  See also Photooxidation of toluene
Butyraldehyde, Ruhrchemie/Rhône–Poulenc's process, 67, 68

C

Cage effects
  chlorine atom, 259
  magnitude in supercritical CO_2, 264
  supercritical fluids, 258–259
  See also Free radical chemistry
Carbohydrate synthesis
  alkylation of derivatives in aqueous solution, 77, 78
  aqueous organometallic reactions, 81
  styryl carbohydrate derivatives, 81, 82f
Carbon–carbon bond formations
  fundamental reactions of organic chemistry, 74, 75
  Knoevenagel condensation alternative, 110
  magnesium reaction in water, 77, 79
  palladium-catalyzed reductive couplings in water, 79, 80
  water as solvent, 65, 66
  See also Coupling reactions in water, indium-promoted; Hydrocarbon functionalization
Carbon dioxide, supercritical
  environmentally benign solvent, 5
  See also Free radical chemistry; Supercritical carbon dioxide; Supercritical fluids as solvent replacements
Carbonyl alkylations, aqueous medium, 77, 78
Carbonyl compounds, borohydride reduction to alcohols, 304–305
Carboxylic esters, debenzylation, microwave irradiation, 294–295
Carriers, volatile methyl siloxanes (VMS) in formulated products, 255

Catalysis
  environmentally benign, green chemistry research, 4
  See also Fluorous biphasic catalysis (FBC); Polymer-facilitated biphasic catalysis
Catalytic pathways, evaluation in supercritical fluids, 281
Caterpillar
  binding affinity of tebufenozide relative to 20-hydroxyecdysone (20E), 16t
  relative acute oral toxicity of tebufenozide, 13t
  See also Tebufenozide
Cathodic delamination
  in-situ phosphatizing layer on metal surface, 54
  plots of control alkyl enamel and water-reducible ISPC coated on bare cold-rolled steel (CRS) coupons, 55f
Cation-exchanged zeolites. See Photooxidation of toluene
Cellulosic biomass
  feedstock for ethanol-fuel production, 144
  See also Genetically engineered Saccharomyces yeasts
Cellulosic fibers, colorants, 18, 19
Chactiacandin, synthesis in water, 81, 84
Chelation
  competitive intramolecular/intermolecular options for indium in coupling reactions, 102–105
  control, stereoselectivity in water, 75
  See also Coupling reactions in water, indium-promoted
Chemical engineering, connections to synthetic organic chemistry and physical chemistry in supercritical fluids, 273, 275f
Chemical environmental impacts, (Score)$_{ij}$ measuring, 236–237
Chemical industry
  hazardous waste, 217–218
  solvent roles in chemistry, 315t
Chemical oxygen demand (COD)
  pollution index, 161
  pollution reduction of vinasse by Aspergillus species, 165, 166t
  See also Sugar cane vinasse, bioconversion
Chemical synthesis
  environmental advantages of photocatalyzed, 225–226
  green chemistry research, 3
  possible advantages in supercritical environment, 271, 273t
  semiconductor catalysis, 221–222
  solvents, 314–316
  water as solvent, 67, 68
  See also Supercritical fluids as solvent replacements
Chemicals, designing safer, green chemistry research, 3
Chemistry
  solvents in, 314–316
  solvent toxicology, 318t
Chemoselectivity, new for aqueous organometallic reactions, 75
Chiral aldehydes, diastereoselective allylation, 101–102
Chlorination
  cyclohexane, 260, 261f
  2,3-dimethylbutane, 260, 262f, 263
  free radical, alkane, 259–260
  neopentane, 263, 264f
  See also Free radical chemistry
Chromium trioxide impregnated wet alumina, oxidation, 301–302
Claisen condensation, aryl/alkyl ketone splitting in reverse, 94
Clean Air Act (CAA), hazardous air pollutants, 322
Cleaning non-polar soils, volatile methyl siloxanes, 247, 248f
Cleaning semi-polar soils, volatile methyl siloxane, 249, 250f
Cleavage reactions
  aldehyde diacetates, 294, 295
  solvent-free of oximes, 297–298
Cloning. See Insect sex pheromone precursors
Coatings
  protective performance, 48, 50
  See also In-situ phosphatizing coatings (ISPC)
Cofermentation. See Genetically engineered Saccharomyces yeasts
Colorants. See Iron-complexed dyes
Color-fastness
  structure relationships in Fe-complexed dyes, 23–26
  See also Iron-complexed dyes
Condensation of acetone
  acidity and basicity by thermoprogrammed desorption (TPD), 197
  acidity and basicity of metal-containing catalysts, 202–203

batch reactor experiments, 201–203
catalyst activity and selectivity, 201t
catalyst characterization, 199–201
catalyst synthesis and characterization, 196–197
CO_2 TPD profiles of hydrotalcite (HTC), Pd/HTC, and Pt/HTC catalysts, 200f
comparing acidity and selectivity of Pd- and Pt-supported HTC, 202
complications by thermodynamic equilibrium limitations, 195
effect of metal loading on selectivity of Pd/HTC and Pt/HTC towards methyl isobutyl ketone (MIBK) and isopropanol (IPA), 203f
effect of reaction temperature on formation of MIBK, 203, 204f
hydrogenation activity with metal loadings, 202
hydrotalcite (HTC) support (Mg-Al hydroxide) in promising catalyst formulation, 196
industrial process for synthesis of commodity chemicals, 194–195
major reaction pathways, 195f
physicochemical properties of catalysts, 199t
reaction system and procedure, 197–198
schematic of reaction system, 198f
solid base catalysts in literature, 195
thermodynamic equilibrium properties, 196t
typical experiment, 197–198
Condensation reactions
synthesis of imines, enamines, and nitroalkenes, 298–300
See also Organic syntheses using microwave irradiation
Conducting polymer, hydrocarbon non-metallic, water as solvent, 69, 71
Conductor band, electron movement in semiconductor, 220
Copolymerization of 3,5-diiodobenzoic acid with acetylene gas, water as solvent, 69, 71
Copper ions–hydrogen peroxide, oxidation, 301
Corrosion resistance
alkyd in-situ phosphatizing coating versus control alkyd formula, 57f
blistering width measured at "x" scribe, 36t
salt spray tests, 54, 56

Coupling aryl halides with acetylene gas, water as solvent, 69, 71
Coupling reactions in water, indium-promoted
1,4-asymmetric induction, 106–108
comparative analysis of additions involving crotylindium and 3-bromoallylindium to 2-hydroxypropanal, 105
competitive indium-promoted allylations in water, 101t
competitive intramolecular/intermolecular chelation options, 102–105
coordination of indium atom to aldehyde carbonyl, 104
coupling involving geometrically biased allylic bromides, 105–106
Cram-like transition state, 102
cyclic transition states with capacity for multi-point steric compression, 109
diastereoselective 1,2-additions to α-oxygenated ketones, 108–109
diastereoselective allylation of chiral aldehydes, 101–102
π-facial discrimination via Felkin–Anh transition states, 102
facial selectivity in nucleophilic additions, 108t
Felkin–Anh paradigm, 106, 107
formation of α-methylene-γ-lactones fused to medium and large rings, 110, 111
indium-promoted alternative to Knoevenagel condensation, 110
indium-promoted C-allylations of β-oxygenated aldehydes in various solvents, 103t
level of syn-1,4-stereocontrol, 107
linking diastereofacial selectivity to R or S configuration, 109
stereochemical course of 1,2-additions involving bromo esters and functionalized aldehydes, 103t
stereoselection in C-allylation of α-oxygenated aldehydes, 101t
thermodynamic stability of Z and E isomers of methyl 2-(bromomethyl)-2-butenoate, 106
Crosslinked polymers
insolubility, 183
See also Polymer-facilitated biphasic catalysis
Crosslinking, polymer film formation, 46–47

Crude hydrolysates. See Genetically engineered Saccharomyces yeasts
5-Cyanovaleramide (5CVAM) production beads (Pseudomonas chlororaphis B23/alginate) as production catalyst in bath reactions with catalyst recycle, 118–119
commercial-scale production, 119, 122
comparing hydrolysis of adiponitrile (ADN) to 5CVAM by Rhodococcus sp. A4/alginate beads and B23/alginate beads, 117f
comparison of A4 and B23, 116, 118
dependence of specific activity of B23/alginate beads on pH, 118
further conversion of 5CVAM to adipamide (ADAM), 119, 121f
high regioselectivity of Pseudomonas putida 3LG-1-5-1A and Rhodococcus sp. A4 catalysts, 115
hydrolysis of ADN to 5CVAM, 115
immobilization of catalysts in alginate beads, 116, 118
light activation of A4 whole cells, 116
microbial catalysts having nitrile hydratase activity, 115
optimization of reaction conditions using immobilized B23 cells, 118–119
ratio of 5CVAM to ADAM with alginate beads, 119, 120f
reaction times and concentrations of 5CVAM, ADN, and ADAM for consecutive 400-gallon batch reactions, 123f
Cyclizations, aqueous organometallic reactions, 79, 80
Cyclohexane
chlorination, 260, 261f
safe range of operation for photocatalytic operation, 226f
See also Free radical chemistry
Cyclohexanone, selective allylation, 75, 76
Cyclohexanone oxime, reaction with dimethylcarbonate (DMC), 95, 96
Cyclohexene, fluorous biphasic oxidation catalysis, 176–177
Cyclopentanoids, preparation by aqueous organometallic reactions, 79, 80

D

Deacylation reactions, microwave irradiation, 294

Debenzylation of carboxylic esters, microwave irradiation, 294–295
Dehydration reactions, microwave-promoted, 298–299
Deoximation reactions, microwave irradiation, 297–298
(+)-3-Deoxy-D-glycero-D-galacto-nonulosonic acid (KDN), synthesis, 67, 68, 81, 82
Designing safer chemicals, green chemistry research, 3
Desilylation reactions, microwave irradiation, 296
Dethioacetalization reaction, microwave irradiation, 296–297
Diastereoselectivity. See Coupling reactions in water, indium-promoted
Dibenzylcarbonate (DBzlC)
benzylation reactions, 97–98
reaction of phenylacetonitrile with DBzlC, 97t
Diels–Alder reactions
supercritical fluids, 276f, 277f, 278t, 279t, 280
water as solvent, 63, 64
yield and selectivity enhancements by fumed silica, 282
3,5-Diiodobenzoic acid, copolymerization with acetylene gas, water as solvent, 69, 71
2,3-Dimethylbutane
chlorination, 260, 262f, 263
See also Free radical chemistry
Dimethylcarbonate (DMC)
environmentally friendly substitute, 3
experimental parameters influencing reactivity, 88
industrial applications, 88
limit in preparation, 87
methylation reactions, 89–93
methylation reactions under continuous-flow and batch conditions, 88–89
proposed mechanism for cleavage of C–CO bond and formation of two esters, 94
proposed mechanism for reaction of DMC with phenylacetonitrile, 90
proposed reaction mechanism of cyclohexanone oxime with DMC, 96
reactions of anilines with DMC, 93t
reactions of aryloxyacetates and aryloxyacetonitriles with DMC, 91t
reactions of benzylaryl- and alkylarylsulfones with DMC, 92t

reactions of benzylic and aliphatic ketones with DMC, 95t
reactions with ketones, 94
reactions with oximes, 95–97
reactivity toward nucleophilic compounds, 88
selective mono-*C*-methylation of alkylarylsulfones, 91–92
selective monomethylation of aroxyacetonitriles and methyl aroxyacetates, 90–91
selective monomethylations of arylacetonitriles and arylacetoesters, 89
selective mono-*N*-methylation of primary aromatic amines, 92–93
use in organic synthesis, 88
Drop-in replacements, solvents for chemical synthesis and processes, 325–326
Drosophila, binding affinity of tebufenozide relative to 20-hydroxyecdysone (20E), 16t
Dyes. *See* Iron-complexed dyes

E

Ecdysonoid mode of action. *See* Tebufenozide
Economic impact, green chemistry, 314
Economics
limitations to use of pheromones for insect control, 34–36
solvent toxicology, 318t
Electrochemical impedance spectroscopy (EIS)
Bode-magnitude plots for cured paint film of polyester-melamine white paint on bare and chromated 2024 T3 Al substrates after soaking in 3% NaCl solution for 72 and 2500 hours, 50, 52f, 53f
Bode-magnitude plots for latex in-situ phosphatizing coatings (ISPC) coated on bare cold rolled steel (CRS), iron phosphated B-1000, and iron phosphated and chromated BD+P60 coupons, 51f
Bode-magnitude plots for latex control coating on bare cold-rolled steel (CRS), phosphatized B-1000, and phosphatized plus chromated BC+P60 panels, 49f
coating performance and under film metallic corrosion, 48, 50

See also In-situ phosphatizing coatings (ISPC)
Enamines, synthesis using microwave irradiation, 298–299
Environment
advantages of pheromone-based pest control, 35
advantages of photocatalyzed chemical synthesis, 225–226
aspects of insect pest control, 34
benign catalysis, 4
green chemistry, 1
Vision 2020 for the Chemical Industry, 1–2
Environmental Index
description, 234–236
methyl ethyl ketone (MEK) and replacement mixture, 241
Environmentally benign catalysis
green chemistry research, 4
See also Condensation of acetone
Environmentally benign solvent substitutes
advantages and disadvantages of categories, 231
case study of methyl ethyl ketone (MEK) replacement, 239–241
chemical environmental impacts, 236–237
classifying research efforts in designing benign solvents, 231
designing mixture replacements, 231
designing new chemicals, 231
environmental indexes, 234–236
infinite dilution activity coefficients of representative solutes in MEK and replacement mixture, 240t
PARIS II algorithm, 237–239
physical properties of MEK and replacement mixture, 240t
(Score)$_{ij}$ measure of chemical environmental impacts, 236–237
screening available solvent databases, 231
substitute solvent design theory, 231–236
technological effectiveness of replacement solvent, 239–241
technological performance: fluid properties, 233–234
technological performance: theory, 231–233
See also Volatile methyl siloxanes (VMS)

Environmental profile, volatile methyl siloxanes (VMS), 246–247
Enzymatic synthesis. *See* Insect sex pheromone precursors
Epidemiology, solvent toxicology, 318*t*
Epoxidation reactions
  fluorous biphasic catalysis approach, 178–180
  product yields and selectivity examples with Ru and Ni catalysts, 179*f*
  structure of perfluorosoluble chiral SALEN ligands, 180*f*
  structures of perfluorosoluble catalysts, 178*f*
  See also Fluorous biphasic catalysis (FBC)
Ethanol
  cofermentation of glucose and xylose present in crude hydrolysates, 157
  design and development of ideal yeast for cofermentation of glucose and xylose to ethanol, 146–147
  environmentally friendly liquid fuel, 143–144
  fermenting glucose-based food crops, 144
  repeated cofermentation of mixture of glucose and xylose, 157, 158*f*
  See also Genetically engineered Saccharomyces yeasts
Ethylbenzene, bioconversion candidate, 135, 138

F

Film formation, polymer chemistry, 46–47
Flavones, synthesis, 306–307
Fluorocarbons
  candidates for environmentally friendly process, 173
  See also Fluorous biphasic catalysis (FBC)
Fluorous biphasic catalysis (FBC)
  alkane and alkene oxidation with perfluorocarbon soluble metalloporphyrins as catalysts, 173–174
  alkane and alkene oxidation with perfluorocarbon soluble non-porphyrin catalysts, 175–177
  basis of process, 173
  epoxidation reactions, 178–180
  examples of product yields and selectivity with Ru and Ni catalysis under FBC conditions, 179*f*
  first generation of perfluoro-modified tetraarylporphyrin (TPP) ligands, 174*f*
  fluorous biphasic oxidation catalysis with cyclohexene as substrate, 177
  inspiring polymer-bound catalysts, 189
  key intermediates in synthesis of perfluorocarbon soluble TPP, 175
  novel two phase homogeneous process, 4
  potential green chemistry concept, 173
  reaction pathway for synthesis of perfluoroponytailed azamacrocycle 6, 176
  structures of perfluorosoluble chiral SALEN ligands, 180*f*
  structures of Ru and Ni complexes of perfluorinated 1,3-diketone, 178*f*
  structures of synthesized perfluoroponytailed azamacrocycles, 175*f*
Formazans
  dye ligands, 24*f*
  dyes for cellulose-based fibers, 20
  See also Iron-complexed dyes
Free radical chemistry
  advantages of supercritical carbon dioxide (SC-CO_2) as solvent, 258
  cage effect in SC-CO_2, 264
  cage effects in supercritical fluids, 258–259
  chlorination of 2,3-dimethylbutane (23DMB), 260, 262*f*, 263
  chlorination of cyclohexane, 260, 261*f*
  chlorination of neopentane, 263, 264*f*
  chlorine atom cage effect, 259
  free radical chlorination of alkane, 259–260
  hydrocarbon functionalization, 265–267
  mechanism of hydrocarbon functionalization and C–C bond formation, 265
  ratio mono- to polychlorides (M/P) as function of viscosity of solvent, 260
  ratio M/P in free radical chlorination of 23DMB in conventional solvents and SC-CO_2 versus inverse viscosity, 262*f*
  ratio M/P in free radical chlorination of cyclohexane in conventional solvents and SC-CO_2 versus inverse viscosity, 261*f*
  ratio M/P in free radical chlorination of cyclohexane in SC-CO_2, 261*f*
  ratio of 1,3- to 1,1-dichloro-2,2-dimethylpropane produced in free radical chlorination of neopentane in SC-CO_2 as function of viscosity, 264*f*
  variation of selectivity with viscosity ob-

served in free radical polymerization of 23DMB in SC-CO_2, 262f
See also Hydrocarbon functionalization
Fruit fly, binding affinity of tebufenozide relative to 20-hydroxyecdysone (20E), 16t

G

Genetically engineered *Saccharomyces* yeasts
cofermentation of glucose and xylose by genetically engineered yeast 1400 (pLNH32), 149f
cofermentation of glucose and xylose by parent yeast strain 1400, 150f
cofermenting glucose and xylose present in crude hydrolysates, 157
comparison of cofermentation of glucose and xylose by yeast strain 1400(pLNH32) and *Pichia stipitis*, 155f
comparison of xylose fermentation by yeast strain 1400(pLNH32) containing xylose reductase gene (XR), xylitol dehydrogenase gene (XD), and xylulokinase (XK) genes, and 1400(pXR-XD) containing XR and XD genes, 153f
design and development of ideal yeast for cofermentation of glucose and xylose to ethanol, 146–147
effective new gene-integration technique for development of superstable, 154, 157
effect of cloning and overexpression of XK gene, 147, 152
effect of replacing original promoters of cloned XR, XD, and XK genes with *Saccharomyces* glycolytic promoters, 152, 154
efforts for finding new yeasts, 144
efforts to express bacterial xylose isomerase gene in *Saccharomyces* yeasts, 144, 146
future perspective, 157, 159
growth using glucose and xylose as sole carbon source, 151t
ideal for industrial production of ethanol, 147
repeated cofermentation of mixture of glucose and xylose to ethanol, 157, 158f
restriction map of pLNH plasmids, 148f
stepwise integration of multiple copies of cloned multiple genes into host chromosomes, 156f
transferring XR and XD for conversion of xylose to xylulose, 146
transforming yeast strain 1400 with plasmids pLNH32 and pLNH33, 147
xylose metabolic pathways in different types of microorganisms, 145f
Global perspectives, conference theme, 2
Glucose
design and development of ideal yeast for cofermentation with xylose to ethanol, 146–147
See also Genetically engineered *Saccharomyces* yeasts
(+)-Goniofufurone, synthesis in water, 81, 83
Green chemistry
definition, 1
drivers affecting impact, 314
environmental consciousness, 313–314
fluorous biphasic catalysis (FBC) concept, 173
synthesis research, 3
value in applicability, 313
view of more chemoselective reaction, 75
Vision 2020 for the Chemical Industry, 2
Green Chemistry and Engineering Conference
organizing committee, 2
themes, 1–2
Green chemistry research
biocatalysis and biosynthesis, 3–4
categorizing approaches, 2–3
designing safer chemicals, 3
environmentally benign catalysis, 4
green chemical synthesis, 3
green solvent systems, 4–5
Green chemistry solvents
appropriateness, 316–317
components of solvent toxicology, 318t
compounds on U.S. EPA 33/50 list, 323t
development of green solvents, 324–325
drop-in replacements, 325–326
evaluating present solvent usage, 323
examples of green solvents, 323–327
future directions in solvents, 326–327
general considerations, 317t
integral part of solution, 316
novel criteria for green solvents: toxicity concerns, 319t
origin of green solvents, 319–320

ozone-depleting solvents restricted under Montreal Protocol, 322t
pollutants by Clean Air Act (CAA), 322
preparation: laws and global regulations governing solvents, 321–322
priority pollutants by EPA (EPA 33/50), 322
representative approaches to evaluating current solvent usage, 324t
solvent toxicology, 317, 319
summary of internal considerations, 319t
summary of solvent toxicology criteria, 321t
systematically selecting green solvents, 320–323
Green solvent systems, green chemistry research, 4
Grubbs' catalyst, transition metal catalyzed polymer synthesis in water, 69, 70

H

Hazardous waste, chemical industry, 217–218
Henry reaction., nitroalkene synthesis, 299
Heterocyclic compounds
2-aroylbenzofurans, 309
2-aryl-1,2,3,4-tetrahydro-4-hydroquinolones, 307
flavone synthesis, 306–307
substituted isoflav-3-enes, 307–308
substituted thiazoles, 308–309
synthesis, 306–309
Heterogeneous catalysis
original concept, 183
selectivity and reactivity versus homogeneous, 183
See also Condensation of acetone
Heterogeneous reactions
use of microwave irradiation techniques, 292–293
See also Organic syntheses using microwave irradiation
Homogeneous catalysis
new approaches using polymers, 182
selectivity and reactivity versus heterogeneous, 183
Human health, green chemistry, 1
Hydrocarbon functionalization
allylation of hydrocarbon via free radical process involving bromine atom, 267
chain lengths for reaction of alkylaromatics with allyl bromides, 266t
concurrent with C–C bond formation, 265–267
direct evidence for intermediacy of bromine radical (Br•) in reaction, 265–266
effect of Z substituent on chain length, 266–267
mechanism of free radical chain process, 265
reactions of alkylaromatics with substituted allyl bromides, 266t
relative rates of addition of $PhCH_2\bullet$ to various allyl bromides, 266
Hydrocarbon non-metallic conducting polymer, water as solvent, 69, 71
Hydrocarbons
photochemical oxidation, 4
photooxidation results, 223t
selective photooxidation in zeolites, 206–207
See also Oxygenation of hydrocarbons; Photooxidation of toluene
Hydroformylation process, water as solvent, 67, 68
Hydrogen peroxide-metal ions, oxidation, 301
Hydrolysis and oxidation of compounds, supercritical fluids, 282–283
Hydroquinolones, synthesis of 2-aryl-1,2,3,4-tetrahydro-4-hydroquinolones, 307
Hydrotalcites (HTC)
acetone condensation, 195
activity and selectivity for acetone condensation, 201t
catalyst characterization, 199–201
catalyst synthesis and characterization, 196–197
physicochemical properties of acetone condensation catalysts, 199t
promising catalyst formulation, 196
See also Condensation of acetone
p-Hydroxybenzoate (HBA), bioconversion of toluene
batch transformation experiments using resting cell suspensions of *Pseudomonas putida* EM 2878, 131
construction and organization of pathway genes, 129, 131

conversion of *p*-cresol to HBA, 133, 134*f*
dependence on level of toluene-4-monooxygenase (T4MO) induction and toluene concentration, 133, 135
generation of non-HBA degrading *P. putida* mutant, 129
HBA profiles with varying levels of *p*-cresol during toluene bioconversion, 133, 134*f*
HPLC analysis of culture supernatants, 137*f*
initial studies of *p*-hydroxybenzoate hydroxylase (PobA) activity in *P. putida* KT2440, 130*f*
Kolbe–Schmitt carboxylation of phenol, 126, 127*f*, 128
maximum yield in batch transformation studies, 135
monomer in liquid crystal polymers (LCP), 126
mono-substituted benzenes with increasing side chain length, 135, 138
mono-substituted toluenes with substitution at C-2 position, 138–139
mono-substituted toluenes with substitution at C-3 position, 139
need for development of cleaner technology for HBA production, 128
other types of mononuclear aromatic molecules of interest in LCP production, 135
plasmid constructions and reactions catalyzed by products of cloned gene expression, 132*f*
PobA enzyme activity, 131*t*
proposed reaction sequence, 130*f*
research to find co-metabolic process to replace Kolbe–Schmitt process, 128–129, 140
results of substrate range experiments, 138*t*
single processing step using proposed bioconversion process, 128
substrate range of *P. putida* EM2878, 135, 138–139
toluene conversion to HBA by resting cells of *P. putida* EM2878, 133, 135
20-Hydroxyecdysone (20E)
concentration in insect's blood, 11
critical events in normal insect molting process relative to changes in 20E concentration, 11*f*
interaction with ecdysone receptor protein (EcR), 10
molting process in insects, 9–11
schematic of interaction with target site in insects, 10*f*
structure, 9
See also Tebufenozide
Hyperecdysonism, tebufenozide causing in caterpillars, 13–14

I

Imines, synthesis using microwave irradiation, 298–299
Indium-catalyzed allylation reaction, organometallic reaction in water, 79, 80
Indium-catalyzed Barbier-type reaction, ring expansions in water, 79, 80
Indium-promoted reactions. *See* Coupling reactions in water, indium-promoted
Industrial chemistry, green solvents, 315*t*
Industry, adoption of green chemistry, 1
Infinite dilution activity coefficients, representative solutes in methyl ethyl ketone (MEK) and replacement mixture, 240*t*
Insect control
economic limitations to use of pheromones, 34–36
environmental advantages of pheromone-based, 35
environmental aspects, 34
mode of action of tebufenozide, 12
need for less toxic strategies, 34–36
suppression of pest populations by mating disruption, 34, 35*f*
See also Insect sex pheromone precursors; Tebufenozide
Insects
molting process, 9–11
See also 20-Hydroxyecdysone (20E)
Insect sex pheromone precursors
biosynthetic pathway leading to major active component of *Trichoplusia ni* sex pheromone, 37*f*
functional expression of cloned pheromone desaturase cDNAs, 38–40
future directions, 40–41
GC/MS total ion spectrum of fatty acid methyl ester derivatives of lipid extract of pYDsTnD11.1-transformed *ole*1 yeast cells, 40*f*
pheromone chemical structures and biosynthetic pathways, 36–37

pheromone desaturase evolution and homology to metabolic acyl-CoA desaturases, 37–38
reason for high cost of synthesis, 36
structure of yeast expression plasmid pYDsTnD11.1, 39f
Insect sex pheromones, methodology for producing nontoxic, 3
In-situ phosphatizing coatings (ISPC)
blistering width measured across "x" scribe, 56t
Bode-magnitude plots for cured paint film of polyester-melamine white paint on bare and chromated 2024 T3 Al substrates after soaking in 3% NaCl solution for 72 and 2500 hours, 50, 52f, 53f
Bode-magnitude plots for latex control coating on bare CRS (cold-rolled steel), iron phosphated B-1000, and iron phosphated plus chromated BD+P60 mild steel panels, 49f
Bode-magnitude plots for latex-ISPCs coated on bare CRS, iron phosphated B-1000, and iron phosphated and chromated BD+P60 coupons, 51f
cathodic delamination, 54
cathodic delamination plots of control alkyd enamel and water-reducible ISPC coated on bare CRS coupons, 55f
cleaner and cheaper chemistry, 58
coating protective performance, 48, 50
commercial baking enamels, 44
disbonding resistance of cured paints on metal by salt water immersion testing, 45
experimental, 44–45
in-situ phosphatizing of metal surface, 47–48
ISPC formulations in study, 44
ISPC formulation stability monitoring method, 44–45
metal phosphate layer on 2024 T3 Al substrate, 50
method of studying formation of interfacial metal phosphate layers, 45
novel surface conversion coating technique, 44
polymer chemistry and paint film quality, 46–47
rheology and storage stability, 45–46
salt spray (fog) test results of corrosion resistance of alkyd ISPC versus control alkyd formula, 57f
smarter chemistry, 56, 58
thermal chemical data for cured paint films of ISPC and control enamels, 47t
water disbonding resistance and ASTM salt spray (fog) tests, 54, 56
Intramolecular pinacol coupling, stereoselectivity in water, 77, 78
In vivo synthesis. *See* Insect sex pheromone precursors
Iodobenzene diacetate (IBD) 'doped' alumina
oxidation, 302–303
oxidation of sulfides, 304
Iron-complexed dyes
aquatic toxicity, 30, 31f
aquatic toxicity using plant protocol involving duckweed (*lemna minor*), 22–23
black Fe-complex, 25
blue Fe-complex, 25
blue-violet Fe-complex, 25
classes of metal complex dyes, 20
colored 1:2 Fe-complexes, 24, 25
evaluating toxicity by Ames test with Prival preincubation modification, 21–22
identifying structures of commercial dyes, 20
lightfastness of black Fe-complex, 26
lightfastness of formazans, 24–25
lightfastness protocol evaluating dyes, 23–24
metal complex dyes examples, 18–19
plots of front count versus concentration for metal salts in dye complexation, 29f
reductive cleavage by reductases producing genotoxic aromatic amines, 22f
red-violet Fe-complex, 25
representative mutagenicity data (number of revertants versus dye concentration), 28f
research approach, 20–23
structure-aquatic toxicity relationships, 28–30
structure-color-fastness relationships, 23–26
structure-mutagenicity relationships, 26–27
unmetallized formazan dye ligands, 24f
unmetallized ligands, 20
unsymmetrical dyes, 26
washfastness of formazans, 24

Irradiation techniques. *See* Organic syntheses using microwave irradiation
Isoflav-3-enes, substituted, synthesis, 307–308

K

Ketones, reactions with dimethylcarbonate (DMC), 94
Ketones, α-oxygenated, diastereoselective 1,2-additions to, 108–109
Knoevenagel condensation, indium-promoted alternative, 110
Kolbe–Schmitt process
  p-hydroxybenzoate (HBA) production, 126, 128
  need for replacement technology for HBA production, 128, 140
  schematic, 127*f*

L

Law, solvent toxicology, 318*t*
Leather, colorant, 18, 19
Lepidoptera, mode of action of tebufenozide, 12
Lewis acid catalyzed reactions, water as solvent, 65, 67
Lightfastness
  formazans, 24–25
  See also Iron-complexed dyes
Liquid crystal polymers (LCP)
  other types of mononuclear aromatic molecules of interest in LCP production, 135
  See also p-Hydroxybenzoate (HBA), bioconversion of toluene
Lower critical solution temperature (LCST)
  definition of behavior, 187
  poly(*N*-alkyl acrylamide)s, 188
  water-soluble polymer-bound catalysts, 186–187

M

Magnesium reaction, carbon-carbon bond formation in water, 77, 79
Mammalian species, relative acute oral toxicity of tebufenozide, 13*t*
Manganese dioxide "doped" silica, oxidation, 301, 302
Mating disruption, suppression of pest populations, 34, 35*f*

Metal complex dyes
  classes, 20
  See also Iron-complexed dyes
Metal ions-hydrogen peroxide, oxidation, 301
Metal-mediated reactions in water. *See* Organometallic reactions
Metal surface, in-situ phosphatization, 47–48
Metalloporphyrins, alkane and alkene oxidation with perfluorocarbon soluble, as catalysts, 173–174
Methyl aroxyacetates, monomethylation with dimethylcarbonate (DMC), 90–91
Methyl arylsulfones, reaction with dimethylcarbonate (DMC), 92
Methyl ethyl ketone (MEK)
  environmental indexes for MEK and replacement, 241
  infinite dilution activity coefficients of representative solutes in MEK and replacement mixture, 240*t*
  physical properties of MEK and replacement mixture, 240*t*
  technological effectiveness of replacement, 239–241
  See also Environmentally benign solvent substitutes
Methyl isobutyl ketone (MIBK)
  effect of metal loading on selectivity of Pd/hydrotalcite (HTC) and Pt/HTC, 202, 203*f*
  effect of reaction temperature on formation, 203, 204*f*
  novel heterogeneous catalyst, 4
  reaction pathway for acetone condensation process, 195*f*
  utility, 194–195
  See also Condensation of acetone
Methyl siloxanes, environmentally benign, 5
Methylation reactions
  selective mono-*C*-methylation of alkylarylsulfones, 91–92
  selective monomethylation of aroxyacetonitriles and methyl aroxyacetates, 90–91
  selective monomethylations of arylacetonitriles and arylacetoesters, 89
  selective mono-*N*-methylation of primary aromatic amines, 92–93
  See also Dimethylcarbonate (DMC)
α-Methylene-γ-lactones, fusion to medium and large rings, 110

Methylene chloride, hydrolysis and oxidative pathways in supercritical fluids, 283, 284f
Microbial biomass. See Sugar cane vinasse, bioconversion
Microwave irradiation
  solvent-free conditions, 293
  solvent-free processes, 5
  See also Organic syntheses using microwave irradiation
Mono-substituted benzenes with increasing side chain length, bioconversion candidates, 135, 138
Mukaiyama Aldol reaction, water as solvent, 65, 67, 68
Mutagenicity
  representative data for Fe-complexed azo dyes, 28f
  structure relationships for Fe-complexed dyes, 26–27
  See also Iron-complexed dyes

N

Nanostructured TiO_2. See Oxygenation of hydrocarbons
Natural products
  (+)-bergenin from arabinose, 81, 84
  (+)-goniofufurone, 81, 83
  papulacandin and chactiacandin, 81, 84
  synthesis using water as solvent, 81
Neopentane
  chlorination, 263, 264f
  See also Free radical chemistry
Nickel complexes, epoxidation reactions by fluorous biphasic catalysis, 178–180
Nitroalkenes, synthesis using microwave irradiation, 299–300
Non-polar soils
  cleaning method, 255
  cleaning with volatile methyl siloxanes, 247, 248f
Non-porphyrin catalysts, alkane and alkene oxidation with perfluorocarbon soluble, 175–177
Nucleophilic additions, facial selectivity to, 108t

O

Olefin metathesis, transition metal catalyzed polymer synthesis in water, 69, 70
Olefins, epoxidation reactions by fluorous biphasic catalysis, 178–180
Organic chemistry
  carbon-carbon bond formation, 74, 75
  fundamental reactions, 74
Organic coatings to metal substrates
  multi-step process including chromates, 43–44
  See also In situ phosphatizing coatings (ISPC)
Organic molecules, photooxidations, 222, 224
Organic syntheses using microwave irradiation
  activated manganese dioxide-silica for oxidation, 301, 302
  borohydride reduction of carbonyl compounds to alcohols, 304–305
  chromium trioxide impregnated wet alumina for oxidation, 301–302
  claycop-hydrogen peroxide for oxidation, 301
  cleavage of aldehyde diacetates, 294, 295
  cleavage of semicarbazones and phenylhydrazones, 298
  condensation of carbonyl compounds with nitroalkanes for nitroalkenes (Henry reaction), 299
  condensation reactions, 298–300
  deacylation reactions, 294
  debenzylation of carboxylic esters, 294–295
  debenzylation of carboxylic esters on alumina surface, 295t
  deoximation reactions, 297–298
  desilylation reactions, 296
  dethioacetalization reaction, 296–297
  impact on heterogeneous reactions, 292–293
  initial experiments with high dielectric solvents, 293
  iodobenzene diacetate (IBD)-alumina for oxidation of sulfides, 304
  nonmetallic oxidants–IBD "doped" alumina, 302–303
  oxidation, reduction, and cycloaddition reactions from a,b-unsaturated nitroalkenes, 299–300
  oxidation of alcohols and sulfides, 300
  oxidation of sulfides, 303–304
  oxidation reactions, 300–303
  practical utility of microwave-assisted solvent-free protocols, 293
  protection/deprotection reactions, 293–298
  reduction reactions, 304–306

reductive alkylation of amines, 305–306
selective and solvent-free oxidation with clayfen (montmorillonite K 10 clay-supported iron(III) nitrate, 300
sodium periodate "doped" silica-oxidation of sulfides to sulfoxides and sulfones, 303–304
solvent-free cleavage of oximes, 297
synthesis of 2-aroylbenzofurans, 309
synthesis of 2-aryl-1,2,3,4-tetrahydro-4-hydroquinolones, 307
synthesis of flavones, 306–307
synthesis of heterocyclic compounds, 306–309
synthesis of imines, enamines, and nitroalkenes, 298–300
synthesis of substituted isoflav-3-enes, 307–308
synthesis of substituted thiazoles, 308–309
typical experimental procedure for oxidation of alcohols, 301
typical experimental procedure for reduction of carbonyl compounds, 305
Organic synthesis
dimethylcarbonate (DMC), 88
water as solvent, 67, 68
See also Dimethylcarbonate (DMC)
Organizations, committee for Green Chemistry and Engineering Conference, 2
Organometallic reactions
cyclizations, 79
magnesium reaction in water, 77, 79
new chemoselectivity in water, 75
new reactivity in water, 77
palladium catalyzed carbon-carbon bond formation in air, 79
ring expansions, 79
stereoselectivity in water, 75, 77
synthesis of polyhydroxylated natural products, 81
water as solvent, 65, 66
Oxidation and hydrolysis of compounds, supercritical fluids, 282–283
Oxidation reactions
activated manganese dioxide-silica, 301, 302
alcohols and sulfides, 300
chromium trioxide impregnated wet alumina, 301–302
claycop-hydrogen peroxide, 301
mechanism of photocatalyzed, 219–221
nonmetallic oxidants–iodobenzene diacetate (IBD) "doped" alumina, 302–303
selective and solvent-free oxidation with clayfen, 300
sulfides, 303–304
typical experimental procedure for oxidation of alcohols, 301
See also Fluorous biphasic catalysis (FBC); Organic syntheses using microwave irradiation; Oxygenation of hydrocarbons
Oximes
deoximation reactions, 297–298
proposed reaction mechanism of cyclohexanone oxime with dimethylcarbonate (DMC), 96
reaction with DMC, 95–97
Oxygenation of hydrocarbons
band position of semiconductor photocatalyst, 221t
conversion and selectivities of photooxidations, 223t
electrons shifting from valence band to conduction band, 220
environmental advantages of photocatalyzed chemical synthesis, 225–226
higher conversions and better selectivities with minimal side reactions, 224
major processes occurring on semiconductor particle following electronic excitation, 221f
mechanism of photocatalyzed oxidations, 219–221
minimizing overoxidation, 219
partial oxidation reactions, 218–219
photocatalysis principles, 219
photocatalytic oxidation processes involving use of semiconductors, 220–221
photooxidations of organic molecules, 222, 224
photoreduction initiated by TiO_2 particles using aqueous HNO_3, 225
photoreductions, 224
process variables, 224–225
recombination of electron/hole pairs, 220
results of photooxidation of hydrocarbons, 223t
safe range of operation for photocatalytic operation of cyclohexane, 226f
safety considerations, 225
semiconductor catalysis for chemical synthesis, 221–222
valence band positions, 220

Ozone-depleting solvents, Montreal Protocol, 322t

P

Paint film
  quality, 46–47
  thermal chemical data, 47t
Palladium catalyzed carbon-carbon bond formation, organometallic reaction in water, 79, 80
Palladium catalyzed coupling between aryl halides with acetylene gas, water as solvent, 69, 71
Papulacandin, synthesis in water, 81, 84
PARIS II software
  properties for solvents, 4–5
  solvent design algorithm, 237–239
Perfluorocarbon soluble catalysts
  alkane and alkene oxidations with metalloporphyrins, 173–174
  alkane and alkene oxidations with nonporphyrins, 175–177
  epoxidation reactions, 178–180
  See also Fluorous biphasic catalysis (FBC)
Performance, coating protective, 48, 50
Pericyclic reactions, water as solvent, 63, 64
Pest control, novel caterpillar control agent, 3
Pharmacology, solvent toxicology, 318t
Phenols, substituted, bioconversion candidates, 138t
Phenylacetonitrile
  monomethylation with dimethylcarbonate (DMC), 89
  proposed mechanism for reaction of DMC, 90
  reaction with dibenzylcarbonate (DBzlC), 97
Phenylhydrazones, cleavage using microwave or ultrasound irradiation, 298
Pheromones, insect sex
  economic limitations, 34–36
  environmental advantages, 35
  methodology for producing nontoxic, 3
  See also Insect sex pheromone precursors
Phosphatization
  in-situ of metal surface, 47–48
  See also In-situ phosphatizing coatings (ISPC)

Photocatalysis. See Oxygenation of hydrocarbons
Photochemical oxidation, selective for hydrocarbons, 4
Photooxidation of toluene
  colorimetric detection in Brønsted acid sites, 211, 213
  correlation of loss of product selectivity and Brønsted acidity in CaY, BaZSM-5, and NaZSM-5, 214–215
  difference spectra following broadband excitation of toluene–O_2 complexes in zeolites BaX and BaY, 209, 210f
  experimental, 207–208
  in-situ FTIR difference spectra after irradiation of toluene and oxygen in CaY, BaZSM-5, and NaZSM-5, 212f
  investigations in BaX and BaY, 208–209
  investigations in CaY, BaZSM-5, and NaZSM-5, 211
  mechanism for formation of benzaldehyde on BaX and BaY, 213–214
  mercury arc lamp as light source, 207
  photoproduct distributions from ex-situ GC analysis, 209t
  zeolite choices for study, 207, 210f
Photoreductions, organic substrates, 224, 225
Physical chemistry
  connections with synthetic organic chemistry and chemical engineering in supercritical fluids, 273, 275f
  effects in supercritical fluids, 283, 286–288
  normalized rate constant as function of density, 287f
  second-order rate constant, 283, 286
Physiology, solvent toxicology, 318t
Pichia stipitis
  comparing cofermentation of glucose and xylose by genetically engineered Saccharomyces yeast strain and P. stipitis, 155f
  expression of XD, XR, and XK genes, 152, 154
  See also Genetically engineered Saccharomyces yeasts
Pinacol coupling
  aromatic aldehydes in presence of aliphatic, 75, 76
  intramolecular, stereoselectivity in water, 77, 78

Plodia, binding affinity of tebufenozide relative to 20-hydroxyecdysone (20E), 16t
Pollutants, U.S. EPA 33/50 program, 322, 323t
Pollution prevention. *See* Environmentally benign solvent substitutes
Poly(alkene oxide)-supported catalysts, catalyst recovery, 183–184
Polyamide and protein fibers, colorants, 18, 19
Polyethylene-bound catalysts
  Ni(0) catalysts for butadiene cyclodimerization and cyclotrimerization, 185f
  PE_{olig}-bound catalysts and catalyst recovery, 186t
  same chemistry as low molecular weight analogs, 185
  synthesis of PE_{olig}-bound ligands, 184f
  See also Polymer-facilitated biphasic catalysis
Poly(*N*-isopropylacrylamide) (PNIPAM)
  catalyst recovery, 183–184
  chemistry for making co- and terpolymers, 188–189
  lower critical solution temperature (LCST) behavior, 188
Polymer, hydrocarbon non-metallic conducting, water as solvent, 69, 71
Polymer-facilitated biphasic catalysis
  ability to make co- or terpolymers from poly(*N*-isopropylacrylamide) (PNIPAM), 188–189
  approaches for carrying out homogeneous catalysis, 182
  approach using catalysts dissolving selectively in lower phase of biphasic mixture, 191–192
  biphasic systems as inspiration, 189
  catalysts, polyethylene-bound (PE_{olig}-bound), and catalyst recovery from analyses of product-containing filtrate from first or second reaction cycle, 186t
  catalysts separating on heating, 183–184
  chemistry for ligand and catalyst preparations, 187
  fluoroacrylate monomers for copolymers for ligand preparation, 190–191
  fluorous phase approach for carrying out homogeneous catalysis, 190f
  generic support substrate for fluorous phase chemistry, 190–191
  insolubility of crosslinked polymers, 183
  lower critical solution temperature (LCST) behavior, 186–187
  Ni(0) catalysts for butadiene cyclodimerization and cyclotrimerization, 185f
  original concept of heterogenized catalysts, 183
  PE_{olig}-bound catalysts, 184–186
  PE_{olig}-bound catalysts with chemistry of low molecular weight analogs, 185
  poly(*N*-alkyl acrylamide)s, 188
  polymer-bound catalysts in biphasic systems, 189–192
  polymer in phase separation and catalyst recovery, 191
  polymers with LCST behavior, 187
  problems using crosslinked polystyrene supports, 183
  recycling of insoluble, crosslinked polystyrene-supported catalyst, 182f
  selectivity and reactivity of homogeneous versus heterogeneous catalysts, 183
  separation of polymer-bound catalyst with polymer having LCST behavior, 187f
  synthesis of PE_{olig}-bound ligands, 184f
  thermomorphic system, 191f
  using soluble polymers, 183
  water-soluble polymer-bound catalysts, 186–189
  work with heterogeneous polymer bound transition metal catalysts, 182
Polymer synthesis, water as solvent, 67, 69
Polymerization reactions, supercritical fluids, 282
Polystyrene, problems of using crosslinked supports, 183
Primary aromatic amines, selective mono-*N*-methylation with dimethylcarbonate (DMC), 92–93
Prival modification, mammalian mutagenicity (Ames) test, 21–22
Properties, software for solvents, 4–5
n-Propylbenzene, bioconversion candidate, 135, 138
4-Propylphenol, bioconversion candidate, 135, 138
Protection/deprotection reactions
  cleavage of aldehyde diacetates, 294, 295
  cleavage of semicarbazones and phenylhydrazones, 298
  deacylation, 294

debenzylation of carboxylic esters, 294–295
deoximation, 297–298
desilylation, 296
dethioacetalization, 296–297
See also Organic syntheses using microwave irradiation
Pseudomonas chlororaphis B23 cells
comparison with *Rhodococcus* sp. A4 cells, 116, 118
nitrile hydratase activity, 115
See also 5-Cyanovaleramide (5CVAM) production
Pseudomonas putida, recombinant
batch transformation experiments using resting cell suspensions of *P. putida* EM2878, 131
construction and organization of pathway genes, 129, 131
generation of non-HBA degrading *P. putida* mutant, 129
p-hydroxybenzoate hydroxylase (PobA) activity in *P. putida* KT 2440, 130f
PobA enzyme activity, 131t
substrate range of *P. putida* EM2878, 135, 138–139
toluene conversion to HBA by resting cells of *P. putida* EM2878, 133, 135
See also p-Hydroxybenzoate (HBA), bioconversion of toluene

Q

Quail, relative acute oral toxicity of tebufenozide, 13t

R

Rat, relative acute oral toxicity of tebufenozide, 13t
Reaction media, environmentally benign, 3
Reactivity
dimethylcarbonate towards nucleophilic compounds, 88
new for aqueous organometallic reactions, 77
Reagents, environmentally benign, 3
Rearrangement of allyl vinyl ether, water as solvent, 63, 64
Recombinant *Aspergillus nidulans. See Aspergillus nidulans*; Sugar cane vinasse, bioconversion

Recombinant *Pseudomonas putida. See* p-Hydroxybenzoate (HBA), bioconversion of toluene
Recombinant *Saccharomyces* yeast, genetically engineered, 3–4
Recombinant yeast strains
expression of desaturase cDNA in desaturase-deficient yeast strain *ole*1, 39
functional expression of cloned pheromone desaturase cDNAs, 38–40
structure of yeast expression plasmid pYDsTnD11.1, 39f
See also Insect sex pheromone precursors
Recycling
catalyst recovery, 182
crosslinked polystyrene-supported catalyst, 182f
Reduction reactions
borohydride reduction of carbonyl compounds to alcohols, 304–305
reductive alkylation of amines, 305–306
typical experimental procedure for reduction of carbonyl compounds, 305
See also Organic syntheses using microwave irradiation
Reductive alkylation, amines, 305–306
Reflux azeotropes, volatile methyl siloxane (VMS) fluids, 247, 249
Regulatory, impact of green chemistry, 314
Replacement solvents
drop-in, 325–326
See also Environmentally benign solvent substitutes; Methyl ethyl ketone (MEK); Supercritical fluids as solvent replacements
Reverse Claisen condensation, aryl/alkyl ketone splitting, 94
Rheology, in-situ phosphatizing coatings (ISPCs), 45–46
Rhodococcus sp. A4 cells
light activation of A4 whole cells, 116
nitrile hydratase activity, 115
See also 5-Cyanovaleramide (5CVAM) production
Ring expansions, aqueous organometallic reactions, 79, 80
Room temperature azeotropes
comparisons of cleaning power for number of RT azeotropes of VMS, 251t
volatile methyl siloxane (VMS) fluids, 247, 249
Ruhrchemie/Rhône–Poulenc's hydrofor-

mylation process, water as solvent, 67, 68
Ruthenium complexes, epoxidation reactions by fluorous biphasic catalysis, 178–180

S

Saccharomyces yeast, genetically engineered, 3–4
Saccharomyces yeasts. *See* Genetically engineered Saccharomyces yeasts
Safer chemicals
  dimethylcarbonate (DMC) alternative to toxic intermediates, 87, 98
  designing, green chemistry research, 3
Safety
  concerns for oxidations, 225
  safe range of operation for photocatalytic operation of cyclohexane, 226f
  tebufenozide to non-target organisms, 12–13
Salt spray (fog) testing
  blistering width measured across "x" scribe, 56t
  corrosion resistance of alkyd ISPC versus control alkyd formula, 57f
  method, 45
  water disbonding resistance for in-situ phosphatizing coatings (ISPCs), 54, 56
Salt water immersion testing
  disbonding resistance of cured paints on metal substrates, 45
  in-situ phosphatizing coatings (ISPCs) on metal panels, 54, 56
  (Score)$_{ij}$, measuring chemical environmental impacts, 236–237
Selectivity
  correlation of loss of product selectivity and Brønsted acidity in photooxidations in zeolites, 214–215
  tebufenozide for caterpillars, 13–14
  See also Methylation reactions
Semicarbazones, cleavage using microwave or ultrasound irradiation, 298
Semiconductors
  band position of semiconductor photocatalyst, 221t
  catalysis for chemical synthesis, 221–222
  electron movement, 220
  illustration of major processes occurring on semiconductor particle following electronic excitation, 221f
  photocatalytic oxidation processes involving use, 220–221
Semi-polar soils, cleaning with volatile methyl siloxane, 249, 250f
Sex pheromones, insect, methodology for producing nontoxic, 3
Silica
  promoting reactions in supercritical carbon dioxide, 281–282
  manganese dioxide "doped", oxidation, 301, 302
  sodium periodate "doped", oxidation of sulfides, 303–304
Siloxanes, methyl, environmentally benign, 5
Siloxanes, volatile methyl. *See* Volatile methyl siloxanes (VMS)
Sodium borohydride, reduction of carbonyl compounds to alcohols, 304–305
Sodium periodate 'doped' silica, oxidation of sulfides, 303–304
Software, properties for solvents, 4–5
Soils, non-polar
  cleaning method, 255
  cleaning with volatile methyl siloxanes, 247, 248f
Soils, semi-polar, cleaning with volatile methyl siloxane, 249, 250f
Soluble polymers, polymer-facilitated catalysis, 183
Solvent-free organic reactions. *See* Organic syntheses using microwave irradiation
Solvent mixtures
  PARIS II algorithm, 237–239
  PARIS II software, 4–5
Solvent systems, green
  green chemistry research, 4
  See also Environmentally benign solvent substitutes; Green chemistry solvents; Volatile methyl siloxanes (VMS)
Solvent usage, evaluating present, 323, 324t
Solvents
  chemistry, 314–316
  roles in chemistry, 315t
  summary of toxicology criteria, 321t
  toxicology, 317–319
  traditional solvation concerns, 316t
  See also Green chemistry solvents; Supercritical fluids as solvent replacements

Southern armyworm, relative acute oral toxicity of tebufenozide, 13t
Spodoptera eridania, relative acute oral toxicity of tebufenozide, 13t
Stability, storage, in-situ phosphatizing coatings (ISPCs), 45–46
Stereoselectivity, new for aqueous organometallic reactions, 75, 77
Storage stability, in-situ phosphatizing coatings (ISPCs), 45–46
Styryl carbohydrate derivatives, synthesis, 81, 82f
Substitute solvent design theory
 environmental indexes, 234–236
 technological performance: fluid properties, 233–234
 technological performance: theory, 231–233
 See also Environmentally benign solvent substitutes
Sugar cane vinasse, bioconversion
 Aspergillus species, 162
 biomass production and pollution reduction, 165
 biomass production and pollution reduction of vinasse by *Aspergillus* species, 166t
 chemical oxygen demand (COD) reduction and biomass production, 167–169
 commercial fertilizer (NPK) solution, 162
 diploids, recombinants, and genealogy, 166
 dry biomass production, 162
 dry biomass production and COD reduction by *A. nidulans* strains, 168f
 effect of temperature on radial growth of *Aspergillus* species, 165t
 effect on radial growth of *A. nidulans* by addition of NPK solution, 164t
 effect on radial growth of *A. nidulans* by addition of sugar can molasses to vinasse, 164t
 effects of addition of NPK, 163
 effects of addition of sugar cane molasses, 163
 growth kinetics, 166
 growth kinetics of *A. nidulans* strains cultivated on liquid vinasse medium, 167f
 material methods, 162–163
 medium, 162
 medium establishment, 163
 parasexual crosses, diploid and recombinant strains, 163
 pollution indexes, 163
 radial growth, 162
 sugar cane molasses and vinasse, 162
 vinasse biodegradation by *Aspergillus* species, 164–165
 vinasse conversion by recombinant strains of *A. nidulans*, 166–169
 vinasse production, 160–161
 vinasse utilization and bioconversion, 161–162
Sulfides
 epoxidation reactions by fluorous biphasic catalysis, 178–180
 oxidation reagents, 300
 oxidation to sulfoxides and sulfones using microwave irradiation, 303–304
Sulfones
 mono-C-methylation with dimethylcarbonate (DMC), 91–92
 oxidation of sulfides using microwave irradiation, 303–304
Sulfoxides, oxidation of sulfides using microwave irradiation, 303–304
Supercritical carbon dioxide
 environmentally benign solvent, 5
 significant change in density, viscosity, and diffusivity near critical point for CO_2, 273, 274f
 See also Free radical chemistry
Supercritical fluids as solvent replacements
 connections among synthetic organic chemistry, physical chemistry, and chemical engineering disciplines, 275f
 defining supercritical region, 272f
 definition, 271
 Diels–Alder reactions, 276f, 277f, 278t, 279t, 280
 evaluating catalytic pathways, 281
 hydrolysis and oxidative pathways for methylene chloride in near-critical and sub-critical water, 283, 284f
 interest in hydrolysis and oxidation of compounds in near-critical and super-critical water, 282–283
 Kolbe–Schmitt synthesis as model of direct carboxylation reaction, 281
 limitation of low solubility of common reagents and reactants in carbon dioxide, 281
 modest rate control by adjusting pressure to alter density, 285f
 normalized rate constant as function of density, 287f
 physical chemistry effects, 283, 286–288

polymerization reactions, 282
possible advantages of supercritical fluids for waste remediation/treatment, and as alternative solvent for waste minimization, 273t
possible synthetic reactions in supercritical CO_2 (SC-CO_2) and co-solvent mixtures, 282
potential environmental and process separation advantages, 281
research and development needs and opportunities, 288–289
second-order rate constant, 283, 286
significant change in density, viscosity, and diffusivity near critical point for CO_2, 273, 274f
use of silica and alumina promoting reactions, 281–282
yield and selectivity of Diels–Alder reactions enhanced by fumed silica, 282
Suzuki reaction, water as solvent, 69, 71
Synthesis
polyhydroxylated natural products, 81
See also Organic syntheses using microwave irradiation
Synthesis, green chemical, green chemistry research, 3
Synthesis, in vivo. See Insect sex pheromone precursors
Synthetic organic chemistry, connections with physical chemistry and chemical engineering in supercritical fluids, 273, 275f

T

Target selectivity. See Tebufenozide
Tebufenozide
binding affinity relative to 20-hydroxyecdysone (20E), 16t
concentrations of 20E in insect's blood, 11
conceptual "gates" determining degree of toxicity in particular insect species, 15f
critical events in insect molting process relative to changes in 20E concentration, 11f
20E and molting process in insects, 9–11
insecticidal mode of action, 12
potent mimic of 20E in caterpillars, 8–9
probably mechanisms for high order of caterpillar selectivity, 13–14
relative acute oral toxicity to representative caterpillar, mammalian, and avian species, 12t
safety to non-target insects/arthropods, 13
safety to non-target organisms, 12–13
schematic of interaction of 20E with target site in insects, 10f
structure, 9
structure of 20E, 9
tradenames, 12
Technical, impact of green chemistry, 314
Tetraarylporhyrin (TPP) ligands, perfluoro-modified, 173–174
Textile dyes
eliminating source of wastewater, 3
See also Iron-complexed dyes
Theory
substitute solvent design, 231–236
See also Environmentally benign solvent substitutes
Thermal properties, polymer paint films, 46–47
Thermomorphic system
catalyst recovery, 184
description and schematic, 191f
Thiazoles, substituted, synthesis, 308–309
Tick, binding affinity of tebufenozide relative to 20-hydroxyecdysone (20E), 16t
Titanium dioxide
photooxidation, 4
See also Oxygenation of hydrocarbons
Toluene
high selectivity of oxidation to benzaldehyde without overoxidation, 222, 223t
photooxidation, 4
See also Photooxidation of toluene
Toluene bioconversion. See p-Hydroxybenzoate (HBA), bioconversion from toluene
Toluenes, substituted
bioconversion, C-2 position substitution, 138–139
bioconversion, C-3 position substitution, 139
bioconversion candidates, 138t
Toxicity
mammalian mutagenicity test (Ames) with Prival modification, 21–22
See also Iron-complexed dyes
Toxicity concerns, green solvents, 319t
Toxicological profile, volatile methyl siloxanes (VMS), 245–246
Toxicology
components of solvent toxicology, 318t

solvent, 317, 319
solvent criteria, 321t
Transition metal catalyzed reactions
  polymer synthesis with water as solvent, 69, 70
  water as solvent, 65, 66
Transition metal oxides and phosphates, acetone condensation, 195
Transition states. *See* Coupling reactions in water, indium-promoted
1,4,7-Triazacyclononane (TACN), highly fluorinated derivatives, 175–176
Trichoplusia ni, model for pheromone biosynthesis, 36–37

U

Uhlmann-type coupling, organometallic reaction in water, 79, 80
U.S. EPA 33/50 program, priority pollutants, 322, 323t

V

Valence band, electron movement in semiconductor, 220
Vinasse. *See* Sugar cane vinasse, bioconversion
Vision 2020 for the Chemical Industry, environment, 1–2
Volatile methyl siloxanes (VMS)
  applications, 247–255
  approval of significant new alternatives program (SNAP), 245
  azeotropes, 247, 249
  azeotropic determinations, 256
  carriers in variety of formulated products, 255
  cleaning of non-polar soils, 247, 248f
  cleaning of semi-polar soils, 249
  comparison of cleaning power for number of room temperature azeotropes of VMS, 251t
  compositions of surfactants in VMS fluids providing high degree of water removal, 254f
  description, 244
  environmental profile, 246–247
  experimental, 255–256
  health and environmental profile, 245–247
  impact of polar solvent addition on ability of VMS fluids to remove solder flux, 250f
  non-polar soil cleaning procedure, 255
  precision water removal, 249, 252
  rosin flux cleaning procedure, 255
  toxicological profile, 245–246
  unusual combination of environmentally benign qualities, 244–245
  water displacement drying (WDD), 252, 253f
WDD method, 256

W

Washfastness
  formazans, 24
  See also Iron-complexed dyes
Waste remediation and minimization, possible advantages of supercritical fluids, 273t
Wastewater, eliminating source in textile dye synthesis, 3
Water as solvent
  chemical synthesis, 67
  Diels–Alder reaction, 63, 64
  hydrocarbon non-metallic conducting polymer, 69, 71
  indium mediated cross-coupling of carbohydrate with allyl bromide derivative following by ozonolysis, 67, 68
  Lewis acid catalyzed reactions, 65, 67
  material synthesis in water, 67, 69
  new materials, 69
  new polymerization processes, 67, 69
  olefin-metathesis for polymer synthesis with Grubbs' catalyst, 69, 70
  organic synthesis, 67
  organometallic reactions, 65
  palladium catalyzed copolymerization of 3,5-diiodobenzoic acid and acetylene gas, 69, 71
  palladium catalyzed coupling between aryl halides and acetylene gas, 69, 71
  pericyclic reactions, 63
  reaction between hydroxymethylanthracene and N-ethylmaleimide, 63, 64
  reaction of carbonyl compounds with allyl halides, 65, 66
  reactions in, 3, 63–67
  rearrangement of allyl vinyl ether, 63, 64
  reasons for choosing, 62–63
  Ru catalyzed ring-opening polymerization for neoglycopolymers, 69, 70
  Ruhrchemie/Rhône–Poulenc's hydroformylation process, 67, 68
  synthesis in water, 67
  transition metal catalyzed reactions, 65, 66

See also Coupling reactions in water, indium-promoted; Organometallic reactions

Water disbonding resistance
  blistering width measured at "x" scribe, 36t
  corrosion resistance of alkyd ISPC versus control alkyd formula, 57f
  salt spray (fog) tests for in-situ phosphatizing coatings (ISPCs), 54, 56

Water displacement drying (WDD) method, 256
  precision water removal by volatile methyl siloxanes (VMS), 252, 253f

Water removal, volatile methyl siloxane, 249, 252

Water-soluble polymer-bound catalysts
  ability to make co- and terpolymers from poly(N-isopropylacrylamide) (PNIPAM), 188–189
  lower critical solution temperature (LCST) behavior, 186–187
  PNIPAM, 188
  separation of polymer-bound catalyst with polymer having LCST behavior, 187f
  typical chemistry for preparation of ligands and catalysts, 187

X

Xylenes, bioconversion candidate, 135, 138

Xylose
  design and development of ideal yeast for cofermentation with glucose to ethanol, 146–147
  See also Genetically engineered *Saccharomyces* yeasts

Xylulokinase gene
  effect of cloning and overexpression, 147, 152
  See also Genetically engineered *Saccharomyces* yeasts

Y

Yeast
  genetically engineered *Saccharomyces*, 3–4
  See also Genetically engineered *Saccharomyces* yeasts; Insect sex pheromone precursors

Z

Zeolites
  acetone condensation, 195
  choices for photooxidation of toluene, 207, 210f
  toluene photooxidation in BaX and BaY, 208–209
  toluene photooxidation in CaY, BaZSM-5, and NaZSM-5, 211
  See also Photooxidation of toluene

Highlights from ACS Books

Desk Reference of Functional Polymers: Syntheses and Applications
Reza Arshady, Editor
832 pages, clothbound, ISBN 0–8412–3469–8

Chemical Engineering for Chemists
Richard G. Griskey
352 pages, clothbound, ISBN 0–8412–2215–0

Controlled Drug Delivery: Challenges and Strategies
Kinam Park, Editor
720 pages, clothbound, ISBN 0–8412–3470–1

Chemistry Today and Tomorrow: The Central, Useful, and Creative Science
Ronald Breslow
144 pages, paperbound, ISBN 0–8412–3460–4

A Practical Guide to Combinatorial Chemistry
Anthony W. Czarnik and Sheila H. DeWitt
462 pages, clothbound, ISBN 0–8412–3485–X

Chiral Separations: Applications and Technology
Satinder Ahuja, Editor
368 pages, clothbound, ISBN 0–8412–3407–8

Molecular Diversity and Combinatorial Chemistry: Libraries and Drug Discovery
Irwin M. Chaiken and Kim D. Janda, Editors
336 pages, clothbound, ISBN 0–8412–3450–7

A Lifetime of Synergy with Theory and Experiment
Andrew Streitwieser, Jr.
320 pages, clothbound, ISBN 0–8412–1836–6

For further information contact:
Order Department
Oxford University Press
2001 Evans Road
Cary, NC 27513
Phone: 1-800-445-9714 or 919-677-0977
Fax: 919-677-1303

Bestsellers from ACS Books

The ACS Style Guide: A Manual for Authors and Editors (2nd Edition)
Edited by Janet S. Dodd
470 pp; clothbound ISBN 0–8412–3461–2; paperback ISBN 0–8412–3462–0

Writing the Laboratory Notebook
By Howard M. Kanare
145 pp; clothbound ISBN 0–8412–0906–5; paperback ISBN 0–8412–0933–2

Career Transitions for Chemists
By Dorothy P. Rodmann, Donald D. Bly, Frederick H. Owens, and Anne-Claire Anderson
240 pp; clothbound ISBN 0–8412–3052–8; paperback ISBN 0–8412–3038–2

Chemical Activities (student and teacher editions)
By Christie L. Borgford and Lee R. Summerlin
330 pp; spiralbound ISBN 0–8412–1417–4; teacher edition, ISBN 0–8412–1416–6

Chemical Demonstrations: A Sourcebook for Teachers, Volumes 1 and 2, Second Edition
Volume 1 by Lee R. Summerlin and James L. Ealy, Jr.
198 pp; spiralbound ISBN 0–8412–1481–6
Volume 2 by Lee R. Summerlin, Christie L. Borgford, and Julie B. Ealy
234 pp; spiralbound ISBN 0–8412–1535–9

The Internet: A Guide for Chemists
Edited by Steven M. Bachrach
360 pp; clothbound ISBN 0–8412–3223–7; paperback ISBN 0–8412–3224–5

Laboratory Waste Management: A Guidebook
ACS Task Force on Laboratory Waste Management
250 pp; clothbound ISBN 0–8412–2735–7; paperback ISBN 0–8412–2849–3

Reagent Chemicals, Ninth Edition
768 pp; clothbound ISBN 0–8412–3671–2

Good Laboratory Practice Standards: Applications for Field and Laboratory Studies
Edited by Willa Y. Garner, Maureen S. Barge, and James P. Ussary
571 pp; clothbound ISBN 0–8412–2192–8

For further information contact:
Order Department
Oxford University Press
2001 Evans Road
Cary, NC 27513
Phone: 1-800-445-9714 or 919-677-0977